Cell Interactions
in Differentiation

ORGANIZING COMMITTEE

Sulo Toivonen
Honorary President

Lauri Saxén
Chairman

Leonard Weiss
Vice-Chairman

S. Nordling   A. Vaheri   J. Wartiovaara

Marketta Karkinen-Jääskeläinen
Editor of the Proceedings

Irma Saxén-Thesleff
Secretary General

Cell Interactions
in Differentiation

Sixth Sigrid Jusélius Foundation Symposium: Helsinki, Finland.
August 1976

Edited by

MARKETTA KARKINEN-JÄÄSKELÄINEN
and LAURI SAXÉN

Third Department of Pathology, University of Helsinki, Finland

and

LEONARD WEISS

*Department of Experimental Pathology, Roswell Park Memorial Institute,
Buffalo, N.Y., USA*

1977

ACADEMIC PRESS
LONDON · NEW YORK · SAN FRANCISCO

A Subsidiary of Harcourt Brace Jovanovich, Publishers

ACADEMIC PRESS INC. (LONDON) LTD
24/28 Oval Road
London NW1

United States Edition published by
ACADEMIC PRESS INC.
111 Fifth Avenue
New York, New York 10003

Library of Congress Catalog Card Number: 77-71827
ISBN: 0-12-398250-2

Printed in Great Britain by Butler and Tanner Ltd
Frome and London

LIST OF PARTICIPANTS

(Full addresses appear on the first page of each chapter)

BENNETT, D. New York, NY, USA
BRIGGS, R. Bloomington, Indiana, USA
BURGER, M. M. Basel, Switzerland
COOKE, J. London, UK
GERISCH, G. Basel, Switzerland
GILULA, N. B. New York, NY, USA
GRAHAM, C. F. Oxford, UK
GROSSBERG, A. L. Buffalo, NY, USA
HUNT, R. K. Baltimore, Md., USA
KALTHOFF, K. Freiburg, West Germany
KARKINEN-JÄÄSKELÄINEN, M. Helsinki, Finland
KIENY, M. Grenoble, France
LASH, J. W. Philadelphia, Pa., USA
LAWRENCE, P. A. Cambridge, UK
LE DOUARIN, N. Nogent-sur-Marne, France

McLAREN, A. London, UK
MÄKELÄ, O. Helsinki, Finland
MARTIN, G. R. San Francisco, Calif., USA
MOSCONA, A. A. Chicago, Illinois, USA
NORDLING, S. Helsinki, Finland
PICTET, R. L. San Francisco, Calif., USA
SAXÉN, L. Helsinki, Finland
SENGEL, P. Grenoble, France
SLAVKIN, H. C. Los Angeles, Calif., USA
TARIN, D. London, UK
THESLEFF, I. Helsinki, Finland
TOIVONEN, S. Helsinki, Finland
VAHERI, A., Helsinki, Finland
WEISS, L. Buffalo, NY, USA
WOLPERT, L. London, UK

PREFACE

"Embryonic induction", the puzzling problem of the thirties, is still a timely problem as the information gained from this field seems to find an increasing number of applications in other biomedical disciplines. Conversely, recent developments in the understanding of intercellular communication between a variety of mature cells, such as those of the CNS and those of the immune system, have provided models to be tested in embryonic systems. The development of both analytical and experimental methods together with many new model systems for morphogenetic tissue interactions developed during recent years have further increased our possibilities of profitably re-examining many classic problems related to "embryonic induction". These recent developments encouraged the organizers of the Sixth International Symposium of the Sigrid Jusélius Foundation to invite a group of representatives of the leading schools in developmental biology and related fields to discuss problems of mutual interest at the Hanasaari Cultural Centre near Helsinki. This volume contains the papers presented at that Symposium in August 1976.

It is the privilege of the organizers of this symposium and the editors of this volume to express their most sincere thanks to the Sigrid Jusélius Foundation and its officials, especially to Professor Nils Oker-Blom, Member and Secretary of the Scientific Advisory Board, and to Mr Marcus Nykopp, Director of the Foundation, who actively contributed to the organization of the symposium. The Sigrid Jusélius Foundation has become the most important source of support for free biomedical research in Finland. While public funds tend to become more and more directed into applied sciences and support projects hopefully believed to solve the problems of today's communities, this foundation has realized the fundamental importance of basic research. The series of the "Sigrid Jusélius Symposia" also reflects the awareness of the foundation and its officials of the necessity for the exchange of scientific data and information at international and interdisciplinary levels.

We also wish to thank our colleagues for their help not only in planning the symposium, but also for their scientific contributions. Drs Anne McLaren, Aron Moscona and Lewis Wolpert planned and organized their sessions in this volume and introduce us to some of the actual problems in their fields. Dr Irma Thesleff, Secretary-General of the Organizing Committee, carried the heaviest burden with never-failing enthusiasm, and Mrs Anja Nykänen gave her expert help throughout our editorial work. We believe the proper way to thank all of our colleagues is to extend our appreciation to the founder of the Finnish School of Developmental Biology, our Honorary President, Professor Sulo Toivonen. The organizers are proud and privileged to have the opportunity to follow in his footsteps.

Hanasaari, Espoo, September 1976   M. K.-J., L. S., L. W.

OPENING REMARKS

Nils Oker-Blom

When the late Mr Fritz Artur Jusélius created the Sigrid Jusélius Foundation half a century ago, soon after the premature death, due to an infectious disease, of his beloved daughter Sigrid, he had two goals in mind: to support scientists in their fight against diseases which are particularly harmful to mankind and to promote international scientific contacts.

As far as the first goal is concerned, emphasis has been put on supporting highly qualified research aimed at elucidating the basic principles of life and the changes in cells and organisms, resulting in what we call disease, with the hope that such knowledge will enable us to prevent or to cure the diseases in question. As a matter of fact most of the post-war biomedical research in this country has been supported by the foundation. The second goal we hope to reach partly by inviting foreign scientists to Finnish institutions and partly by arranging international symposia in which scientists can discuss their problems, and draw the guide-lines for further work.

The first symposium was held in 1965 on the "Control of Cellular Growth in Adult Organism". The themes of the four following symposia have been "Regulatory Functions of Biological Membranes". "Cell Interactions and Receptor Antibodies in Immune Response", "Biology of Fibroblast" and finally "Amyloidosis". Thus, the present symposium on "Cell Interactions in Differentiation" complements very well the subject matter of the other symposia.

I wish to congratulate the organizers, Professor Lauri Saxén and his co-workers and staff, who have succeeded in planning an interesting scientific program, and particularly that they have succeeded in collecting a great number of distinguished scientists.

It is a special pleasure for me also to mention the Honorary President of the symposium, Professor Sulo Toivonen. As is well known Sulo Toivonen is a pioneer in the many fields of the puzzling problem of embryonic induction. In addition to his personal contribution to the field Professor Toivonen has introduced a large group of younger scientists into this field in Finland, many of whom have presented results of their recent work in this symposium. Thus, it was a privilege to have him with us as, if he may excuse the words, a bridge between classic embryology and its concepts, and the present, ambitious approaches of the younger generation.

At a recent seminar at our institute crystallization of virus proteins was discussed and a hanging drop micro method was presented in which the milieu, the amount of reacting molecules etc. can be well controlled, which resulted in the most beautiful crystals. Perhaps a symposium like this can be compared to such a hanging drop. The right amount of the right ideas in the right milieu may crystallize into one new, great and beautiful leading principle. Let us hope that it has happened here.

INTRODUCTION

The classic work of Hans Spemann and his school demonstrated two fundamental properties of embryonic cells: (1) they are not necessarily predetermined and can be experimentally diverted from their normal developmental
pathway, and (2) cells within an embryonic organism require extrinsic
messages in order to express their developmental capacities. The aim of our
work today is still to unravel the mechanisms behind these phenomena, and
even if the terminology has changed and the goals can be more clearly specified, the basic problem has remained the same: how does a cell become
adjusted to the synchronized development of an entire multicellular organism
so as to express its genetic information at a strictly controlled time and place?
To do this, a cell must be able to respond to (and to emit) different kinds
of signals:

> At the organismal and organ level, each cell must sense its position and
> respond to a control system; this is termed *positional information*.
> At the tissue and cell level, each cell must exchange messages with both
> like and unlike adjacent cells. Such *homo-* and *heterotypic cell interactions* con-
> stitute a central guiding mechanism for cytodifferentiation and morpho-
> genesis.

None of these communicative systems has yet been fully understood, and it has
become increasingly evident that we should not look for unifying, simplified
concepts and for general mechanisms, but rather to analyse each process separately until their individual basic mechanisms are better understood. This, however, does not mean that those working on these probably somewhat different
interactive events should not communicate and search for common features
in these processes of fundamental biological significance.

CONTENTS

I. Early Determinative Events in Embryogenesis

II. Positional Information and Morphogenetic Signals

III. Morphogenetic Tissue Interactions

IV. Molecular Mechanisms of Cell Contact Interactions

V. Cell Recognition

I
Early Determinative Events in Embryogenesis

Early Determinative Events in Embryogenesis: An Introduction

ANNE McLAREN

MRC Mammalian Development Unit,
Wolfson House, London, England

In discussing determinative events in embryogenesis, it is customary to distinguish between endogenous factors, such as the presence of intracellular organelles or cytoplasmic microheterogeneities, and outside influences, whether from the external environment or from neighbouring cells. An elegant example of the identification and analysis of a pre-localized cytoplasmic determinant in an insect egg will be presented by Dr Kalthoff.

In mammals it seems that localized cytoplasmic factors have little if any determinative influence in early development: cells of early cleavage stage embryos are totipotent, and only become differentiated as a result of differences in their environment, arising from the location of the cells within the embryo. The geometry of cleavage, responsible for cell location in normal mouse embryos, will be discussed by Dr Graham. The distinction between insect and mammal may be less absolute than appears at first sight: development is a continuous process, the dividing line of fertilization is taken only for our convenience, and endogenous determinants present at fertilization may have formed in response to external factors during oogenesis.

The role of the nucleus in cell type determination will be discussed by Dr Briggs. Nuclear transfer experiments in amphibia have suggested that no irreversible gene loss occurs in the course of differentiation, but the degree to which the epigenotype of the differentiated cell nucleus can be set back to its starting value by re-exposure to egg cytoplasm is still a matter of debate. So the reversibility of nuclear changes during differentiation may also be less absolute than is sometimes assumed.

Dr Martin will take up the story of mammalian development, with reference to teratocarcinoma cell lines and their differentiation under various culture conditions, as well as after retransplantation to a host blastocyst. The *in vitro* teratocarcinoma system provides a remarkably close parallel to early development in the normal embryo, and may in the future offer a means of identifying some of the events controlling cell allocation and determination.

Analysis of a Cytoplasmic Determinant in an Insect Egg

Klaus Kalthoff

Biologisches Institut I (Zoologie) der Universität, Albertstrasse 21a, 7800 Freiburg, West Germany

THE CONCEPT OF PRELOCALIZED CYTOPLASMIC DETERMINANTS

The development of increasingly complex structures from apparently homogeneous layers of cells is a most fascinating phenomenon in developmental biology. The realization of the emerging spatial patterns depends upon recognizable differences, in structure or alignment, which develop between cells or groups of cells. Each of these prospective pattern elements takes up specific cell activities and thereby assumes its distinct features. This process of cellular *differentiation* is thought to be programmed by earlier events referred to as *determination*, a term which is usually defined by operational criteria such as the capability of the cells to differentiate autonomously upon transplantation or in culture (Gehring, 1973). The process of determination may be visualized as a stepwise "instruction" and subsequent "commitment" of cells to embark on specific developmental pathways (Sander, 1976). In a higher organism, different instructing and committing signals must be handed out to different cell groups in proper *spatial arrangement*, so that the various developmental pathways on which the cells embark give rise to the spatial pattern characteristic of that organism.

The study of spatial pattern formation during embryogenesis therefore proceeds along two lines of questions and experiments which are interrelated but basically distinct: One line is along the *time axis* and is mainly concerned with the processes that channel a given group of cells into special activities (cellular differentiation). The other line is mainly concerned with the organization of such activities along the *spatial axes* of the embryo (pattern specification). In contrast to considerable progress in the molecular analysis of cellular differentiation, the molecular basis of pattern specification is almost unknown. This is reflected in a multiplicity of conceptual frameworks for pattern specification at the cellular and supracellular levels. Some of these involve the idea of cytoplasmic determinants, which are thought to be prelocalized in the egg cell in a spatial arrangement. If different determinants become incorporated into different cells during cleavage, they could serve to specify a coordinated pattern of different cell activities.

The development of insect embryos is especially suggestive of cytoplasmic determinants because the periplasm ("cortex") of the egg remains undisturbed by early mitotic divisions. Only after an extended period of intravitelline cleavage, i.e. after a series of nuclear divisions within the yolk rich endoplasm of the egg, most of the nuclei, with jackets of cytoplasm around them, migrate towards the egg surface. After further nuclear divisions, the blastoderm cells are formed by infoldings of the oolemma. Theoretically, each of these cells might pick up a particular morphogenetic instruction which has been laid down during oogenesis or had resulted from epigenetic processes during the plasmodial period of the egg. In the following sections, three cases which illuminate the role of cytoplasmic determinants in insect embryogenesis will be discussed. They had been selected to draw attention to experimental systems which may allow the analysis of cytoplasmic determinants to be taken to the molecular level. A broad review including a thorough re-evaluation of "classical" work on pattern specification in insect embryogenesis has recently been presented by Sander (1976). For a comprehensive account of the causal analysis of insect embryogenesis, the reader is referred to the review of Counce (1973).

THE OOSOME

The posterior pole of the eggs of many *Coleoptera*, *Diptera*, and *Hymenoptera* carries basophilic, granular material referred to as polar granules or the oosome. It has long been known that the "pole cells" which originate at this site and incorporate the oosome material produce the primary germ cells. These observations suggest that the oosome acts as a cytoplasmic determinant for the formation of germ cells, although the pole cells may give rise to other cell types as well (see Counce, 1973). The idea of germ cell determination by oosome material was supported by the loss of fertility which follows selective destruction or removal of posterior pole material. However, it is notoriously hard to show why parts of an embryo, in this case the germ cells, are *not* formed after experimental interference. In fact, fertile animals *can* develop after removal or X irradiation of the oosome as Günter (1971) has observed in the ichneumonid *Pimpla turionellae*. Positive proof of the determination of prospective germ cells by oosome material came from transplantation experiments. Okada *et al.* (1974) performed a "rescue" experiment in which they sterilized *Drosophila* eggs by UV irradiation of the posterior pole, and then restored the fertility by transplanting cytoplasm from the posterior pole of unirradiated eggs. Illmensee and Mahowald (1974) transferred posterior polar plasm from *Drosophila* eggs at early cleavage stages to the anterior tip of host eggs of the same age. In the presumptive somatic region of the first host "pole cells" were formed which, after transplantation into the posterior region of a second host, gave rise to progeny with the genetic label of the first host (Fig. 1). This experiment has now opened the way for a biochemical characterization of the active fraction.

The germ cell case, somewhat unfortunately, has been taken as a paradigm for cytoplasmic determinants in general. Thus the "cortex" of insect eggs, in particular dipteran eggs, has been visualized as a "mosaic" of qualitatively different cytoplasmic determinants (see Kalthoff, 1976). The germ cells, however, are peculiar in several respects (see Sander, 1975). They tend to segregate early and preserve a full

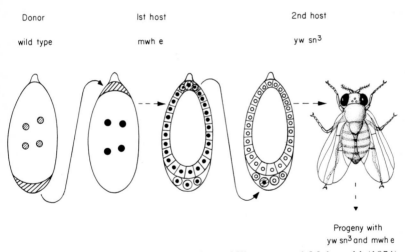

Donor   1st host   2nd host

wild type   mwh e   yw sn^3

Progeny with
yw sn^3 and mwh e

Fig. 1. Schematic illustration of the procedure of Illmensee and Mahowald (1974) to demonstrate the determinative action of the oosome in *Drosophila*. Oosome material was transplanted to the anterior pole region of a first host. The anterior "pole cells" formed there were then transferred to the posterior region of a second host, the progeny of which included individuals with the genetic marker of the first host.

chromosome complement in species which eliminate chromosomes from their somatic cells. Whereas pattern specification in the somatic part of the body initiates specialized cell activities, the germ cells have to be kept in a nonspecialized state as they will give rise to all types of cells in the future embryo. It is consistent with this fundamental difference, that the determination of the germ cells is clearly dissociated from the specification of the somatic body pattern. Mutations or experimental treatment which interfere with fertility frequently leave the somatic part of the body unaffected (Counce, 1973; Gehring, 1973). Conversely, monster embryos in which the head and thorax are replaced by a second abdomen carry germ cells only in the original abdomen (Bull, 1966; Yajima, 1970; see also Fig. 3 of this chapter). There is no indication whatsoever that the oosome case is paradigmatic of pattern specification in general.

Besides the oosome, the ultrastructure of oocytes does not reveal any signs of cytoplasmic determinants. With biochemical and immunological techniques, regional differences in the protein contents of cricket egg fragments have been found (see Sander, 1976, for references). None of these unevenly distributed proteins has been tested for morphogenetic activity.

LONG-RANGE PATTERN SPECIFICATION BY POSTERIOR POLE MATERIAL IN THE *EUSCELIS* EGG

The lack of ultrastructural or direct biochemical evidence for cytoplasmic determinants other than the oosome means that the case for such determinants is almost entirely based upon the results of experimental interference with the development of insect eggs. Experiments which demonstrate positively that posterior determinants are involved in the specification of the longitudinal body pattern were carried out by Sander (1960, 1975) with eggs of the leaf hopper, *Euscelis plebejus*. The posterior pole region of these eggs carries a cluster of symbiotic bacteria, and this can be used as a visible marker to indicate the position of posterior pole material which can be pushed anteriorly by invaginating the egg with a blunt needle. In combination with transverse fragmentation of the egg, such translocation experiments have clearly demonstrated that posterior pole material plays a key role in the specification of the basic body pattern of *Euscelis*.

Anterior fragments of *Euscelis* eggs, after fragmentation during cleavage stages, either produce only head structures or no embryonic parts at all (Fig. 2b). They become capable, however, of forming complete embryos if posterior pole material is shifted anteriorly before fragmentation (Fig. 2c$_2$). This dramatic increase in morphogenetic capacity must be ascribed to the presence of the posterior pole material moved to the anterior fragment. Similar results have been obtained from analogous experiments with eggs of the bean weevil, *Bruchidius obtectus*, and the ichneumonid *Pimpla* (see Sander, 1976, for references). In larger posterior fragments of *Euscelis* eggs containing the translocated posterior pole material (Fig. 2d), partial germ bands with reversed polarity (EDC) or aberrant patterns with two sets of posterior segments joined in mirror-image symmetry (EDCDE) were found.

The action of posterior pole material (PPM) has two important

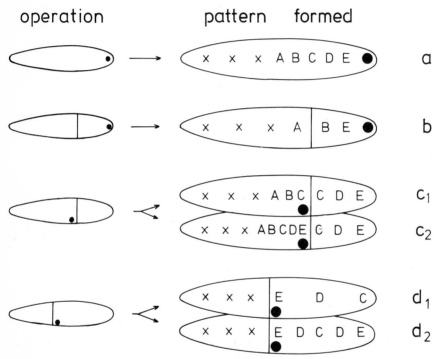

Fig. 2. Combined fragmentation and translocation of posterior pole material in *Euscelis* eggs (after Sander, 1960, 1975). Horizontal bar in outline of egg indicates fragmentation at the respective level, black disc represents the symbiont ball indicating the position of posterior pole material before and after translocation. Both operations were carried out at late intravitelline cleavage. The segment patterns identified some days later are symbolized by letters: crosses represent "extraembryonic" material; A, procephalon; B, gnathocephalon; C, thorax; D and E, abdomen. The symbols at the upper and lower ends of each series indicate the approximate location of the precursor cells for the respective pattern elements in the blastoderm, as inferred from direct observation of the developing germ band. The remaining symbols have been distributed evenly in between.

characteristics. First, the addition of PPM to an anterior fragment does not always cause this fragment to form a complete germ band (Fig. $2c_2$), but sometimes the increase in the morphogenetic potential is only submaximal. In such cases, the PPM does not cause the formation of the terminal abdominal segments that are normally formed in its vicinity. Instead, segments of the middle region are added (Fig. $2c_1$). Second, the length of a given set of segments (e.g. CDE) may vary by a factor of more than 2 (compare anterior fragment in Fig. $2c_2$ to posterior fragment in Fig. $2d_1$). These characteristics suggest that the PPM does not produce a sequence of qualitatively different determinants but instead initiates a gradient of one determinant, the local level of which could cause the respective blastoderm cells to form different segments. The results diagrammed in Fig. 2 could then be ascribed to changes in the level and shape of such a gradient (see Sander, 1975, 1976). Whatever the nature of this gradient may be, it is conceptually important that different quantities of one morphogenetic determinant may release qualitatively different cell activites. Lewis *et al.* (1977) have recently considered the precision with which a spatial gradient of a diffusible morphogen can specify a pattern of discrete and stable cell states by threshold mechanisms in the reacting cells.

Experiments to characterize the active fraction of the PPM in *Euscelis* eggs are now under way. To exert its influence in the anterior fragment, it must become located close to the oolemma, and it is apparently *not* contained in the symbiotic bacteria (Sander, personal communication).

ANTERIOR DETERMINANTS IN DIPTERAN EGGS

The involvement of anterior egg components in the specification of the longitudinal body pattern is most evident in dipteran eggs. Eggs of chironomid midges have an apparent predisposition to produce, upon various types of experimental interference, abnormal embryos with longitudinal pattern duplications. Embryos of this kind were first described by Yajima (1960, 1964) who observed, after centrifugation or partial UV irradiation of *Chironomus dorsalis* eggs, mirror image duplications of the abdomen without head and thoracic segments ("double abdomens") and, conversely, Janus type heads without thoracic and abdominal segments ("double cephalons"). In each of these aberrant segment patterns, the polarity of a considerable part of the antero-posterior axis is reversed, and pattern elements are replaced by elements that are normally formed only in the other egg half. After appropriate treatment of the eggs, both types of aberrant pattern are perfectly symmetrical in their external and internal morphology except that the germ cells are found only in the posterior parts of the double abdomens and double cephalons (Yajima, 1970; Gollub and Sander, personal communication). A "genocopy" of the double abdomens in *Chironomus dorsalis* has been found in the *bicaudal* mutant syndrome of *Drosophila melanogaster* (Bull, 1966). With respect to the concept of prelocalized cytoplasmic determinants, it is most important that the *maternal* genotype is the controlling factor in the *bicaudal* mutant. The abnormalities must therefore be ascribed to a defective oogenetic condition. A "dicephalic" *Drosophila* embryo which apparently represents a longitudinal mirror duplication of head and thoracic segments has recently been observed by Lohs-Schardin and Sander (1976).

In a chironomid midge of the genus *Smittia*, double abdomens can be produced by several unrelated types of experimental interference including UV irradiation of the anterior pole region (Kalthoff and Sander, 1968), centrifugation (Kalthoff *et al.*, 1977), puncture of the egg at the anterior pole (Schmidt *et al.*, 1975), and application of RNase to the anterior pole region (Kandler-Singer and Kalthoff, 1976). It is hard to conceive that all these different procedures could *de novo* gener-

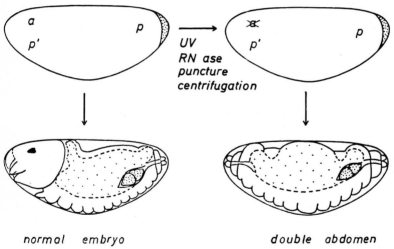

normal   embryo   double   abdomen

Fig. 3. Production of the aberrant pattern "double abdomen" in *Smittia* eggs by different types of experimental interference, all of which are thought to inactivate or displace anterior determinants designated as *a*. These are thought to cooperate with other factors abbreviated as *p'*, in the anterior pole region, so as to allow the formation of head and thorax. Upon inactivation or displacement of *a*, *p'* is assumed to cause abdomen formation in the anterior egg half, as the formation of the normal abdomen is ascribed to factors *p* not further analysed. Note that the germ cells (shaded) are present only in the normal abdomen.

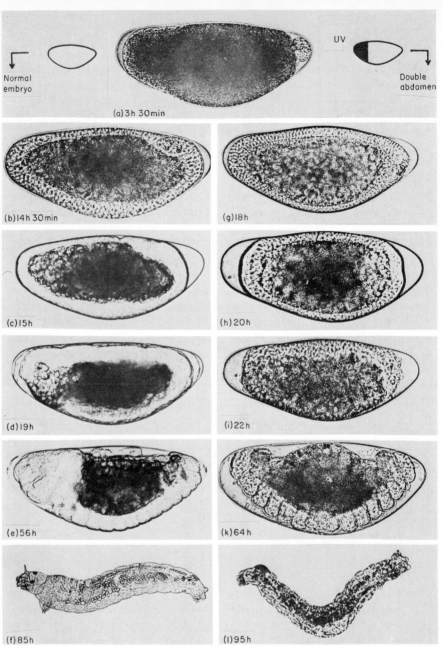

Fig. 4. Development of a normal larva (b–f) and a "double abdomen" (g–l) in *Smittia* eggs. The aberrant pattern was produced in this case by UV irradiation of the anterior egg quarter during intravitelline cleavage (a). The two developmental pathways differ markedly during germ anlage formation (b–d versus g–i), whereas segmentation and histological differentiation appear similar in both cases (e, k). The double abdomen germ anlage develops by fusion of two layers of thick blastoderm formed near the pole regions (h, i). Thus pattern specification may occur independently in each half of the double abdomen. Besides the abnormal germ anlage, the rotations of the embryo around the long egg axis are irregular in the double abdomen, and amnion and serosa are not formed. The two parts of the double abdomen develop in strict symmetry and synchrony unless one partner is handicapped by the lack of space in the egg shell. The egg age is given in hours after deposition at 19°C.

ate specific determinants for the formation of an abdominal end. It is much more likely that the different methods have in common the displacement or inactivation of crucial anterior egg components (designated as *a* in Fig. 3). This view is strongly supported by the fact that the UV induction of double abdomens is photoreversible, photoreversal being commonly ascribed to a light-dependent, enzymatic *repair* of UV damage to nucleic acids (see below). It seems therefore that double abdomen formation does not result from the release of unspecific UV photoproducts, but from the inactivation of specific anterior determinants (Kalthoff, 1976).

The development of double abdomen in *Smittia* eggs does not involve cell death to any extent that could be observed in a high-quality time-lapse film. The anterior abdomen is apparently made from cells that would normally form head and probably thorax (Fig. 4). The formation of the anterior abdomen does not even require interaction with the posterior egg half as can be proved by combined fragmentation and UV irradiation (Sander, unpublished). These results demonstrate that potentially the conditions (designated as *p* and *p'* in Fig. 3) which allow the formation of an abdomen exist not only in the posterior, but potentially also in the anterior egg half. The determinant *p'*, however, is not expressed so long as the anterior determinant *a* is active at the proper place and time. This interpretation implies that the elimination of *a* does not cause the concomitant elimination of *p'*. A speculative scheme showing the putative interaction of the anterior determinant *a*, the posterior determinants *p* and *p'*, and the reacting cells in the egg has been outlined elsewhere (Kalthoff, 1976).

POSSIBLE NATURE OF THE ANTERIOR DETERMINANT IN THE *SMITTIA* EGG

An effort is being made in our laboratory to characterize the origin, localization, and biochemical nature of the anterior determinant(s) in the *Smittia* egg. This is greatly facilitated, because UV irradiation under appropriate conditions causes double abdomen formation in virtually every treated egg (Fig. 5). The dependence of the double abdomen yield upon the irradiated egg area indicates that most of the effective targets for UV are localized in the anterior pole region. They are apparently deposited there during oogenesis since double abdomens can be induced with full yield immediately after egg deposition (Kalthoff, 1971). These results suggest that the anterior determinants may originate from the nurse cell adjacent to the anterior pole of the oocyte during oogenesis.

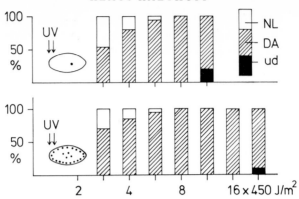

Fig. 5. Experiment demonstrating the extranuclear localization of the effective targets for UV induction of double abdomens. During the first hour after deposition the anterior egg half is entirely devoid of nuclei (upper graph). After nuclear migration, the periplasm is populated with nuclei (lower graph). The double abdomen yield after UV irradiation is independent of the presence or absence of nuclei in the target area (anterior eighth of the egg). NL, normal larvae; DA, double abdomens; ud, undifferentiated eggs. Number of eggs per column ≈ 80. Note that over a certain UV dose range (abscissa) the double abdomen yield is 100%.

The localization of the effective targets is extranuclear, because the yield of double abdomens after UV irradiation is independent of the presence or absence of nuclei in the irradiated egg region (Fig. 5). Moreover, UV irradiation of centrifuged eggs showed that the targets were contained in the clear cytoplasmic fraction (Fig. 6). Within this fraction, the targets were apparently *not* stratified under conditions causing stratification of both endoplasmic reticulum (ER) and mitochondria (Fig. 6). It is concluded that the anterior determinants in the *Smittia* egg are contained in the clear cytoplasm but apparently not associated with mitochondria or ER (Kalthoff *et al.*, 1977).

To obtain a clue to the biochemical nature of the anterior determinants, an action spectrum for the UV induction of double abdomen was determined (Fig. 7). The maximum efficiency per incident quantum was found between 280 and 285 nm, indicating a protein moiety in the effective targets. A minor peak was also found at 265 nm, suggesting a nucleic acid (see Kalthoff, 1973). Further independent evidence for the involvement of a nucleic acid is the photoreversibility of the UV induction of double abdomen (Fig. 8). Photoreversal is defined as the mitigation of UV effects by subsequent irradiation with light of longer wavelength. The latter is effective only after but not before UV, and has to be received by the same egg area. These results, together with the wavelength dependence, temperature dependence, and dose

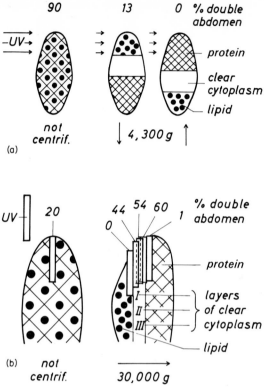

Fig. 6. Analysis of target localization for UV induction of double abdomens by irradiation of centrifuged eggs. Accumulation of protein or lipid in the irradiated anterior quarter caused a considerably decrease in the double abdomen yield (a). Removal of protein and lipid from the irradiated anterior egg region enhanced double abdomen formation (b). Stratification of mitochondria in layer II and endoplasmic reticulum (ER) in layers I and II was not parallelled by corresponding differences in the double abdomen yields after irradiation of the respective layer. The maximum yield of double abdomens was found after UV irradiation of layer III which apparently contained little if any organelles except ribosomes.

rate saturation of photoreversal after UV induction of double abdomens, indicate that the underlying molecular mechanism belongs to the well-known type of "direct photoreactivation" (Kalthoff, 1973). This is commonly ascribed to light-dependent, enzymatic splitting of UV induced pyrimidine dimers in nucleic acids (Cook, 1970). In fact we have recently observed that light after UV removes uridine dimers in the RNA of the egg, and that the photoreactivable sector, after UV inactivation of eggs at different wavelengths, is correlated with the

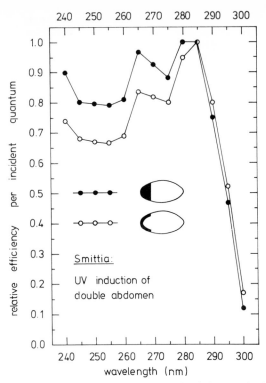

Fig. 7. Action spectrum for UV induction of double abdomens in *Smittia* eggs. The relative efficiency per incident quantum was determined by dose-effect curves at each wavelength. Wavelength dependent shielding was taken into account assuming target localization throughout the irradiated anterior egg quarter (●) or only within a superficial layer of periplasm (○). Peaks at 285 and 265 nm indicate a nucleic acid–protein complex as the effective targets.

amount of uridine dimers produced at that wavelength (Jäckle and Kalthoff, unpublished).

Application of RNase to the anterior pole of *Smittia* eggs causes the same switch in the developmental program as UV irradiation (Kandler-Singer and Kalthoff, 1976). By puncturing eggs at the anterior pole during submersion in RNase, the aberrant segment pattern "double abdomen" can be produced with a maximum yield of 29% (Fig. 9). As controls, eggs were punctured in water, denatured RNase, and oxidized RNase; none of these produced double abdomens. Fragments of RNase S (subtilisin modified RNase A) were inactive, whereas recombination of the fragments resulted in a double abdomen yield comparable with that of the native enzyme (Table 1). This is

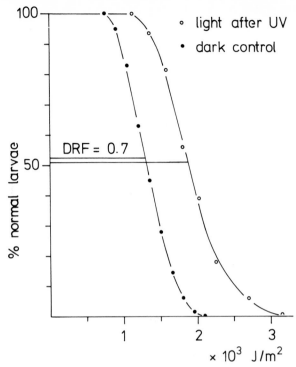

Fig. 8. Dose-effect curves for UV induction of double abdomens with and without sub-sequent exposure of eggs to photoreverting light. The effect of a given UV dose plus light was equivalent to the effect of the 0.7 fold UV dose in the dark control, i.e. the dose reduction factor (DRF) was 0.7. Number of eggs per point ≈ 100.

regarded as reliable proof that the switch in pattern formation resulted from RNase activity introduced into the egg. Neither application of other enzymes to the anterior pole nor application of RNase to other egg regions produced double abdomens in significant yields.

The data obtained so far are compatible with the idea that ribonu-cleoprotein (RNP) particles act as common targets for both UV and RNase. One might speculate that "masked" maternal messenger RNA acts as an anterior determinant in the egg of *Smittia*. Protein com-ponents in such ribonucleoprotein particles may serve to inhibit both premature translation and RNase digestion prior to translation (Spirin, 1969). On the other hand, UV damage to the RNA component might be enhanced rather than prevented by the protein components. This interpretation would help to explain the poor double abdomen yield resulting from RNase application prior to nuclear migration (Fig. 9)

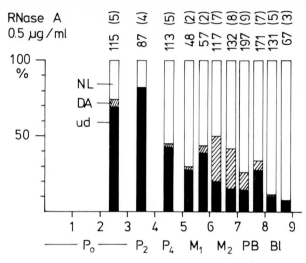

hours after deposition / stages at 20° C

Fig. 9. Production of the aberrant pattern "double abdomen" in *Smittia* by puncturing eggs at the anterior pole during submersion in RNase. Within a small concentration range (0.5 to 0.8 µg/ml), and after puncturing at late nuclear migration (M_2), the double abdomen yield reached a maximum of 29%. The yield decreased rapidly around preblastoderm stage and was also poor prior to nuclear migration. NL, normal larvae; DA, double abdomens; ud, undifferentiated eggs. Figures on top of each column indicate the number of analysed eggs and, in brackets, the number of clusters from which the eggs were derived.

TABLE I

Production of the aberrant pattern "double abdomen" in *Smittia* eggs (5.5 to 7.5 h after deposition) by puncturing at the anterior pole during submersion in RNase. NL, normal larvae; ud, undifferentiated eggs; DA, double abdomens

Agent	Conc. (µg/ml)	Batches	Eggs (total)	NL	ud (numbers)	DA	DA (% tot.)	DA (% surv.)
RNase A	0.8	19	382	95	203	84	22	47
RNase A denatured	0.5–1.0	23	476	436	40	—	0	0
RNase A oxidized	10.0	4	93	93	—	—	0	0
H_2O		8	158	151	7	—	0	0
RNase S	1.0	9	189	76	63	50	26	40
S-Peptid	0.16	9	205	195	10	—	0	0
S-Protein	0.84	4	81	77	3	1	1	1
	1.68	7	170	151	15	4	2	3
S-Peptid plus S-Protein	0.16 + 0.84	8	184	69	77	38	21	36

as opposed to the almost constant production of double abdomens by UV, from egg deposition until late nuclear migration (Kalthoff, 1971). It is also in accordance with this speculation that the light-dependent respecification of the normal segment pattern after UV irradiation at early stages can be delayed until nuclear migration (Kalthoff *et al.*, 1975).

CONCLUDING REMARK

The gross change in the morphogenetic program of the *Smittia* embryo, which is apparently caused by function or disfunction of stored RNP particles, is probably not as unique as it may appear. Homeotic mutations which switch morphogenetic pathways from, for example, antenna to leg or haltere to wing in *Drosophila* may be regarded as related phenomena. Such switches in morphogenetic pathways are apparently caused by disfunction of single genes (Lewis,1963; Garcia-Bellido, 1975). As such genes might become activated by RNAs or their translational products (Davidson and Britten, 1971), and our data suggest that stored RNP particles act as anterior determinants in *Smittia* eggs, the molecular mechanisms underlying these phenomena may be similar.

ACKNOWLEDGEMENTS

I wish to thank Prof. K. Sander and Dr C. M. Bate for reading the manuscript critically. Work in the author's laboratory is supported by the Deutsche Forschungsgemeinschaft, SFB 46.

REFERENCES

Bull, A. L. A genetic factor which affects the polarity of the embryo in *Drosophila melanogaster. J. exp. Zool.* **161**, 221–242 (1966).

Cook, J. S. Photoreactivation in animal cells, in A. C. Giese (ed.), Photophysiology, Vol. 5, pp. 191–223, Academic Press, London, New York (1970).

Counce, S. J. Causal analysis of insect embryogenesis, in S. J. Counce and C. H. Waddington (ed.), Developmental Systems, Vol. 2, pp. 1–156, Academic Press, London, New York (1973).

Davidson E. H. and Britten, R. J. Note on the control of gene expression during development. *J. theor. Biol.* **32**, 123–130 (1971).

Garcia-Bellido, A. Genetic control of wing disc development in Drosophila, in Cell Patterning, Ciba Foundation Symposium 29, pp. 241–263, Associated Scientific Publishers, Amsterdam (1975).

Gehring, W. J. Genetic control of determination in the Drosophila embryo, in F. H.

Ruddle (ed.), Genetic Mechanisms of Development, pp. 103–128, Academic Press, London, New York (1973).

Günther, J. Entwicklungsfähigkeit, Geschlechtsverhältnis und Fertilität von Pimpla turionellae (Hymenoptera, Ichneumonidae) nach Röntgenbestrahlung oder Abschnürung des Eihinterpols. Zool. Jb. Anat. **88**, 1–46 (1971).

Illmensee, K. and Mahowald, A.P. Transplantation of posterior polar plasm in Drosophila. Induction of germ cells at the anterior pole of the egg. Proc. nat. Acid. Sci. (Wash.) **71**, 1016–1020 (1974).

Kalthoff, K. Position of targets and period of competence for UV induction of the malformation "double abdomen" in the egg of Smittia spec. (Diptera, Chironomidae). Wilhelm Roux' Arch. Entwickl.-Mech. Org. **168**, 63–84 (1971).

Kalthoff, K. Action spectra for UV induction and photoreversal of a switch in the developmental program of the egg of an insect (Smittia). Photochem. Photobiol. **18**, 355–364 (1973).

Kalthoff, K. Specification of the antero-posterior body pattern in insect eggs, in P. A. Lawrence (ed.), Insect Development, pp. 53–75, Blackwell, Oxford (1976).

Kalthoff, K., Hanel, P. and Zissler, D. A morphogenetic determinant in the anterior pole of an insect egg (Smittia spec., Chironomidae, Diptera). Localization by combined centrifugation and UV irradiation, Develop. Biol. **55**, 285–305 (1977).

Kalthoff, K., Kandler-Singer, I., Schmidt, O., Zissler, D. and Versen, G. Mitochondria and polarity in the egg of Smittia spec. (Diptera, Chironomidae): UV irradiation, respiration measurements, ATP determinations and application of inhibitors, Wilhelm Roux' Arch. Entwickl.-Mech. Org. **178**, 99–121 (1975).

Kandler-Singer, I. and Kalthoff, K. RNase sensitivity of an anterior morphogenetic determinant in an insect egg (Smittia spec., Chironomidae, Diptera). Proc. nat. Acad. Sci. (Wash.) **73**, 3739–3743 (1976).

Kalthoff, K. and Sander, K. Der Entwicklungsgang der Missbildung "Doppelabdomen" im partiell UV-bestrahlten Ei von Smittia parthenogenetica (Dipt., Chironomidae). Wilhelm Roux' Arch. Entwickl.-Mech. Org. **161**, 129–146 (1968).

Kauffman, S. Control circuits for determination and transdetermination. Science **181**, 310–318 (1973).

Okada, M., Kleinman, I. A. and Schneiderman, H. A. Restoration of fertility in sterilized Drosophila eggs by transplantation of polar cytoplasm. Develop. Biol. **37**, 43–54 (1974).

Lewis, E. B. Genes and developmental pathways. Amer.Zool. **3**, 33–56 (1963).

Lewis, J., Slack, J. M. W. and Wolpert, L. Thresholds in development. J.theor.Biol. **65**, in press (1977).

Lohs-Schardin, M. and Sander, K. A dicephalic monster embryo of Drosophila melanogaster. Wilhem Roux' Arch. Entwickl.-Mech. Org. **179**, 159–162 (1976).

Sander, K. Analyse des ooplasmatischen Reaktionssystems von Euscelis plebejus Fall. (Cicadina) durch Isolieren und Kombinieren von Keimteilen. II. Mitteilung: Die Differenzierungsleistungen nach Verlagern von Hinterpolmaterial. Wilhelm Roux' Arch. Entwickl.-Mech. Org. **151**, 660–707 (1960).

Sander, K. Pattern specification in the insect embryo, in Cell Patterning, Ciba Foundation Symposium 29, pp. 161–182, Associated Scientific Publishers, Amsterdam (1975).

Sander, K. Specification of the basic body pattern in insect embryogenesis. Advanc. Insect Physiol. **12**, 125–238 (1976).

Schmidt, O., Zissler, D., Sander, K. and Kalthoff, K. Switch in pattern formation after puncturing the anterior pole of Smittia eggs (Chironomidae, Diptera). Develop. Biol. **46**, 216–221 (1975).

Spirin, A. S. Informosomes, *Europ. J. Biochem.* **10**, 20–35 (1969).

Yajima, H. Studies on embryonic determination of the harlequin-fly, *Chironomus dorsalis*. I. Effects of centrifugation and of its combination with constriction and puncturing, *J. Embryol. exp. Morph.* **8**, 198–215 (1960).

Yajima, H. Studies on embryonic determination of the harlequin-fly, *Chironomus dorsalis*. II. Effects of partial irradiation of the egg by ultraviolet light. *J. Embryol. exp. Morph.* **12**, 89–100 (1964).

Yajima, H. Study of the development of the internal organs of the double malformations of *Chironomus dorsalis* by fixed and sectioned materials. *J. Embryol. exp. Morph.* **24**, 287–303 (1970).

Genetics of Cell Type Determination

Robert Briggs

Department of Zoology, Indiana University, Bloomington, Indiana 47401, USA

INTRODUCTION

During development the cells of the various regions of the embryo first become determined as to type, and later express this determination in the form of differentiated morphological and molecular phenotypes. It can be assumed on general grounds that this determination of the parts of the embryo must be endowed with exceptional stability—otherwise development would be more error-prone than we observe it to be. The most conclusive evidence that determination is a highly stable, heritable, property of individual cells comes from studies of cells in clonal culture. For example, cartilage cells inherit their specific determination over many cell generations, whether it is overtly expressed or not (Coon, 1966). The same is true of myoblasts (Konigsberg, 1963; Richler and Yaffe, 1970), retinal pigment cells (Cahn and Cahn, 1966), neuroblastoma cells (Schubert *et. al.*, 1971), and melanoma cells (Davidson *et. al.*, 1966). Other lines of evidence, particularly from studies of imaginal disc determination in *Drosophila* (Hadorn, 1966; Gehring, 1972), all point to the same conclusion—that the various cell types making up the organism are highly stabilized in their determined states, and that this determination is a heritable property of individual cells. There are well-known exceptions to this rule, cases in which cells do change from one determined state to another, but these are limited in number and in the types of changes exhibited.

The search for molecular-genetic mechanisms of cell type determination is now on in earnest, and there is no shortage of theoretical and experimental approaches to the problem. All of these approaches depend upon the position taken with respect to one fundamental question—the question whether the genome is identical in different types of somatic cells. It appears that the position taken by most investigators is that this issue is now settled—that the evidence is convincing that there are no irreversible genetic changes accompanying cell type determination. Thus the problem becomes entirely one of gene regulation. It is certainly so that the evidence pointing to this conclusion is very persuasive; yet it will do no harm to scrutinize it to see how complete it is. That is one of the purposes of this chapter. The second purpose is to review some recent experiments bearing on this problem—experiments dealing with gene control in somatic cell hybrids and in somatic nuclei transplanted into oocytes.

NUCLEAR TRANSPLANTATION AND OTHER STUDIES OF GENE CONTENT OF SOMATIC CELLS

Four types of evidence bearing on the problem of the gene content of somatic cells will be considered. Three of these, from cytological, molecular, and somatic cell hybridization studies, will be dealt with rather briefly. The fourth, from nuclear transplantation studies, will be presented in greater detail.

Cytological and molecular studies

The cytological studies of Beerman (1952, 1956) provide some of the most impressive evidence indicating that different cell types have the same genes, but exhibit different patterns of gene activity. Beerman carried out a detailed comparison of the banding patterns exhibited by the polytene chromosomes of several cell types in *Chironomus* larvae. The patterns were shown to be the same in cells of the salivary gland, midgut, Malpighian tubules, and rectum. Since it now appears that, in some cases at least, individual bands can be equated to individual genes (Judd *et. al.*, 1972), this result indicates that the same genes are present in all of the cell types Beerman analysed. Beerman also analysed "puffing" patterns. Puffing is now known to be an indicator of gene activation and to involve a spinning out of the chromosomal fibres and an activation of RNA synthesis. Beerman's analysis showed that while the same bands are present, these bands show characteristically different puffing patterns in different cell types.

There are some reports in the cytological literature that tell a different story, indicating that the genome may be changed as cells differentiate. For example, polytene nuclei of brain ganglia cells in *Drosophila hydei* show differential replication of euchromatic DNA relative to the DNA of the chromocentre. Since the chromocentre consists of the heterochromatic portions of the X and the Y, this result indicates that brain ganglia cells contain a genome unbalanced in favour of autosomal genes. Other cases of differential replication of portions of the genome are also on record (Pavan, 1965; Endow and Gall, 1975). However, the main reservation in the interpretation of banding patterns is that changes in genes at the molecular level would not be detectable cytologically. The vast majority of mutations in *Drosophila* do not lead to any visible changes in the bands of the polytene chromosomes, and since this is so we can assume that molecular changes associated with cell type determination might also be undetectable. In view of this reservation the most we can conclude from the cytological evidence is that much of it is consistent with, but does not prove, the identity of the genome in different cell types.

The molecular studies of DNA's of different tissues have led to a conclusion like that just given above for the cytological work. These studies have included determinations of the total amounts of DNA per chromosome set, and of the characteristics of these DNA's. Of particular interest are the attempts to determine, by means of DNA–DNA hybridization, whether DNA's of different cell types contain different nucleotide sequences. Differences have been reported to exist between somatic and germ line cells in *Ascaris* (Tobler *et. al.*, 1972), but none were found among the DNA's of different types of somatic cells of the mouse (McCarthy and Hoyer, 1964). However, the sensitivity of the hybridization methods is such that significant differences could go undetected. As with the cytological work, the results of the DNA studies are consistent with, but do not prove, the identity of the genome in different cell types.

Nuclear transplantation

The most rigorous test of gene content and function is that provided by nuclear transplantation. To the author's knowledge, the idea for this type of experiment was first proposed by Spemann (1938). Spemann had shown earlier that when a newt's egg is constricted into two connected halves prior to first cleavage the zygote nucleus is at first limited to one half, which may cleave into 16 to 32 cells before one of the daughter nuclei migrates into the other half. The important result is that the half experiencing the delayed nucleation can develop into a complete normal embryo, showing that there is no segregation or elimination of genes during the early cleavages. Spemann then addressed the question whether changes in the genetic make up of somatic nuclei might occur at later stages of development, and he explicitly proposed that decisive information on this question might be obtained were it possible to transfer the nuclei into enucleated eggs. For if the nucleus of a differentiated somatic cell were shown to be capable of promoting complete and normal development, then it would be demonstrated conclusively that all of the genes required in ontogeny must be present and must be capable of being activated in the normal ways by the egg cytoplasm.

To actually carry out the experiment proposed by Spemann required the development of methods for activating and enucleating recipient eggs, isolating donor cells, and transferring their undamaged nuclei into the recipient eggs. The methods were worked out and tested first on nuclei of blastula cells of the frog, *Rana pipiens* (Briggs and King, 1952). Blastula cells were chosen because they are known to be undetermined, and on this basis it was assumed that their nuclei might be capable of promoting normal development if properly transferred into enucleated eggs. The assumption proved correct. In the first series of experiments approximately one-third of the eggs receiving transplanted blastula nuclei cleaved normally, and the majority (60%) of these developed to advanced embryonic or larval stages. Subsequently this result was improved upon and it was clearly established that blastula and early gastrula nuclei are capable of promoting normal development when transplanted back into enucleated eggs. This having been established, the way was open to transplant nuclei from older embryos to test whether they undergo changes in the course of cell type determination. The results of such tests, carried out in several laboratories on a variety of amphibians, agree in showing that as development proceeds beyond gastrulation the nuclei of the various parts of the embryo show an ever decreasing capacity to elicit normal development when transplanted into enucleated eggs. However, a minority of nuclei from cells of advanced embryos, or even nuclei from cells of adult frogs, promote extensive development. In what follows we wish to examine two questions. The first of these concerns the extent of development promoted by nuclei from cells known to be fully determined—i.e. to conform to a well-defined morphological and molecular phenotype. If any of these nuclei can be shown to promote complete and normal development, then the case for the genetic integrity of the determined cells is made, regardless of the reasons why the majority of the transplanted nuclei may fail to promote normal development. Of course, the second question does concern these failures and the problem of their relevance to the processes of cell change during development.

Development capacity of nuclei of determined cells

The early work on this problem suggested that as cells become determined their nuclei lose the capacity to promote normal development when transplanted into enucleated eggs. Nuclear transfers were done from ectoderm, central nervous system, optic vesicle, notochord, somite, lateral plate mesoderm, and endoderm; and from species of *Rana, Bufo, Ambystoma, Triturus, Pleurodeles*, and *Xenopus* (see reviews by Briggs and King, 1959; Gallien, 1966; Gurdon, 1974). With the exceptions to be noted below, all of this work agreed in showing that once these embryonic regions were clearly set aside their nuclei showed reduced developmental capacities. Analysis of the patterns of abnormalities exhibited by nuclear-transplant embryos suggested that the patterns might be related to the cell type providing the nuclei, but the specificity of the relationship could not be established. The other main points emerging from this early work were as follows: (1) The abnormal patterns of development exhibited by nuclear-transplant embryos were stable—i.e. they were exhibited unchanged in clones of embryos produced by serial nuclear transplantation (King and Briggs, 1956). (2) The abnormalities could not be corrected by parabiosing nuclear-transplant embryos with normal ones (Briggs *et. al*, 1960). (3) Most interestingly, the abnormalities produced by a somatic (endoderm) nucleus in the development of the recipient egg were *not* corrected when a haploid set of egg chromosomes was combined with the diploid set from the endoderm (Subtelny, 1965). (4) Most of the nuclear-transplant embryos exhibited changes in karyotype (Briggs *et al.*, 1960). Some of these points will be discussed below in this chapter.

An important exception to the generality of results summarized above was reported for *Xenopus* by Gurdon (1962). While *Xenopos*, along with the other amphibia, shows a steady decrease in the developmental capacity of the nuclei with advancing embryonic age, this decline occurs less rapidly in *Xenopus* than in the other species (Fischberg *et. al.*, 1958). Most importantly, it appears that even in young tadpoles some nuclei retain the capacity to promote complete development of recipient eggs. This has been definitely established for intestinal nuclei from stage 46–48 tadpoles—tadpoles which are some 4 to 7 days old, just beginning to feed, with intestinal cells columnar in shape, mostly free of yolk, and with a striated border. Nuclei from intestinal cells of this type were transplanted into 726 eggs (Gurdon, 1962). Ten of these eggs (1.5%) developed into feeding tadpoles. Furthermore, it was shown that some of the partially cleaved recipient eggs, which cannot themselves develop normally, nonetheless contain nuclei which, if transferred into a second set of eggs, promote development to feeding tadpole stages. Altogether, some 7% of intestinal cell nuclei were calculated to retain the capacity to promote development to larval stages, and some of these larvae were reared to sexual maturity and were shown to be fertile (Gurdon and Uehlinger, 1966).

In order to eliminate the possibility that a minority of undetermined cells might be providing the nuclei that promote normal development, Gurdon and his colleagues have recently transplanted nuclei from cultured skin cells of *Xenopus* (Gurdon *et. al.*, 1975). Under appropriate conditions more than 99.9% of the cells growing out from skin explants can be shown to contain immunoreactive keratin, and therefore to be of a well-defined, determined cell type (Reeves and Laskey, 1975). Nuclei transplanted from these cells elicit cleavage in about 30% of the recipient eggs. None of these original recipients develop beyond neurulation, but if their nuclei are again transferred into enucleated eggs some of these eggs do develop to advanced embryonic stages. For example, in one series of experiments 129 initial nuclear transfers were made, leading to the formation of 28 partial and 12 complete blastulae—none of which developed beyond gastrulation. Eleven of the partial blastulae provided nuclei for transfer into a new series of eggs. Eleven clones, totalling 371 eggs, were produced. Twenty of these eggs, distributed among 6 clones, developed to advanced embryonic stages approximating the Nieuwkoop–Faber stage 42. These embryos possessed functional muscle, nerve, a heartbeat and blood circulation, eyes with lenses, and other types of differentiated cells. However, all embryos stopped developing at this stage, became oedematous, and died without feeding. That these advanced embryos had developed with the transplanted skin cell nuclei, and not with the egg nuclei, was demonstrated through the use of the nucleolar marker. Donor cell nuclei were 1-nu, and were injected into eggs from 2-nu females which had been UV-irradiated to inactivate the egg nucleus. The advanced embryos in 4 of the 6 clones mentioned above were shown to be 1-nu diploid, or 2-nu tetraploid. Given this result the only possibility of error would occur if the UV inactivation of the recipient egg nucleus were to fail, and a new mutation eliminating one nucleolus were to occur in combination with a doubling of the egg chromosomes.[1] Such a mutation might occur prior to maturation and be combined with a chromosomal doubling through retention of a polar body, or less probably the egg might complete maturation, undergo a replication of the haploid set of chromosomes, and then a mutation to give a 1-nu diploid embryo.

---

[1] Gurdon and Laskey (1970) have considered this possibility and have carried out a test, the results of which indicate that the error being discussed here does not occur in their experiments.

The mutation leading to the elimination of the nucleolus is not particularly rare, occurring in 1.5% of three populations of frogs sampled by Blackler (1968). However, the chance that it would be combined with the other factors needed to give errors of the type mentioned above can be regarded as insignificant.

Another set of experiments on nuclei of a well-defined cell type has been reported recently. Wabl et al. (1975) isolated lymphocytes from immunized adult frogs by adsorbing them onto a nylon grid bearing an appropriate hapten. At least 98% of the cells sticking to the grid were shown to be immunoglobulin-bearing lymphocytes. Nuclei from these lymphocytes promoted cleavage in about 20% of the enucleated eggs into which they were transplanted. None of these initial transfers led to development beyond gastrulation, but retransfers to a second series of eggs led in some cases to more advanced development. The most advanced embryos reached stage 43–44 and possessed a variety of different cell types, but all died at this stage. These results are almost exactly like those obtained by Gurdon et al. in their experiments on skin cell nuclei.

We now review briefly the results of several experiments, in addition to the ones described above, in which the developmental capacity of nuclei from cells of adult frogs has been tested in the usual way, by transplantation into enucleated eggs. The first experiments on nuclei of adult animals were those of King and McKinnell (1960) and King and DiBerardino (1965). In the latter paper, particularly, it was shown that nuclei from the kidney adenocarcinoma of the frog, *Rana pipiens*, promote development to advanced embryonic stages (stage 24–25, Shumway). None of the embryos started feeding. Similar results were reported by McKinnell et al. (1969) on the developmental capacity of nuclei from triploid tumour cells. Somewhat more advanced development was observed on the part of eggs into which had been transplanted nuclei of spermatogonia, this also with *Rana pipiens* (DiBerardino and Hoffner, 1971). The most advanced nuclear transplant in this series was a young tadpole which fed for a few days before dying. This is the only instance on record, so far as I am aware, in which a nucleus from a cell of adult origin has promoted development to a functional larval stage—i.e. to the feeding stage. The nucleus in this case came from a germ line cell.

Nuclei of a variety of cultured cells have been tested by Laskey and Gurdon (1970) and by Kobel *et al.* (1973). Cultures providing the donor cells were derived from adult kidney, lung, heart, skin, and from established cell line A-8, derived originally from liver. All of the cultures were from *Xenopus laevis*, and all experiments involved serial nuclear transfers. That is to say, the original transfers of nuclei from the cultured cells gave limited development. Retransfers from blastulae of the first series gave more advanced development on the part of some of the recipient eggs, to a late tailbud stage for A-8 nuclei, and to an advanced embryonic stage (stage 41) for all of the cultured cell nuclei. None of the embryos developed to the feeding larval stage. It should be mentioned that this result holds for all tests so far reported for nuclei of normal cells of adult frogs. Comparable tests of nuclei from cultured epithelial cells of advanced embryos (stage 40) gave similar results except that in three cases, out of 3546 transfers, the recipient eggs developed into adult frogs (Gurdon and Laskey, 1970).[2]

INTERPRETATION OF NUCLEAR TRANSPLANTATION STUDIES

Do nuclei of determined cells retain the complete unaltered genome?

The best evidence on this problem comes from studies of nuclei from cells which individually display unmistakable phenotypes. Among those so far tested, melanophores, ciliated ectodermal cells (Kobel et al., 1973), keratinized skin cells (Gurdon et al., 1975), and immunoglobulin-bearing lymphocytes (Wabl et al., 1976) seem to satisfy best this requirement for a clear-cut phenotype. The most extensive development was promoted by lymphocyte nuclei and by nuclei of cultured skin cells (see above). Some of the eggs into which these nuclei were transplanted developed into advanced embryos, possessing a variety of differentiated cell types. However, none of the embryos transformed into feeding larvae. The fact that the development went as far as it did has been taken to mean that nuclei of specialized cell types retain a full complement of genes capable of functioning normally in development (Gurdon et al., 1975). While this is likely to be correct, there is another interpretation that has not yet been excluded. According to this alternative view, the development of the test eggs may depend to some degree on RNA's produced by the maternal nucleus and stored in the egg cytoplasm. If this is so it would mean that some of the genes of the trans-

---

[2] I have restricted this review to nuclear transplantation studies done with amphibians, for the reason that most of the work in the field has been done with these organisms. However, it should be mentioned that nuclear transfers have been done successfully in *Drosophila*. In particular, Ilmensee (1973) has shown that nuclei from various regions of the early gastrula are capable of promoting the development of recipient eggs to one of the three larval instars, or rarely to the pupal stage. Pole cells or gonads from these recipients, grafted into normal hosts, give gametes which, in suitable crosses, produce embryos that develop normally to adulthood. The nuclei of the various regions of the early gastrula are thus demonstrated to the totipotent. The gastrula in *Drosophila* is known to exhibit regional determination, but the state of determination of the individual cells is not known.

planted nucleus might remain inactive and development still go to advanced stages. An example of such a situation is provided by the homozygous anucleolate mutant which lacks the genes for rRNA synthesis but inherits enough rRNA in the oocyte cytoplasm to carry it through to an advanced embryonic stage (Brown and Gurdon, 1964). Are there enough of other kinds of stored RNA's to permit extensive development in the absence of the function of the genes coding for them? Specific answers cannot be given to this question, but we can say that transcripts of a large number of different genes are known to be present in the mature oocyte and to persist throughout embryonic development (Hough et al., 1973). Their role is unknown, but the fact that they persist to stages at which nuclear-transplant embryos stop developing means that there is an unresolved problem, and that is to unravel the relative contributions of stored (maternal) and newly synthesized RNA's to the development of the nuclear-transplant embryos.

Two considerations may argue against the idea that stored RNA's play an important role in later stages of embryonic development (Gurdon et al., 1975). One of these derives from the observation that development beyond cleavage in amphibians is known to require RNA synthesis on the part of the zygotic nuclei. The same can be said for nuclear-transplant embryos. They also will not develop beyond gastrulation unless the transplanted nuclei synthesize RNA, but this does not mean necessarily that all genes must be present or active. It is possible that certain classes of genes in the transplanted nuclei may be capable of functioning and others may not, and this could depend on the types of controls to which the genes were exposed in the donor cells prior to transplantation.

A second consideration against the idea of cell differentiation being dependent on regulation of translation of stored mRNA's comes from the work of Gurdon et al. (1974) and Woodland et al. (1974) on injected globin mRNA's. Briefly, globin mRNA's injected into fertilized Xenopus eggs persist throughout embryonic development and are translated at all stages of development and in both erythropoietic and non-erythropoietic portions of the embryo. The fact that the translational machinery is generally present argues against a translational and for a transcriptional control of cell differentiation. This argument is based on the assumption that injected mRNA's are handled in the same way as are the RNA's produced in the cell. This may or may not be so. RNA's produced during oogenesis and stored in the oocyte might be specially packaged, or complexed with macromolecules that regulate their accessibility for translation. Thus the experiments on injected mRNA's do not exclude an important role for stored mRNA's in amphibian development. In this connection it may be mentioned that some cases are known in which translational control is important. In tunicates it appears that development to the tadpole stage occurs entirely on the basis of stored RNA's (Lambert, 1971). And in some embryos which require new RNA synthesis to proceed beyond gastrulation, stored (oocyte) RNA may still be involved in the specification of structures which differentiate during post-gastrula development. Thus, the polar lobe of mollusc embryos contains either specific stored mRNA's, or regulators of the translation of stored RNA's involved in the development of the post-trochal region of the embryo (Newrock and Raff, 1975).

Why do transplanted nuclei of determined cells fail to promote the complete development of recipient eggs?

As we have seen in an earlier section of this chapter, a large number of tests have been done on nuclei of adult origin with a consistent result. In the best cases, development of the test eggs proceeds to the pre-larval stage and stops. Nuclear transplantation is a rather difficult technique and it may be that nuclear damage, or failure of integration of chromosomal and cytoplasmic division cycles, accounts for the cessation of development. In fact, it is known that most of the abnormal nuclear-transplant embryos display chromosomal abnormalities, and that these abnormalities arise during the first cleavages following nuclear transfer. Apparently some of the heterochromatic regions of the transplanted nuclei fail to replicate in time for the first division of the recipient egg, and this leads to the formation of chromosome bridges and breaks (DiBerardino and Hoffner, 1970). However, the embryos showing the most advanced development do not show detectable changes in karyotype (Wabl et al., 1976), yet they fail to continue their development. Chromosomal changes not visible cytologically could of course have arisen as a consequence of the nuclear transfer. An alternative interpretation, not excluded by the evidence at hand, is that nuclei of determined cells have undergone some change as a part of the determination process—a change which is not reversed when the nuclei are transplanted, and which accounts for the incompleteness of development.

On general grounds we would expect cell type determination to involve an extremely reliable mechanism permitting some genes to act and ensuring that others are kept inactive. The mechanism for keeping genes inactive must be especially important. For example, genes specifying pituitary hormones must act only in the relatively small number of cells in the pituitary itself, and be kept inactive elsewhere.[3] Otherwise even a very low level of synthesis in the great mass of cells of the body as a whole should give detectable hormone levels in hypophysectomized animals, and this is not the case. More specific support for a highly stabilized heritable mechanism of nuclear differentiation is provided by the work of Sonneborn (1954). In *Paramecium* a nucleus genetically capable of determining two mating types is restricted to determine only one. This occurs at only one point in the life cycle, shortly after fertilization, as a result of exposure to a specific cytoplasmic factor. Once established this nuclear determination is faithfully inherited and cannot be changed, even by a subsequent exposure to the cytoplasmic factor controlling the opposite mating type. Comparable nuclear determinations also occur with respect to the function of genes controlling mating type in *Tetrahymena* (Nanney, 1956) and trichocyst discharge in *Paramecium* (Sonneborn, unpublished). Recently, heritable cytoplasmically induced changes in nuclear function have been found also in the axolotl (Brothers, 1976). Exposure of nuclei to a cytoplasmic factor (o^+ substance) at a specific time late in cleavage induces a heritable activation of genes required for gastrulation and later development. This activation persists in the absence of the factor inducing it. Conversely, if nuclei go through the critical period without being exposed to o^+ substance they lose the capacity to be induced when the substance is provided to them later. The main point of all of this is to indicate that nuclei may be restricted in their function in ways so stabilized as to be irreversible or very difficult to reverse, and to suggest that this may be a factor limiting the development of nuclear-transplant embryos.

Can restrictions in the developmental capacity of transplanted somatic nuclei be reversed?

Two types of experiments will be described. The first tests whether restrictions in the developmental capacity of transplanted somatic nuclei can be corrected by the addition of a normal haploid set of chromosomes from the egg. If the restrictions are due to aneuploidy originating as a consequence of the nuclear transfer (see above), or to simple deficiencies whatever their origin, then the addition of an intact set of chromosomes should bring about an improvement. Experiments to test this point have been reported by Subtelny and more recently by Aimar (1972). In the first of the papers dealing with this problem Subtelny and Bradt (1960, 1961) showed that it is possible to produce "hybrids"

---

[3] This particularly cogent example was drawn to my attention by my colleague, Dr Arthur Koch.

between somatic cells and germ cells by transplanting somatic nuclei into eggs which are allowed to retain their own nuclei. Following the transfer the diploid somatic nucleus and the haploid egg nucleus move to the epicentre of the egg and fuse, just as do the sperm and egg nuclei in normal fertilization. If the transplanted nucleus is from an undetermined blastula cell the resulting triploid hybrid develops normally. In a later paper Subtelny (1965) reports on an extensive set of experiments in which he compares two groups of eggs—one group developing with chromosomes derived solely from transplanted post-gastrula endoderm nuclei and the second group developing with "hybrid" nuclei containing endoderm and egg chromosomes. The experiments were designed so that the only difference between the two groups was in the presence or absence of the egg chromosomes. This was accompiished by using cloned endoderm nuclei and transplanting them into eggs of a single frog, some eggs being enucleated and some not. The important result from these experiments was that the two groups of eggs exhibited the same abnormalities and restrictions in their development. In other words, the addition of the haploid set of chromosomes from the egg did *not* correct the restrictions in the developmental capacity of the endoderm nuclei, as would have happened had these restrictions been simply a consequence of aneuploidy or other deficiencies. Rather it appears that endodern nucleus may be exerting a negative control over the function of the egg nucleus in the hybrid—a result similar to that seen in certain somatic cell hybrids (see below).

Another approach to this problem of reversibility stems from a model provided by nature. One of the best examples of reversal of specialization of nuclear function is seen in spermatogenesis (Bloch and Hew, 1960; Bloch and Brack, 1964). Prior to the beginning of the meiotic divisions the genome of the spermatocyte is synthesizing RNA's for a variety of sperm-specific products (acrosomal enzymes, protamines, etc.). This synthesis of RNA is then shut down as "protamines", which are now being synthesized in the cytoplasm, move into the nucleus displacing the proteins previously present. One result of this shift in nuclear proteins is a very tight packing of transcriptionally inert chromosomes in the sperm head. Another result is the transformation of a nucleus previously specialized for the synthesis of sperm-specific products into one potentially capable of engaging in any kind of synthesis. The completion of this transformation requires the removal of protamines from their association with the sperm chromosomes. The egg must possess a very efficient mechanism for doing this; protamine is lost from the sperm nucleus shortly after it enters the egg in fertilization, and the nucleus then becomes totipotent again.

With this model in mind some attempts have been made to reverse the restrictions on the developmental capacity of endoderm nuclei by replacing their proteins with protamine or spermine. Preliminary experiments indicate that exposure of donor endoderm cells to 0.01 or 0.02% protamine prior to and during nuclear transfer may lead to a two- to three-fold increase in the number of recipient eggs developing to larval stages (Briggs, unpublished). In a more extensive set of experiments Hennen (1970) has shown that spermine treatment of donor cells gives a significant (118%) increase in the developmental capacity of endoderm nuclei. As Hennen points out, the beneficial effects of spermine or protamine could be non-specific, for example through binding of enzymes that might otherwise be activated during nuclear transfer and lead to nuclear damage. The more interesting possibility, also considered by Hennen, is that spermine (or protamine) acts by displacing histones and other polycations from DNA, and is in turn removed by the egg cytoplasm following nuclear transfer, leaving the nucleus in a condition to function normally in development.

POSITIVE AND NEGATIVE GENE CONTROL IN CELL HYBRIDS AND IN OOCYTES

Cell hybrids

We are concerned here with the regulation of genes that account for the differences between cell types; i.e. with genes that control "luxury" functions and together represent the epigenotypes of determined cells (Abercrombie, 1967; Ephrussi, 1972). The discovery that these genes are subject to negative control was reported first by Davidson et al. (1966). Hybrids made between Syrian hamster melanoma cells (which contain Dopa oxidase and produce melanin) and mouse fibroblasts (which do not) were found consistently to lack the enzyme and to be unpigmented. This negative effect on the expression of the melanoma epigenotype was referred to by Ephrussi as "extinction". In the next several years extinction of luxury functions was found to occur in several other cell hybrids and to exhibit some very important properties, of which three will be mentioned here (see reviews by Ephrussi, 1972; and Davidson, 1974). First, extinction is specific in the sense that hybrids which have lost the luxury function of one of the parental cell types still express the household functions of both. In other words, extinction is not the result of a generalized inhibition of one of the two genomes by the other. Second, extinction is not an all or none phenomenon but rather is subject to a gene dosage effect. Hybrids containing

one genome (roughly speaking) from each of the parental cell types
usually show extinction, but if the "differentiated" parent contributes
an extra genome then its luxury functions may be expressed in the
hybrid. Third, extinction depends upon the continued presence of certain chromosomes from the non-differentiated parent. Cell hybrids
may, in successive mitoses, lose preferentially some of the chromosomes
of the non-differentiated parent. Extinguished luxury functions may
then be re-expressed, even if they had been totally extinguished over
many cell generations previously. This is a particularly important
observation because it shows: (1) that extinction is not a consequence
of a loss of genes specifying luxury functions, (2) that the mechanisms
of regulation (extinction) involves the chromosomes and interchromo-
somal communication, and (3) that determination at the gene level
is highly stabilized in the sense that the determination of which genes
can function and which cannot is inherited over many cell generations
whether the genes are expressed or not.

We turn now to the important question of positive control of genes
specifying luxury functions. As has been mentioned in an earlier section
of this chapter, cell type determination must involve a reliable mechanism permitting some genes to act and ensuring that others are kept inactive. Are the genes that are never expressed still present? And if they
are can they still be brought to activity? On the question of presence
or absence we should mention the work of Suzuki et al. (1972), who
have shown by molecular hybridization that the gene for silk fibroin
is present in the same number of copies in the silk gland (where the
gene is active) and in other parts of the silkworm (where it is not active).
Similar results have been obtained for the gene specifying haemoglobin
(de Jiménez et al., 1971). Thus, in a few cases we have direct evidence
of the presence of genes in cells in which they are silent.

Can silent genes be activated? Recent cell hybridization studies indicate that the answer to this question is "yes", under some circumstances. An example of activation is provided by the recent work of
Brown and Weiss (1975) on hybrids formed between rat hepatoma cells,
which synthesize several liver-specific enzymes, and mouse lymphoid
cells, which of course do not. Hybrids can be produced with varying
numbers of chromosomes from each parent, ranging from a hepatoma/
lymphoid ratio of 1 to a ratio close to 6—i.e. greatly favouring the hepatoma parent. Hybrids with approximately equal numbers of chromosomes from each parent show partial or complete extinction of the liver
enzymes; hybrids with a hepatoma/lymphoid chromosome ratio of 2
or greater do not show this extinction—i.e. they continue to produce
liver-specific enzymes. Since the rat and mouse enzymes can be distin-

guished from each other it is possible to determine which of the parental genomes is responsible for the synthesis seen in the hybrid. The analyses show that both rat and mouse enzymes are produced, indicating that genes for these liver-specific enzymes, previously silent in the lymphoma cells, have been activated. It is important to add these genes become inactive again if, in the course of successive cell generations, the ratio of hepatoma chromosomes to lymphoid chromosomes is reduced.

The cell hybridization studies have established the important points that cell type determination does not involve losses of genes normally never expressed, and does involve a highly stabilized inherited chromosomal system of gene control. In a given cell type certain genes are somehow selected to be capable of functioning while others are maintained in an inactive state. The normally active genes can be turned off by the genome of another cell type, but inherit the capacity to become active again when the "foreign" chromosomes are lost. Conversely, normally inactive genes can be activated when exposed to an excess of chromosomes from a cell in which those genes are normally active. Again the change is temporary—the activated genes revert to their normal inactive state when a more normal chromosome balance is restored.

A note of caution may be interjected here. All of the hybridization studies summarized above were done with established cell lines, generally aneuploid and malignant in character. As Ephrussi (1972) has pointed out, it is not certain that the gene control mechanisms operating in these cells are the same as those in normal somatic cells. The fact that some of these cell lines exhibit many of the characteristics of the normal cells from which they were derived leads one to think that the underlying mechanisms of control may also be the same, but this remains to be demonstrated.

Oocytes

Through the work of Gurdon and associates it has been shown that oocytes, expecially amphibian oocytes, offer unique opportunities for the study of gene regulation. Gurdon (1968) showed first that when somatic nuclei are transplanted into oocytes they increase greatly in volume, come to resemble in many ways the oocyte nucleus (germinal vesicle), and at the same time show a pronounced increase in RNA synthesis. The nature of the RNA's, and whether they can be translated or not, were not established in this early work. Subsequently, Gurdon et al. (1971) made the remarkable discovery that rabbit haemoglobin mRNA is translated following injection into *Xenopus* oocytes, the

efficiency of translation being one-tenth to half that of cultured intact rabbit reticulocytes and far superior to that of cell-free systems.
Messenger RNA's for several other proteins were found also be translated in *Xenopus* oocytes (review, Gurdon, 1974).

The fact that injected mRNA's are translated suggested that RNA's
produced by transplanted nuclei ought to be translated also. This in
turn would open the way to the study of gene regulation by analysis
of the proteins produced by nuclei transplanted into oocytes. The
central question, from the point of view of cell type determination, is
whether the epigenotype of transplanted somatic cell nuclei is retained,
or altered to conform to that of the oocyte. This question may be
answered if experiments are designed which permit recognition of the
synthesis of proteins normally found only in the donor somatic cell, or
only in the oocyte, or in both. Some encouraging beginnings toward
an analysis of this type have been reported recently. Gurdon *et al.* (1976)
have shown that nuclei from HeLa (human) cells persist and synthesize
RNA for many days following their transfer into *Xenopus* oocytes. These
RNA's are translated into proteins, 3 of which can be shown to be of
HeLa origin. Some 13 additional HeLa proteins also differ enough from
Xenopus oocyte proteins to be recognizable. These proteins are *not*
present in the host oocytes. Gurdon *et al.* (1976) present evidence to
indicate that the synthesis of some, but not all, HeLa proteins results
from a selective transcription of HeLa genes in *Xenopus* oocyte cytoplasm.

In another set of experiments Etkin (1976) has approached the problem of gene regulation by transplanting liver nuclei into oocytes, and
then examining the oocytes for synthesis of: (1) an enzyme normally
found in liver but not in oocytes, and (2) an enzyme normally found
in both. The enzymes chosen were alcohol dehydrogenase (ADH) for
the liver enzyme and LDH for the ubiquitous enzyme. The experiment
required that enzymes produced by the transplanted nuclei be dis-
tinguishable from those produced by the oocyte. This was accomplished
by transplanting liver nuclei from one species of amphibian (*Ambystoma
texanum*) into oocytes of another (*A. mexicanum* = axolotl), these species
being chosen because they produce distinct forms of both enzymes.
When the nuclear transfers were done, and the recipient oocytes later
analysed, it was found that the transplanted liver nuclei continued to
produce the ubiquitous enzyme (LDH) but stopped producing the liver
enzyme (ADH). This indicates that the transcriptional pattern of the
liver nuclei is being modified to correspond in some respects at least
to that of the oocyte. The crucial experiment, not yet on record so far
as the author is aware, is one in which a nucleus from a clearly defined

somatic cell type is tested for its capacity to direct the synthesis of oocyte-specific products.[4]

Work utilizing oocytes in the manner just described is still in its earliest stages, but it would appear already that it offers a realistic possibility of exploring the extent to which patterns of gene expression (epigenotypes) associated with cell type determination can be altered by the cytoplasmic environment, and possibilities also of analysing the molecular mechanisms of determination at the gene level. (See Gurdon *et al.*, 1976, for further discussion on this subject.)

SUMMARY

The fundamental cellular event in embryonic development is that of determination—the restriction of cells to specific functions. Determination is known to be a highly stabilized, heritable state of individual cells; a state which must depend upon some as yet unknown mechanism which selects certain combinations of genes to be capable of functioning while others are kept inactive. It is usually assumed that this mechanism must be one of gene regulation, with all genes being present in all cell types. We examine several lines of evidence and find that taken together they strongly support the idea that different cell types contain the same genes, even though no single type of evidence is completely conclusive by itself.

The most rigorous test of gene content and function is that of nuclear transplantation, for if the nucleus of a well-defined cell type promotes complete development when transplanted into an enucleated egg it is demonstrated conclusively that the nucleus must retain a complete genome. Nuclei from well-defined cell types of adult frogs have been shown to be capable in some cases of promoting the development of test eggs into advanced embryos. These embryos have been found to possess a variety of well-differentiated cell types, indicating that the genome of the transplanted nucleus retains many sets of genes in addition to those that were functioning in the donor cell. The interpretation of these experiments is complicated somewhat by the fact that the test embryos die prior to the feeding larval stage, and by the presence in the test eggs of stored (maternal) RNA's of unknown function in development.

Other lines of evidence also point to, but do not prove, the equivalance of genomes in different cell types. Studies of banding patterns

---

[4] An experiment by De Robertis *et al.* (in press, *Biochem. Soc. Symp.* 1976) indicates that nuclei of an established cell line, derived originally from *Xenopus* kidney, when transferred into *Pleurodeles* oocytes, are reprogrammed to produce certain proteins which are normally produced in the oocyte and not in the donor cell line.

in polytene chromosomes, and of the amounts and types of DNA's in various cells, generally fail to reveal differences associated with cell type. Of course, some differences might exist and still escape detection in these studies. Recent studies on specific genes indicate that genes which are normally never expressed except in one cell type are retained in other cells and can sometimes be brought to expression. For example, studies of gene amplification indicate that the gene for silk fibroin is present in the same number of copies in the silk gland (where it is active) and in other parts of the silkworm (where it is not). Also, recent studies of hepatoma-lymphoid cell hybrids show that genes for liver-specific functions are present in potentially functional form in the lymphoid genome.

While determination apparently does not usually involve losses or irreversible changes in genes, it does involve the establishment of highly stabilized epigenotypes—i.e. cell type specific combinations of functional and non-functional genes. In cells with balanced genomes the epigenotype appears extraordinarily stable, and is inherited over many cell generations whether it is expressed or not. In cell hybrids the pattern of expression of a given epigenotype may be altered by effects exerted on it by the chromosomes of a different cell type, but reverts to its original pattern when the "foreign" chromosomes are lost. The expression of the epigenotype may also be modified by the cytoplasm. Somatic nuclei transplanted into oocytes retain some functions and apparently lose others. It is not known as yet whether they may take on functions normally restricted to the oocyte nucleus.

ACKNOWLEDGEMENTS

Work from the author's laboratory, referred to in this paper, has been supported by research grants from the Research Grants Division, National Institutes of Health, USPHS (R01 GM05850), and from the National Science Foundation (GB 8671).

This paper is a contribution 1045 from the Department of Zoology, Indiana University.

REFERENCES

Abercrombie, M. General review of the nature of differentiation, in A. V. S. DeReuk and J. Knight (ed.), Cell Differentiation, Churchill Ltd, London (1967).
Aimar, C. Analyse par la Greffe nucléaire des propriétés morphogenetiques des noyaux embryonnaires chez Pleurodeles waltlii (amphibien urodele). Application a l'étude de la gémellarité expérimentale. *Ann. Embryol. Morphol.* **5**, 5–42 (1972).
Beermann, W. Chromomerenkonstanz and spezifische Modifikationen der Chromoso-

menstruktur in des Entwicklung und Organdifferenzierung von Chironomus tentans. *Chromosama* **5**, 139–198 (1952).

Beermann, W. Nuclear differentiation and functional morphology of chromosomes. *Cold Spr. Harb. Symp. quant. Biol.* **21**, 217–232 (1956).

Berendes, H. D. and Keyl, H. G. Distribution of DNA in heterochromatin and euchromatin of polytene nuclei of Drosophila hydei. *Genetics* **57**, 1–13 (1967).

Blackler, A. W. New cases of the Oxford nucleolar marker in the South African clawed toad. *Rev. suisse Zool.* **75**, 506–509 (1968).

Bloch, D. P. and Brack, S. D. Evidence for the cytoplasmic synthesis of nuclear histone during spermiogenesis in the grasshopper *Chortophaga viridifasciata* (DeGeer). *J. Cell Biol.* **22**, 327–340 (1964).

Bloch, D. P. and Hew H. Y. C. Changes in nuclear histones during fertilization and early embryonic development in the pulmonate snail, *Helix aspersa. J. biophys. biochem. Cytol.* **8**, 69–81 (1960).

Briggs, R. and King, T. J. Transplantation of living nuclei from blastula cells into enucleated frogs' eggs. *Proc. nat. Acad. Sci. (Wash.)* **38**, 455–563 (1952).

Briggs, R., and King, T. J. Nucleocytoplasmic interactions in eggs and embryos, in J. Brachet and A. Mirsky (ed.), The Cell, Vol. 1, pp. 537–617, Academic Press, New York (1959).

Briggs, R., King T. J. and DiBerardino, M. A. Development of nuclear-transplant embryos of known chromosome complement following parabiosis with normal embryos, in Symposium on the Germ Cells and Earliest Stages of Development, pp. 441–447, Inst. Intern. d'Embryol. Foundazione A. Baselli, Milano (1960).

Brothers, A. J. Stable nuclear activation dependent on a protein synthesized during oogenesis. *Nature (Lond.)* **260**, 112–115 (1976).

Brown, D. D. and Gurdon J. B. Absence of ribosomal RNA synthesis in the anucleolate mutant of *Xenopus laevis. Proc. nat. Acad. Sci. (Wash.)* **51**, 139–146 (1964).

Brown, J. E. and Weiss, M. C. Activation of production of mouse liver enzymes in rat hepatoma—mouse lymphoid cell hybrids. *Cell* **6**, 418–494 (1975).

Cahn, R. D. and Cahn, M. B. Heritability of cellular differentiation: clonal growth and expression of differentiation in retinal pigment cells *in vitro. Proc. nat. Acad. Sci. (Wash.)* **55**, 106–114 (1966).

Coon, H. G. Clonal stability and phenotypic expression of chick cartilage cells *in vitro. Proc. nat. Acad. Sci. (Wash.)* **55**, 66–73 (1966).

Davidson, R. L. Gene expression in somatic cell hybrids. *Ann. Rev. Genet.* **8**, 195–218 (1974).

Davidson, R. L., Ephrussi, B. and Yamamoto, K. Regulation of pigment synthesis in mammalian cells, as studied by somatic hybridization. *Proc. nat. Acad. Sci. (Wash.)* **56**, 1437–1440 (1966).

de Jiménez, E. S., Domínguez, J. L., Webb, F. H. and Bock, R. M. Comparative DNA-RNA hybridization in differentiated cells. *J. mol. Biol.* **61**, 59–71 (1971).

DiBerardino, M. A. and Hoffner, N. Origin of chromosomal abnormalities in nuclear transplants—A reevaluation of nuclear differentiation and nuclear equivalence in amphibians. *Develop. Biol* **23**, 185–209 (1970).

DiBerardino, M. A. and Hoffner, H. Development and chromosomal constitution of nuclear transplants derived from male germ cells. *J. exp. Zool.* **176**, 61–72 (1971).

Endow, S. A. and Gall, J. G. Differential replication of satellite DNA in polyploid tissues of *Drosophila virilis. Chromosoma* **50**, 175–192 (1975).

Ephrussi, B. Hybridization of Somatic Cells. Princeton Univ. Press, Princeton, N.J. (1972).

Etkin, L. D. Regulation of lactate dehydrogenase (LDH) and alcohol dehydrogenase (ADH) synthesis in liver nuclei following their transfer into oocytes, *Develop. Biol.* **52**, 201–209 (1976).

Fischberg, M., Gurdon, J. B. and Elsdale, T. R. Nuclear transfer in amphibia and the problem of the potentialities of the nuclei of differentiating tissues. *Exp. Cell Res., Suppl.* **6**, 161–178 (1958).

Gallien, L. La greffe nucléaire chez les amphibiens. *Ann. Biol.* **5–6**, 241–269 (1966).

Gehring, W. The stability of the determined state in cultures of imaginal discs in Drosophila, in H. Ursprung and R. Nöthiger (ed.), The Biology of Imaginal Discs, Springer-Verlag, Heidelberg (1972).

Gurdon, J. B. The developmental capacity of nuclei taken from intestinal epithelium cells of deeding tadpoles. *J. Embryol. exp. Morph.* **10**, 622–640 (1962).

Gurdon, J. B. Changes in somatic cell nuclei inserted into growing and maturing amphibian oocytes. *J. Embryol. exp. Morph.* **20**, 401–414 (1968).

Gurdon, J. B. The Control of Gene Expression in Animal Development, Harvard Univ. Press, Cambridge, Mass. (1974).

Gurdon, J. B., De Robertis, E. M. and Partington, G. Injected nuclei in frog oocytes provide a living cell system for the study of transcriptional control. *Nature (Lond.)* **260**, 116–120 (1976).

Gurdon, J. B., Lane, C. D., Woodland, H. R. and Marbaix, G. Use of frog eggs and oocytes for the study of messenger RNA and its translation in living cells. *Nature (Lond.)* **233**, 177–182 (1971).

Gurdon, J. B. and Laskey, R. A. The transplantation of nuclei from single cultured cells into enucleate frogs' eggs. *J. Embryol. exp. Morph.* **24**, 227–248 (1970).

Gurdon, J. B., Laskey, R. A. and Reeves, O. R. The developmental capacity of nuclei transplanted from keratinized skin cells of adult frogs. *J. Embryol. exp. Morph.* **34**, 93–112 (1975).

Gurdon, J. B. and Uehlinger, V. "Fertile" intestine nuclei. *Nature (Lond.)* **210**, 1240–1241 (1966).

Gurdon, J. B., Woodland H. R. and Lingrel, J. B. The translation of mammalian globin mRNA injected into fertilized eggs of *Xenopus laevis*. I. Message stability in development. *Develop. Biol.* **39**, 125–133 (1974).

Hadorn, E. Dynamics of determination, in M. Locke (ed.), Major Problems in Developmental Biology, pp. 85–104, Academic Press, New York (1966).

Hennen, S. Influence of spermine and reduced temperature on the ability of transplanted nuclei to promote normal development in eggs of *Rana pipiens*. *Proc. nat. Acad. Sci. (Wash.)* **66**, 630–637 (1970).

Hough, B. R., Yancey, P. H. and Davidson, E. H. Persistence of maternal RNA in *Engystomops* embryos. *J. exp. Zool.* **185**, 357–368 (1973).

Illmensee, K. The potentialities of transplanted early gastrula nuclei of *Drosophila melanogaster*. Production of their imago descendants by germ-line transplantation. *Wilhelm Roux' Arch. Entwickl.-Mech. Org.* **171**, 331–343 (1973).

Judd, B. H., Shen, M. W. and Kaufman, T. D. The anatomy and function of a segment of the X chromosome of *Drosophila melanogaster*. *Genetics* **71**, 139–156 (1972).

King, T. J. and Briggs, R. Serial transplantation of embryonic nuclei. *Cold Spr. Harb. Symp. quant. Biol.* **21**, 271–290 (1956).

King, T. J. and DiBerardino, M. A. Transplantation of nuclei from the frog renal adenocarcinomia. I. Development of tumor nuclear-transplant embryos. *Ann. N.Y. Acad. Sci.* **126**, 115–126 (1965).

King, T. J. and McKinnel, R. G. An attempt to determine the developmental potentialities of the cancer cell nucleus by means of transplantation, in Cell Physiology
of Neoplasia, pp. 591–617, Univ. of Texas Press, Texas (1960).

Kobel, H. R., Brun, R. B. and Fischberg, M. Nuclear transplantation with melanophores, ciliated epidermal cells, and the established cell line A-8 in Xenopus laevis.
J. Embryol. exp. Morph. **29**, 539–547 (1973).

Konisberg, J. Clonal analysis of myogenesis. Science **140**, 1273–1284 (1963).

Lambert, C. C. Genetic transcription during the development and metamorphosis of
the tunicate Ascidia callosa. Exp. Cell Res. **66**, 401–409 (1971).

Laskey, R. A. and Gurdon, J. B. Genetic content of adult somatic cells tested by nuclear
transplantation from cultured cells. Nature (Lond.) **228**, 1332–1334 (1970).

McKinnell, R. G., Deggins, B. A. and Labat, D. D. Transplantation of pluripotential
nuclei from triploid frog tumors. Science **165**, 394–396 (1969).

McCarthy, B. J., and Hoyer, B. H. Identity of DNA and diversity of messenger RNA
molecules in normal mouse tissues. Proc. nat. Acad. Sci. (Wash.) **53**, 915–922 (1964).

Nanney, D. L. Caryonidal inheritance and nuclear differentiation. Amer. Nat. **90**, 291–
307 (1956).

Newrock, K. M. and Raff, R. A. Polar lobe specific regulation of translation in embryos
of Ilyanassa obsoleta. Develop. Biol. **42**, 242–259 (1975).

Pavan, A. Nucleic acid metabolism in polytene chromosomes and the problem of differentiation, in Brookhaven Symposia in Biology. No. 18, pp. 222–241 (1965).

Reeves, O. R. and Laskey, R. A. In vitro differentiation of a homogeneous cell population—the epidermis of Xenopus laevis. J. Embryol. exp. Morph. **34**, 75–92 (1975).

Richler, C. and Yaffe, D. The in vitro cultivation and differentiation capacities of myo-
genic cell lines. Develop. Biol. **23**, 1–18 (1970).

Schubert, D., Humphreys, S., deVitry, F. and Jacob, F. Induced differentiation of
a neuroblastoma. Develop. Biol. **25**, 514–546 (1971).

Sonneborn, T. M. Patterns of nucleocytoplasmic integration in Paramecium. Caryo-
logia **6**, Suppl., 307–325 (1954).

Spemann, H. Embryonic Development and Induction, Hafner Publishing Co., New
York (1938).

Subtelny, S. On the nature of the restricted differentiation-promoting ability of transplanted Rana pipiens nuclei from differentiating endoderm cells. J. exp. Zool. **159**,
59–92 (1965).

Subtelny, S. and Bradt, C. Transplantations of blastula nuclei into activated eggs from
the body cavity and from the uterus of Rana pipiens. I. Evidence for fusion between
the transferred nucleus and the female nucleus of the recipient eggs. Develop. Biol.
2, 393–407 (1960).

Subtelny, S and Bradt, C. Transplantations of blastula nuclei into activated eggs from
the body cavity and from the uterus of Rana pipiens. II. Development of the recipient
body cavity eggs. Develop. Biol. **3**, 96–114 (1961).

Suzuki, Y., Gage, L. P. and Brown, D. D. The genes for silk fibroin in Bombyx mori.
J. molec. Biol. **70**, 637–649 (1972).

Tobler, H., Smith, K. D. and Ursprung, H. Molecular aspects of chromatin elimination
in Ascaris lumbricoides. Develop. Biol. **27**, 190–203 (1972).

Wabl, M. R., Brun, R. R. and DuPasquier, L. Lymphocytes of the toad, Xenopus laevis,
have the gene set for promoting tadpole development. Science **190**, 1310–1312 (1976).

Woodland, H. R., Gurdon, J. B. and Lingrel, J. B. The translation of mammalian
globin mRNA injected into fertilized eggs of Xenopus laevis. II. The distribution of
globin synthesis in different tissues. Develop. Biol. **39**, 134–140 (1974).

Interactions between Embryonic Cells during the Early Development of the Mouse

C. F. GRAHAM and S. J. KELLY

Zoology Department, South Parks Road, Oxford, England

INTRODUCTION

During development to the blastocyst stage (40 to 120 cells), the mouse embryo is not in direct contact with maternal tissue and is surrounded by a jelly coat, the zona pellucida; operations can be conducted on these preimplantation stages and the embryo subsequently returned to the uterus to continue development. Most of the information about cell interactions in the early embryo is therefore restricted to studies on embryos developing to the blastocyst stage.

The blastomeres of the mouse remain totipotent up to the 8-cell stage (see: Progressive restriction, p. 46, and subsequently their environment determines that they should form either trophectoderm or inner cell mass (ICM) (see: Environmental causes of restriction, p. 47. In order to develop, the cells of the embryo must be able to do several things:

(1) Create a structure in which cells are exposed to different environments.
(2) Detect their environment (e.g. with receptors).
(3) Respond to the environment with a change in behaviour (e.g. expression of new gene products).

The information which is available about these three processes is discussed below.

RESTRICTION OF DEVELOPMENTAL POTENTIAL

Progressive restriction

One-cell stage

Eggs which contain localized morphogenetic factors or reference points do not develop normally when parts of the egg containing these specialized regions are ablated (review Reverberi, 1971; Graham, 1976).

The following experiment suggests that the one-cell mouse egg does not contain such regions. Fertilized eggs were bisected so that one half contained the female pronucleus and the other half contained the male pronucleus (Tarkowski and Rossant, 1976). One quarter of the operated embryos would be expected to die before the blastocyst stage because they lack an X-chromosome (Morris, 1968), and we are only concerned with the haploid embryos which contained this chromosome. Seven out of the twelve operated embryos which were recovered had formed blastocysts, suggesting that all X-chromosome-containing embryos could develop to this stage. If we assume that many different regions of the egg were ablated from the X-chromosome-containing halves in these experiments, then these observations prove that the fertilized one-cell egg does not contain specialized morphogenetic regions whose developmental effect can be detected by the blastocyst stage.

Two- to eight-cell embryos

There is good evidence that each blastomere of 2-, 4-, and 8-cell stage embryos is totipotent; that is, they are able to form all the tissues of foetal and adult mice (review Kelly, 1975):

(1) Whole mice can be formed by one cell of a 2-cell embryo.
(2) Each cell of a 4-cell embryo when combined with carrier blasto-meres can form parts of the trophoblast plus parietal endoderm, of the visceral yolk sac, and of the foetus half way through pregnancy. At least three of the four blastomeres can form most of the tissues of an adult mouse including the germ cells.
(3) Each 8-cell blastomere of a pair derived from a single 4-cell stage blastomere regularly form parts of the foetus, of the visceral yolk sac, and of the trophoblast plus parietal endoderm half way through pregnancy when combined with carrier blastomeres. Single blastomeres of 8-cell stage embryos can form the majority of tissues of an adult mouse including the germ cells.

Although the experiments have not proved that each blastomere of a single 8-cell stage embryo is capable of forming a complete mouse, they strongly suggest that this is the case.

Blastocyst stage

Totipotency is lost by the blastocyst stage (reviewed Gardner and Papaioannou, 1975; Gardner and Rossant, 1976). The cells on the outside of the blastocyst can only form trophoblast giant cells and parts of the chorion and ectoplacental cone. The cells on the inside of the blastocyst (ICM cells) can form all the tissues of the embryo and adult except the trophoblast. Models of mouse development then have to explain how these two restrictions of developmental potential occur between the 8-cell stage and blastocyst. Since blastocyst formation can start at the 21-cell stage (McLaren and Bowman, 1973) it may be that these restrictions come into effect during an 18 hour period or less.

Environmental causes of restriction

Inside and outside environments

Isolated 8-cell stage blastomeres mainly form trophectoderm (Tarkowski, 1959a, b; Tarkowski and Wroblewska, 1967; Sherman, 1975; Rossant, 1976), and each blastomere of a 4-cell stage embryo tends to form this tissue when placed on the outside of aggregated blastomeres (Hillman *et al.*, 1972). The exposure of a large part of the cell surface to the external medium or lack of contact with other cells over most of the cell surface are then the environmental conditions which are required for trophectoderm formation.

In contrast, ICM cells develop from blastomeres which have been surrounded by other cells. All the blastomeres of 4- and 8-cell stage embryos which have been surrounded by other blastomeres tend to form ICM. The environmental condition of cell contact over most of the cell surface is then required for ICM development.

Time of action of environment

Up to the 8-cell stage, all blastomeres are exposed to the external medium and are outside cells; since these cells remain totipotent (see: Two- to eight-cell embryos, p. 46, they cannot have been acted on by the environment. The outside cells probably start to pump inwards by the 21-cell stage and are developmentally restricted by the mid-blastocyst and so the embryo's cells respond to outside conditions by forming trophectoderm sometime between 8- and 21-cell stages. By the late blastocyst, the ICM cells respond to outside conditions by forming primitive endoderm which confirms that embryonic cells respond to out-side conditions by forming trophectoderm only during a limited stage of development (Rossant, 1975a, b).

The inside cells must also respond to their environment in a relatively short time interval. It is rare to find inside cells at the 8-cell stage, while nearly all 16-cell stage embryos have one to three inside cells (Herbert and Graham, 1974). There is evidence that some more cells can be recruited to the inside cell population after the 16-cell stage (Barlow *et al.*, 1972), and the developmental capacity of these cells is restricted by the mid-blastocyst (Rossant, 1975a, b).

CREATION OF A STRUCTURE WITH INSIDE AND OUTSIDE CELLS

It is often found that in many animal phyla, cleavage produces a solid ball of cells of which some are inside and others are outside. This arrangement is not trivial because numerous other embryos form hollow spheres by cleavage and such embryos occur in the Mammalia (e.g. marsupial native cat: Hill, 1911). The creation of the mouse morula with inside and outside cells is the consequence of at least two processes.

Compaction

The process

In the early 8-cell stage of most mouse strains, the blastomeres are nearly spherical and only touch their neighbours over a small area of the cell surface. If the cells remained spherical during subsequent development then the external medium would penetrate the interstices and there would be no inside cells. This situation is avoided by the blastomeres compacting in the late 8-cell stage; the surface of the embryo becomes smooth and the cells pack close together so that neighbouring cells are in contact over large areas. Compaction is reversibly inhibited by calcium-depleted medium and by Cytochalasin B, but it is not affected by Colchicine and Colcemid. These observations suggest that microtubules are not involved in compaction and that calcium depletion and Cytochalasin B may act on motile systems in the cell surface (Ducibella and Anderson, 1975), although Cytochalasin B also acts as a metabolic inhibitor.

Mechanism

Something like the compacted state of the 8-cell stage would be produced by forces which minimized the surface area of the embryo (Ducibella and Anderson, 1975), providing that cells which have structured surface membranes are considered in this analysis to be fluid droplets without such structural surfaces.

(1) If the blastomeres were covered by adhesion factors which reached a lower free energy state after interaction with factors on the surface of other blastomeres, then the cells might maximize their surface area of contact; such a process may be called "minimizing the total adhesive free energy" (Steinberg, 1963). Such forces would not automatically create the compacted state as it is observed because factors which maximized cell contact area would tend to produce interdigitating membranes between the cells and these are rarely seen.

(2) The microvilli on the exposed outside surfaces of compacted embryo overlap between cells (Calarco and Epstein, 1973). This suggests that the surface of compacted embryos may be minimized and held under tension by the microvilli of adjacent cells adhering and pulling against each other; similar mechanisms have been proposed for the locomotion of tissue culture cells (Ciba Foundation, 1973). The surface trophectoderm layer of the blastocyst appears to be under tension (Cole, 1967), but we do not know if this is the case in the compacted 8-cell embryo.

It is clear that mechanisms (1) and (2) may depend on cell adhesion, but there is no direct evidence for this. There is no information about the molecules involved, but it is possible that the factor which holds the embryonal carcinoma (EC) cells of teratocarcinomas together may have a similar function in the early embryo (Oppenheimer, 1973, 1975; Oppenheimer et al., 1969; Oppenheimer and Humphreys, 1971). There are also a large array of antigens on the surface of 8-cell embryos which might hold the cells together (see: Antigens, p. 52).

Formation of inside cells

Progenitors of inside cells

The one to three inside cells of 16-cell stage embryos are often derived from the first cell to divide away from the 2-cell stage. The evidence for this view is as follows:

(1) The first blastomere to divide from the 2-cell stage usually produces amongst its daughter cells the first pair of blastomeres to reach the 8-cell stage (observed in 11/12 embryos: Graham, unpublished; Lewis and Wright, 1935).

(2) There is then a tendency for the first cells to divide away from the

8-cell stage to form the inside cells of the 16-cell stage (Barlow *et al.*, 1972).

(3) The first pair of blastomeres to reach the 8-cell stage contribute significantly more cells to the ICM than do the last pair of blastomeres to reach the 8-cell stage (S. J. Kelly, unpublished).

We must next enquire why this might be the case.

Orientation of mitosis

Oscar Hertwig stated the rule that mitotic spindles are formed along the longest axis of the cell, and this rule seems to hold for sea urchins during cleavage (Hörstadius, 1973; Rappaport, 1974). Spindles of 8-cell mouse embryos with single blastomeres in mitosis are often arranged radially and this is often the longest axis of early dividing blastomeres in the 8- to 11-cell embryo (Graham, unpublished). This situation continues until there are one to three inside blastomeres (9- to 11-cell stage). There are not unvarying orientations of the spindles; in some embryos there are no inside cells by the 16-cell stage and these are recruited later by division of the outside cells (Barlow *et al.*, 1972).

If this analysis of cleavage is correct, then it is only necessary to understand the molecular basis of two processes in order to understand how inside and outside environments are created in the mouse embryo: these processes are compaction and orientation of the products of mitosis.

CONSEQUENCES OF COMPACTION

Limited movement inside the embryo

Tight and gap junctions are infrequent in the uncompacted 8-cell stage embryo (Ducibella and Anderson, 1975), but 8 hours later, in the morula, the blastomere surfaces exposed to the environment are joined by tight junctions which prevent the penetration of lanthanum (Ducibella *et al.*, 1975). At this stage (16-cell stage) and later it is unlikely that cells move in and out of the surface layer of the embryo although inside cells can still be produced by cell division. When pairs of labelled 8-cell stage blastomeres were each placed on the surface of forty compacted 8- to 16-cell embryos, then 92% of their daughter cells remained on the outside of the embryo and formed part of the trophectoderm (Hillman *et al.*, 1972).

Cell movement within the embryo has been studied by combining blastomeres, labelled with tritiated thymidine, with unlabelled blastomeres (Garner and McLaren, 1974; S. J. Kelly, unpublished). These combinations have been made in uncompacted embryos at the 8-cell stage, and they show that there is little cell mixing between the labelled and unlabelled blastomeres during development to the blastocyst. Similarly a study of the distribution of oil droplets inside the embryo from the 2- and 4-cell stages to the blastocyst suggests that the cells do not move about the embryo (Wilson *et al.*, 1972). It is possible that the junctions between compacted cells are partly responsible for limiting the movement of cells.

Inside cells are sealed from the outside environment

In the morula (16-cell stage), the focal tight junctions between the outside cells provides the morphological basis for these cells to form a functional epithelium and they might therefore delimit the interior of the embryo from the external environment (Ducibella *et al.*, 1975). These authors argue that the interior environment is likely to contain different ions for the following reasons:

(1) Sodium-dependent amino acid transport changes at the morula stage (Borland and Tasca, 1974).
(2) Fluid from cytoplasmic vesicles is released into intercellular spaces before blastocoele formation properly begins (Calarco and Brown, 1969).
(3) The trophectoderm layer of the blastocyst of the rabbit engages in active transport of sodium and chloride (Cross and Brinster, 1970; Cross, 1973).

Certainly the ionic composition of the mouse blastocoele is different from that of the external environment of the embryo (Borland *et al.*, 1977), but it is not certain that ionic composition is the feature of the interior of the morula which determines the inside blastomeres to form ICM cells.

So far nobody has discovered a substitute for the effect of the outside cells on the inside cells. When a glass slide is pressed onto the embryo, from the 4- or 8-cell stage onwards, so that the embryo is forced to develop as a monolayer, then the ICM does not develop normally. All the cells may accumulate fluid and contact with the zona pellucida cannot substitute for the effect of outside cells (Stern, 1973). It is of great importance to discover the special features of the interior environment of the morula.

TROPHECTODERM AND INNER CELL MASS GENE PRODUCTS

Differences between the inside and outside cells develop rapidly, and are clear-cut by the blastocyst stage.

Protein synthesis

The ICM has a different pattern of protein synthesis when compared to the mural trophoblast (Van Blerkom *et al.*, 1976). On two-dimensional gels it is found that there are eight proteins which are synthesized in large quantities by the ICM but are produced in only trace amount by the mural trophoblast; conversely there are three proteins produced in large quantities in mural trophoblast which are barely detectable amongst the proteins synthesized by ICM.

Antigens

The early mouse embryo cross-reacts with antibodies prepared against mouse embryos and embryonal carcinoma (EC) cells of teratocarcinomas (see Wiley and Calarco, 1975; Searle *et al.*, 1976; Solter and Schachner, 1976). The most interesting of these antibodies are those which react specifically with either the ICM cells or the trophectoderm cells. Thus anti-PCC4 (an EC cell line) only reacts with ICM cells and does not react with the trophoblast or pre-blastocyst embryos (Gachelin, 1976); this observation demonstrates the appearance of new gene products on the cell surface of the ICM. Similarly it has been shown that the ICM possess different exposed antigenic sites when compared to the trophectoderm, using antisera prepared against another teratoma line (Edidin, 1976).

Alkaline phosphatase

There are at least two types of alkaline phosphatase activity in the pre-implantation mouse embryo. One type, which is detected by the Burstone histochemical technique, is found in all the blastomeres of the embryo from the 4-cell stage through to the blastocyst (Solter *et al.*, 1973; Izquierdo and Marticorena, 1975). The other activity is detected by the Gomori and Gomori-Takamatsu techniques; this activity is present in the late morula (17-cell stage) and is localized in the ICM of the blastocyst (Izquierdo and Ortiz, 1975; Mulnard, 1965). It is not clear whether these techniques detect different enzymes or the same enzyme which has a different stability in different parts of the embryo when exposed to histochemical abuse.

During subsequent development, the ICM cells are known to stimulate trophectoderm proliferation (review Gardner and Johnson, 1975) and the outside cells of the ICM form endoderm in response to an outside environment (Rossant, 1975a, b). The cell interaction responsible for these changes provides rich material for analysis.

THE USE OF TERATOCARCINOMAS IN MOUSE EMBRYOLOGY

It is clear that our present information about mouse embryos gives a good idea of how the embryo might create a structure in which cells are exposed to different environments. However, we do not know how these environments are detected or how the detection system acts to change the pattern of gene products in different cells (Trophectoderm and inner cell mass gene products, p. 52. It is probable that answers to these questions will only be obtained if it is possible to work with bulk living material which mimics normal embryogenesis and which can be used for detailed biochemical studies; such material is provided by the EC cells of teratocarcinomas. Here we will be concerned with the ability of these EC cells to interact with normal cells in the embryo. Such interactions are necessary if a case is to be made that EC cells will provide an accurate analogue of normal mouse embryogenesis (see also Martin, this volume).

EC cells will incorporate into most of the tissues of an adult mouse when injected into the blastocyst (Brinster, 1974, 1975; Mintz et al., 1975; Mintz and Illmensee, 1975; Illmensee and Mintz, 1976; Papaioannou et al., 1975). In most of these experiments between 2 and 40 EC cells were injected, but the results obtained with single EC cell injections were similar (Illmensee and Mintz, 1976) and the experiments will be discussed together.

It has been shown that the EC cells can develop into melanocytes of the skin and retina, the dermal elements of the skin follicle, erythrocytes, immunoglobulin-producing cells, and liver cells by the direct recognition of differentiated characters coded by EC cells. These cells can also be traced in the chimaeras by making use of electrophoretic variants of the enzyme glucose phosphate isomerase (Carter and Parr, 1967). Using this marker it has been shown that the EC cells can form a substantial part of the following tissues: brain, heart, kidney, liver, salivary gland, thymus, lung, intestine, pancreas, stomach, reproductive tract and gonad. In at least one case it has been shown that the EC cells can form viable sperm (Mintz et al., 1975) and this observation provides a complete proof of their genetic totipotency. The success of the cell injection experiments suggests that EC cells can participate in all the tissue interactions of normal embryogenesis. We are particularly concerned to establish that they can interact with the cells of the early embryo so that they can be used to follow the events of early embryogenesis.

ACKNOWLEDGEMENTS

The MRC kindly supported our own studies reported here. Wish to thank Richard Gardner and Janet Rossant for checking the manuscript.

REFERENCES

Barlow, P. W., Owen, D. A. J. and Graham, C. F. DNA synthesis in the preimplantation mouse embryo. *J. Embryol, exp. Morph.* **27**, 431–445 (1972).

Borland, R. M., Biggers, J. D. and Lechene, C. P. Studies on the composition and formation of mouse blastocoele fluid using electron probe microanalysis. *Develop. Biol.* **55**, 1–8 (1977).

Borland, R. M. and Tasca, R. J. Activation of Na$^+$-dependent amino acid transport system in preimplantation mouse embryos. *Develop. Biol.* **30**, 169–182 (1974).

Brinster, R. L. The effect of cells transferred into the mouse blastocyst on subsequent development. *J. exp. Med.* **140**, 1049 (1974).

Brinster, R. L. Can teratocarcinoma cells colonize the mouse embryo? in M. I. Sherman and D. Solter (ed.), Teratomas and Differentiation, pp. 51–58, Academic Press, New York (1975).

Calarco, P. G. and Brown, E. H. An ultrastructural and cytological study of preimplantation development of the mouse. *J. exp Zool.* **171**, 253–284 (1969).

Calarco, P. G. and Epstein, C. J. Cell surface changes during preimplantation development in the mouse. *Develop. Biol.* **32**, 208–213 (1973).

Carter, N. D. and Parr, C. W. Isoenzymes of phosphoglucose isomerase in mice. *Nature (Lond.)* **216**, 511 (1967).

Ciba Foundation, Locomotion of Tissue Cells, Ciba Foundation Symposium 14, Elsevier, Amsterdam (1973).

Cole, R. J. Cinemicrographic observations on the trophoblast and zone pellucida of the mouse blastocyst. *J. Embryol. exp. Morph.* **17**, 481–490 (1967).

Cross, M. Active sodium and chloride transport across the rabbit blastocoele wall. *Biol. Reprod.* **8**, 556–575 (1973).

Cross, M. and Brinster, R. L. Influence of ions, inhibitors, and anoxia on transtrophoblast potential of rabbit blastocyst. *Exp. Cell Res.* **62**, 303–309 (1970).

Ducibella, T., Albertini, D. F., Anderson, E. and Biggers, J. D. The preimplantation mammalian embryo: characterization of intercellular junctions and their appearance during development. *Develop. Biol.* **45**, 231–250 (1975).

Ducibella, T. and Anderson, E. Cell shape and membrane changes in the eight-cell mouse embryo: prerequisite for morphogenesis of the blastocyst. *Develop. Biol.* **47**, 45–58 (1975).

Edidin, M. The appearance of cell-surface antigens in the development of the mouse embryo: a study of cell surface differentiation, in Embryogenesis in Mammals, Ciba Foundation Symposium 40, pp. 177–194, Associated Scientific Publishers, Amsterdam (1976).

Gachelin, G. Le teratocarcinome expérimental de la souris: une système modèle pour l'étude des relations entre antigenes des surfaces cellulaires et differenciation embryonnaire. *Bull. Cancer* **63**, 95–110 (1976).

Gardner, R. L. and Johnson, M. H. Investigation of cellular interaction and deployment in the early mammalian embryo using intra- and inter-specific chimaeras, in Cell Patterning, Ciba Foundation Symposium 29, pp. 183–200, Associated Scientific Publishers, Amsterdam (1975).

Gardner, R. L. and Papaioannou, V. E. Differentiation in the trophectoderm and inner cell mass, in M. Balls and A. E. Wild (ed.), The Early Development of Mammals, pp. 107–132, Cambridge Univ. Press (1975).

Gardner, R. L. and Rossant, J. Determination during embryogenesis, in Embryogenesis in Mammals, Ciba Foundation Symposium 40, pp. 5–18, Associated Scientific Publishers, Amsterdam (1976).

Garner, W. and McLaren, A. Cell distribution in chimaeric mouse embryos before implantation. *J. Embryol. exp. Morph.* **32**, 495–503 (1974).

Graham, C. F. The formation of different cell types in animal embryos, in C. F. Graham and P. F. Wareing (ed.), The Developmental Biology of Plants and Animals, pp. 14–28, Blackwell's Scientific Publications, Oxford (1976).

Herbert, M. and Graham, C. F. Cell determination and biochemical differentiation of the early mammalian embryo. *Curr. Top. Develop. Biol.* **8**, 151–178 (1974).

Hill, J. P. The early development of the Marsupalia, with special reference to the native cat. *Quart. J. micr. Sci.* **56**, 1–134 (1910).

Hillman, M., Sherman, M. I. and Graham, C. F. The effect of spatial arrangement on cell determination during mouse development. *J. Embryol. exp. Morph.* **28**, 263–278 (1972).

Hörstadius, S. Experimental Embryology of Echinoderms, Oxford Univ. Press (1973).

Illmensee, K. and Mintz, B. Totipotency and normal differentiation of single teratocarcinoma cells cloned by injection into blastocysts. *Proc. nat. Acad. Sci. (Wash.)* **73**, 549–553 (1976).

Izquierdo, L. and Marticorena, P. Alkaline phosphatase in preimplantation mouse embryos. *Exp. Cell Res.* **92**, 399–402 (1975).

Izquierdo, L. and Ortiz, M. E. Differentiation in mouse morulae. *Wilhelm Roux' Arch. Entwickl.-Mech. Org.* **177**, 67–74 (1975).

Kelly, S. J. Studies of the potency of the early cleavage blastomeres of the mouse, in M. Balls and A. E. Wild (ed.), The Early Development of Mammals, pp. 97–105, Cambridge Univ. Press (1975).

Lewis, W. H. and Wright, E. S. On the early development of the mouse egg. *Contr. Embryol. Carneg. Instn.* **148**, 115–143 (1935).

McLaren, A. and Bowman, P. Genetic effects on the timing of early development in the mouse. *J. Embryol. exp. Morph.* **30**, 491–498 (1973).

Mintz, B. and Illmensee, K. Normal genetically mosaic mice produced from malignant teratocarcinoma cells. *Proc. nat. Acad. Sci. (Wash.)* **72**, 3585–3589 (1975).

Mintz, B., Illmensee, K. and Gearhart, J. D. Developmental and experimental potentialities of mouse teratocarcinoma cells from embryoid body cores, in M. I. Sherman and D. Solter (ed.), Teratomas and Differentiation, pp. 59–82, Academic Press, New York (1975).

Morris, T. The XO and OY chromosome constitution in the mouse. *Genet. Res.* **12**, 125–136 (1968).

Mulnard, J. G. Studies on the regulation of mouse ova *in vitro*, in G. E. W. Wolsten-holme and M. O'Connor (ed.), Preimplantation States of Pregnancy, Ciba Foundation Symposium, pp. 123–144, Churchill Ltd, London (1965).

Oppenheimer, S. B. Utilization of L-glutamine in intercellular adhesion: ascites tumor and embryonic cells. *Exp. Cell Res.* **77**, 175–182 (1973).

Oppenheimer, S. B. Functional involvement of specific carbohydrate in teratoma cell adhesion factor. *Exp. Cell Res.* **92**, 122–126 (1975).

Oppenheimer, S. B., Edidin, M., Orr, C. W. and Roseman, S. An L-glutamine requirement for intercellular adhesion. *Proc. nat. Acad. Sci. (Wash.)* **63**, 1395–1402 (1969).

Oppenheimer, S. B. and Humphreys, T. Isolation of specific macromolecules required for adhesion of mouse tumour cells. *Nature (Lond.)* **232**, 125–127 (1971).

Papaioannou, V. E., McBurney, M. W., Gardner, R. L. and Evans, M. J. Fate of teratocarcinoma cells injected into early mouse embryos. *Nature (Lond.)* **258**, 70–73 (1975).

Rappaport, R. Cleavage, in J. Lash and J. R. Whittaker (ed.), Concepts of Development, pp. 76–98, Sinauer Associates, Stamford (1974).

Reverberi, G. (ed.). Experimental Embryology of Marine and Fresh Water Invertebrates, North-Holland Publishing Co., Amsterdam (1971).

Rossant, J. Investigation of the determinative state of mouse inner cell mass. I. Aggregation of isolated inner cell masses with morulae. *J. Embryol. exp. Morph.* **33**, 979–990 (1975a).

Rossant, J. Investigation of the determinative state of mouse inner cell mass. II. The fate of isolated inner cell masses transferred to the oviduct. *J. Embryol. exp. Morph.* **33**, 991–1001 (1975b).

Rossant, J. Postimplantation development of blastomeres isolated from 4- and 8-cell mouse eggs. *J. Embryol. exp. Morph.* **36**, 283–290 (1976).

Searle, R. F., Sellens, M. H., Elson, J., Jenkinson, E. J. and Billington, W. D. Detection of alloantigens during preimplantation development and early trophoblast differentiation in the mouse by immunoperoxidase labeling. *J. exp. Med.* **143**, 348–359 (1976).

Sherman, M. I. The role of cell–cell interactions during early mouse embryogenesis, in M. Balls and A. E. Wild (ed.), The Early Development of Mammals, pp. 145–166, Cambridge Univ. Press (1975).

Solter, D., Damjanov, I. and Skreb, N. Distribution of hydrolytic enzymes in early rat and mouse embryos—A reappraisal. *Z. Anat. Entwickl.-Gesch.* **39**, 119–126 (1973).

Solter, D. and Schachner, M. Brain and sperm cell surface antigens (NS-4) on preimplantation mouse embryos. *Develop. Biol.* **52**, 283–290 (1976).

Steinberg, M. S. Reconstruction of tissues by dissociated cells. *Science* **141**, 401–408 (1963).

Stern, S. M. Development of cleaving mouse embryos under pressure. *Differentiation* **1**, 407–412 (1973).

Tarkowski, A. K. Experiments on the development of isolated blastomeres of mouse eggs. *Nature (Lond.)* **184**, 1286–1287 (1959a).

Tarkowski, A. K. Experimental studies on regulation in the development of isolated blastomeres of mouse eggs. *Acta Theriol.* **3**, 191–267 (1959b).

Tarkowski, A. K. and Rossant, J. Haploid blastocysts from bisected zygotes. *Nature (Lond.)* **259**, 664–665 (1976).

Tarkowski, A. K., and Wroblewska, J. Development of blastomeres of mouse eggs isolated at the 4- and 8-cell stage. *J. Embryol. exp. Morph.* **18**, 155–180 (1967).

Van Blerkom, J., Barton, S. C. and Johnson, M. H. Molecular differentiation in the preimplantation mouse embryo. *Nature (Lond.)* **259**, 319–321 (1976).

Wiley, L. M. and Calarco, P. G. The effects of anti-embryo sera and their localization on the cell surface during mouse pre-implantation development. *Develop. Biol.* **47**, 407-418 (1975).

Wilson, I. B., Bolton, E. and Cuttler, R. H. Preimplantation differentiation in the mouse egg as revealed by microinjection of vital markers. *J. Embryol. exp. Morph.* **27**, 467–479 (1972).

The Differentiation of Teratocarcinoma Stem Cells *in vitro:* Parallels to Normal Embryogenesis

GAIL R. MARTIN

Department of Anatomy and Cancer Research Institute, School of Medicine, University of California, San Francisco, California 94143, USA

INTRODUCTION

It has been suggested by several authors that pluripotent mouse embryonal carcinoma cells, which are the stem cells of the tumours known as teratomas, can be used as an alternative to the cells of the early embryo for studies of mammalian embryonic cell differentiation. This suggestion is based on the observation that there are numerous similarities between the two cell types, but that unlike the pluripotent cells of the embryo, the tumour stem cells are readily obtainable and large numbers of them can easily be grown *in vitro*. In the discussion which follows the author will briefly review the evidence that the two cell types are similar and present some new data on the parallels between normal embryogenesis and the differentiation of embryonal carcinoma cells *in vitro*.

Description of the tumours

Solid mouse teratoma tumours are characterized by their content of a variety of differentiated cell types including derivatives of all three primary germ layers, endoderm, mesoderm and ectoderm, as well as a distinctive cell type known as embryonal carcinoma (reviewed by Stevens, 1967). The tumour cells may also be found as aggregates in the ascitic fluid of animals bearing solid intraperitoneal teratocarcinomas. Such aggregates are termed embryoid bodies because of their resemblance to certain stages of normal embryogenesis (Stevens, 1959 1960; Pierce *et al.*, 1960).

Two types of embryoid body can be obtained, simple and cystic. Simple embryoid bodies consist of an inner core of embryonal carcinoma cells surrounded by a single layer of endodermal cells. They resemble the foetal portion of the 5-day mouse embryo, showing the inversion of germ layers which is typical of rodent embryos. The dotted line in Fig. 1 shows the morphological relationship of these embryoid bodies to the normal early mouse embryo. Cystic embryoid bodies are more complex, many of them containing embryonal carcinoma cells, a variety of differentiated tissues and a fluid-filled cyst. These structures bear striking similarities to the foetal portion of older mouse embryos, but clearly are disorganized in comparison with them (Hsu and Baskar, 1974).

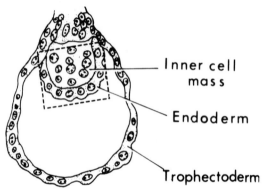

Fig. 1. Mouse embryo at approximately 5 days of development. The endoderm has formed on the free surface of the inner cell mass. The dotted line delineates the portion of the embryo to which simple embryoid bodies are analogous.

Methods of obtaining tumours

In mice, teratoma tumours can be obtained in the following ways:

(1) They arise spontaneously in the testis or the ovary; the incidence of tumour occurrence is strain-dependent, and the introduction of certain genes into the genetic background of a mouse strain can increase the incidence 10- to 30-fold (Stevens, 1973). In male mice the tumours arise through the abnormal proliferation of primordial germ cells early in foetal development (Stevens, 1962) while in female mice the tumours arise through parthenogenetic development of ovarian oocytes (Stevens and Varnum, 1974).

(2) They can be obtained experimentally by transplanting the foetal genital ridge of male embryos of certain inbred strains to the testis of an adult mouse (Stevens, 1964).

(3) They can be obtained experimentally in many inbred strains by

implanting a normal embryo of up to $7\frac{1}{2}$ days of development in an extrauterine site (Stevens, 1968, 1970; Solter *et al.*, 1970; reviewed by Damjanov and Solter, 1974a).

These data are summarized in Table I. No significant differences between teratoma tumours obtained in these different ways have as yet been found, except in the tumours described by Artzt and Bennett (1972), which were initiated by ectopic implantation of embryos homozygous for the *T*-locus mutant gene t^{w18}. Such tumours did not contain any mesodermal derivatives, presumably because the embryos from which they were derived were incapable of forming mesodermal cells, and would consequently have died *in utero*.

TABLE I

Methods of obtaining teratomas in mice

	Sex, strain of host	Source of tumour	Time tumours first detectable	Sex of tumour cells
Spontaneous tumours	♂ 129	Primordial germ cells in the foetal genital ridge	Approximately 15th day of foetal development	♂
	♀ LT	Parthenogenetically activated ovarian eggs	30 days post natal or older	♀ (haploids possible)
Experimentally induced tumours	♂ 129, A	Primordial germ cells	Shortly after transplant of genital ridge of a 12–13 day embryo to the testis of an adult	♂
	♂ or ♀ Many strains	Embryo (2 cell to $7\frac{1}{2}$ days of development)	Shortly after transfer of embryo to an ectopic site	♂ or ♀

Benign vs. malignant teratoma tumours

Teratomas can be either benign or malignant. Histological studies indicate that when they are first detectable all teratoma tumours contain the distinctive cell type known as embryonal carcinoma (Stevens, 1959; reviewed by Damjanov and Solter, 1974a). Only those tumours in which some of these cells continue to multiply as undifferentiated cells are malignant. Such tumours, known as teratocarcinomas, are characterized by rapid growth and are retransplantable. The embryonal carcinoma cells are found haphazardly distributed throughout the tumour in the form of small groups or nests. Those tumours in which the embryonal carcinoma cells do not continue to multiply as undifferentiated cells and consist of only differentiated cell types usually cease growth within 6 weeks of initiation. They are not retransplantable. Such tumours are known as benign teratomas or simply teratomas, although this latter term is often used as a generic term to designate both the benign and malignant tumours.

In addition to histological evidence that the embryonal carcinoma cells are the pluripotent stem cells of these tumours, cloning experiments performed *in vivo* (Kleinsmith and Pierce, 1964) and *in vitro* have now provided definitive evidence that a single embryonal carcinoma cell can give rise to the tumour and all of the differentiated cell types found in it. There is also some evidence, not as yet conclusive, to support the idea that the embryonal carcinoma cells are the only malignant cells found in the tumours (reviewed by Martin, 1975).

GROWTH AND DIFFERENTIATION OF EMBRYONAL CARCINOMA CELLS *IN VITRO*

Recently, methods have been developed for establishing teratocarcinoma stem cell lines *in vitro*. Clonal lines of mouse embryonal carcinoma cells have been isolated (Kahan and Ephrussi, 1970; Rosenthal *et al.*, 1970; and many others, as reviewed by Martin, 1975), and have been shown to retain their pluripotency during growth *in vitro*; upon reinjection of the cells into histocompatible hosts, the cells form tumours containing a wide variety of differentiated cell types.

Although many attempts were made to obtain differentiation of clonal embryonal carcinoma cells *in vitro*, the early workers in this field reported little success, and therefore attention was focussed on embryoid bodies, the ascitic form of the tumour. It was reasoned that since simple embryoid bodies consisted of only two cell types, the outer endodermal cells and the pluripotent embryonal carcinoma cells of the core, it might be possible to study the differentiation of the stem cells *in vitro* by explanting such embryoid bodies from the animal and culturing them *in vitro*. On the basis of morphological and biochemical criteria, differentiation of the cells can be obtained *in vitro* providing two requirements are met. Firstly the cells must be attached to the substratum, and secondly they must be grown without subculturing over a period of several weeks (Levine *et al.*, 1974; Teresky *et al.*, 1974; Gearhart and Mintz, 1974, 1975). While such studies with explanted embryoid bodies and non-clonal cell lines derived from them (Lehman *et al.*, 1974) and have been useful in defining the conditions necessary for teratocarcinoma cell differentiation *in vitro*, it may be that some of the differentiation observed is the terminal expression of cells which have previously become determined (i.e. committed to some developmental pathway) *in vivo*.

More recently, several groups have reported that clonal embryonal carcinoma cells can differentiate *in vitro* under the appropriate conditions (Martin and Evans, 1975a; Nicolas *et al.*, 1975; Sherman, 1975a). One of the most striking observations to emerge from these studies was that the differentiation of some embryonal carcinoma cells *in vitro* parallels the normal development of the embryo (Martin and Evans, 1975a, b). The earliest stage in the differentiation of such embryonal carcinoma cells *in vitro* is the formation of simple embryoid bodies, which are identical to those found in animals bearing intraperitoneal teratocarcinomas. While some of the pluripotent embryonal carcinoma cell lines which have been isolated can form only simple embryoid bodies *in vitro*, others are capable of forming simple embryoid bodies which, when kept in suspension, continue to differentiate to form cystic embryoid bodies. The development of cystic embryoid bodies from simple ones has many features in common with normal mouse embryogenesis, including the formation of what appears to be a primitive streak and the third germ layer, mesoderm. The morphological features of cystic embryoid body formation *in vitro* and the similarities between this process and normal embryogenesis are described below.

Once either simple or cystic embryoid bodies are formed *in vitro*, the differentiation to a wide variety of cell types, including pigmented epithelium, cartilage and muscle, can occur, providing that the embryoid bodies are allowed to attach to a substratum (Martin and Evans, 1975c; Evans and Martin, 1975).

The formation of embryoid bodies *per se* does not, however, appear to be a requirement for differentiation, since there are several pluripotent embryonal carcinoma cell lines which can differentiate *in vitro* without first undergoing embryoid body formation (Nicolas *et al.*, 1975; Sherman, 1975a).

THE FORMATION OF CYSTIC EMBRYOID BODIES *IN VITRO* BY CLONAL EMBRYONAL CARCINOMA CELLS

As we have previously reported, clonal teratocarcinoma stem cell lines can be maintained in the undifferentiated state by frequent subculture (every 3–4 days) in the presence of feeder cells (Martin and Evans, 1975a, b). To obtain differentiation, the pluripotent embryonal carcinoma cells are subcultured in the absence of feeder cells. Although the cells attach to the surface of the tissue culture dish, they quickly form aggregates which are rather loosely attached to the substratum. Within 2–4 days after plating under these conditions, such aggregates become much rounder and more loosely attached. At this time, the cell clumps can easily be detached from the substratum by gently pipetting medium over the surface of the dish. In the phase-contrast microscope the individual clumps appear to be composed of only one cell type at the time of detachment (Fig. 2a). If these clumps are subsequently prevented from attaching to a substratum (by transferring them to a non-adhesive bacteriological petri dish) an outer layer of endodermal cells forms over their whole surface (Fig. 2b). The apparent increase in clump size is probably due to both clump aggregation and cell multiplication.

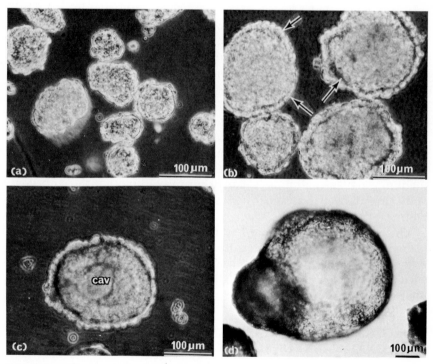

Fig. 2. Cystic embryoid body formation *in vitro*: phase contrast photomicroscopy. (a) Clumps of clonal embryonal carcinoma cells at the time of detachment from the substratum; (b) after 2 days in suspension the cell clumps become simple embryoid bodies showing a clear outer layer of endoderm (arrows); (c) an embryoid body after 5 days in suspension—a cavity is apparent; (d) a cystic embryoid body, after 9 days in suspension, has expanded into a balloon-like structure.

Over the next 4–6 days, changes within the interior of the clumps become apparent in the phase-contrast microscope; a cyst appears to form (Fig. 2c). Between 7–10 days in suspension, some of the clumps undergo a dramatic change in size and structure. They transform into large (1–3 mm diam.) eccentric, thin-walled, fluid-filled balloons (Fig. 2d). In most cases there is a solid core of cells on one side of the balloon.

In order to examine the internal changes which are occurring during this process of cystic embryoid body formation, aliquots of the cell population were taken at various times after detachment. The cell clumps were fixed in glutaraldehyde, embedded in Epon, and the samples sectioned. A detailed report of the changes in the structure of the clumps as the cells form fully cystic embryoid bodies is currently in preparation (Martin *et al.*, MS submitted). The results are reviewed below.

At the time of detachment, the cell clumps were found to be less homogeneous than was apparent in the phase-contrast microscope. While some of the clumps appeared to consist only of embryonal carcinoma cells, in most there were clearly some endodermal cells present. In some cases there were only one or two endodermal cells on the surface of the clump, while in other cases there was a crescent of endodermal cells (Fig. 3a). However, in none was there a complete layer of endodermal cells encircling the embryonal carcinoma cells.

Within 1 or 2 days after detachment, the endodermal cell layer was found to be complete, and entirely surrounded the core of embryonal carcinoma cells (Fig. 3b). At this time such clumps are identical to the simple embryoid bodies which are found in the peritoneal cavity of mice carrying the ascitic form of the tumour (Martin and Evans, 1975a, b). These structures resemble the foetal portion of the 5-day mouse embryo, as noted above.

By 3–4 days in suspension, internal changes in the embryonal carcinoma cell core of the embryoid bodies could be detected; a small eccentrically placed focus of necrosis became apparent (Fig. 3c). By approximately 4–5 days in suspension, this incipient cavity had enlarged, possibly by increased cell death, and the core cells lining one side of the cavity had elongated to form an epithelium (Fig. 3d). Such core cells have a striking morphological similarity to the embryonic ectoderm of the normal 5–6-day mouse embryo (egg-cylinder stage).

During the culture period the development of the embryoid bodies becomes distinctly asynchronous. Between 6–8 days in suspension, a proportion of the embryoid bodies have enlarged, presumably by the influx of fluid into the cavity, and the "ectodermal" cell layer had become attenuated. It is not yet clear whether or not such embryoid bodies continue to develop. However, in other embryoid bodies, a

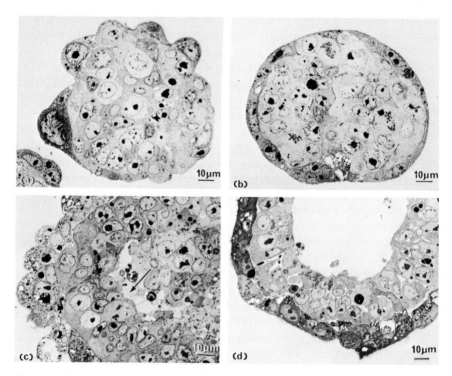

Fig. 3. Cystic embryoid body formation *in vitro*: Epon semi-thin sections. (a) A clump of clonal embryonal carcinoma cells at the time of detachment from the substratum. These are 5 endodermal cells present on the outer surface of the clump. (b) An embryoid body formed after the clumps had been in suspension for 1 day. There is a complete outer layer of unspecialized endoderm. (c) An embryoid body which has been in suspension for 3 days. The beginnings of a cavity (arrow) which is analogous to the proamniotic cavity of the normal 5–6-day mouse embryo is visible. The outer cells have the characteristics of typical visceral endoderm. (d) An embryoid body which has been kept in suspension for 4 days. The cavity has enlarged and the core cells have columnarized so that they now resemble typical embryonic ectoderm.

process which is analogous to normal primitive streak and mesoderm formation occurs: interposed between the outer endodermal cell layer and the inner ectodermal cells, some loosely attached cells which resemble mesoderm appeared (Fig. 4).

After this time, that is by approximately 8–10 days of development in suspension, a striking change occurred in many of the embryoid bodies in the culture. One side of the cell clumps ballooned out, as shown in Fig. 2 (above). A typical section of one such balloon is shown in Fig. 5a. The thin wall of the balloon is a two-layered structure, with

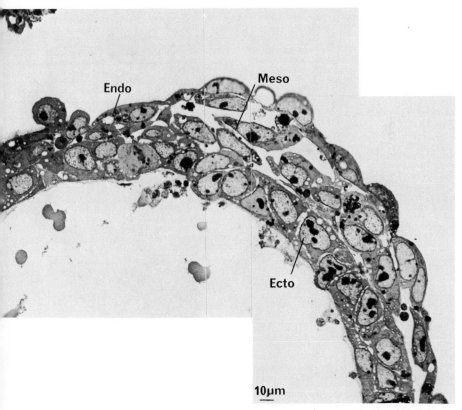

Fig. 4. The formation of mesoderm. A cystic embryoid body after 8 days in suspension. The mesodermal cells are apparent between the inner ectodermal and outer endo-dermal cell layers.

an outer layer of endodermal cells and an inner layer of what appear to be mesodermal cells. The solid portion of these cystic embryoid bodies contains an ectodermal cyst (Fig. 5). Although as yet there is no conclusive evidence, one possibility is that these balloon-like structures might form by proliferation of the mesodermal cell layer shown in Fig. 4, which then splits to form an ectodermal vesicle covered with mesoderm and an endodermal cell layer lined with mesoderm which then balloons out. If this is the case, then the two-layered wall of the balloon is analogous to the yolk sac of the normal 7–8-day mouse embryo. Consistent with this idea is our observation that within the mesodermal cell lining of the balloon wall there were structures resembling capillaries filled with blood cells (Fig. 6). In the normal embryo, the primitive blood cells form in the mesodermal lining of the yolk sac.

From these results it is quite clear that the development of cystic

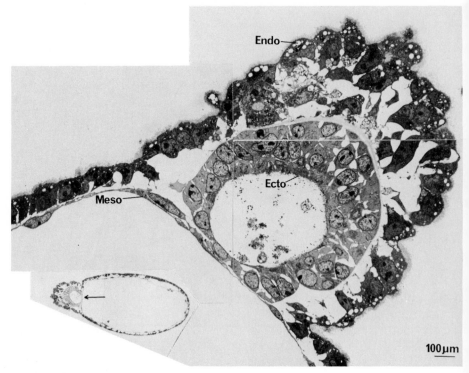

Fig. 5. An expanded cystic embryoid body balloon present in the culture after 10 days in suspension. Inset shows the complete structure, while the remainder of the figure is an enlargement of the area designated by the arrow. The whole embryoid body is approximately 0.5 mm in length.

Two cavities are visible. The smaller of the two is analogous to the proamniotic cavity of the mouse embryo and is surrounded by cells which resemble embryonic ectoderm. The larger cavity is lined by a continuous layer of mesoderm, and is probably analogous in some respects to the exocoelom of the normal 7-day mouse embryo. The two-layered wall of the balloon, consisting of endoderm and mesoderm, is analogous to the wall of the yolk sac of the normal $7\frac{1}{2}$-day embryo.

embryoid bodies *in vitro*, from homogeneous, clonal populations of embryonal carcinoma cells, bears a striking resemblance to the normal development of the early mouse embryo. That is, embryoid body formation occurs in an orderly sequence of developmental events. First a layer of endoderm is formed over the surface of clumps of undifferentiated cells. This is followed by cavitation and ectodermalization of the embryonal carcinoma core cells. Finally, mesoderm appears between the ectodermal and endodermal cell layers.

Admittedly, the events described above occur more slowly in the cultures of teratocarcinoma cells, and there is a lesser degree of

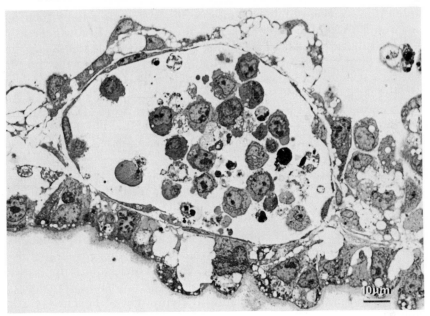

Fig. 6. A capillary-like structure filled with cells resembling blood cells which is located in the mesodermal lining of the wall of a balloon approximately 11 days after detachment of the undifferentiated cell clumps.

synchrony and organization than in the embryo developed *in vivo* or *in vitro* (Wiley and Pederson, in press). This could be due to genetic or epigenetic differences between teratocarcinoma stem cells and normal embryonic cells. However, it should be pointed out that the culture medium we have used (Dulbecco's modified Eagle's + 10% calf serum) may not be optimal. It should also be noted that in such cultures of teratocarcinoma cells there are none of the normal trophoblast-derived tissues such as extra-embryonic ectoderm which are present in normal embryos; such cells may well be required to provide inductive influences which affect the normal course of development of the embryo. Perhaps in assessing the ability of the teratocarcinoma stem cells to mimic normal embryonic cell behaviour, it would be more relevant to compare them with isolated inner cell masses cultured under identical conditions. Solter and Knowles (1975) have demonstrated that such isolated inner cell masses do develop into embryoid bodies, and a more detailed analysis of this phenomenon, in direct comparison with the teratocarcinoma cells, would be useful.

It seems likely that the differentiation of teratocarcinoma stem cells is controlled by mechanisms similar to those which operate during the expression of normal embryonic developmental programs. The first

differentiative event in normal mouse embryogenesis is the formation
of trophoblast on the outer surface of the morula. It is now generally
believed that the outer cells become trophoblast while the inner cells
remain pluripotent as a consequence of their position in the morula
(Tarkowski and Wroblewska, 1967; reviewed by Herbert and Graham,
1974). The second differentiated cell type to form in the mouse embryo
is the endodermal cell layer which appears only on the free surface of
the pluripotent inner cell mass (ICM) of the blastocyst. There is now
some evidence that this differentiation too is stimulated by "positional
information". It has been shown that when ICMs are freed from their
surrounding trophoblast and cultured *in vitro* (Solter and Knowles,
1975) or *in vivo* (Rossant, 1975) an endodermal cell layer is formed over
the whole surface of the ICM, including that portion which was previously covered by trophoblast.

    The fact that clumps of embryonal carcinoma cells form endoderm
rather than trophoblast over their whole surface suggests an analogy
between them and the cells of the ICM. Our observations support the
hypothesis that the position of the cells is the most important factor
in determining whether or not this differentiation will occur (Martin
and Evans, 1975a, b). Thus, at the time of detachment the clumps of
embryonal carcinoma cells are morphologically homogeneous, or have
some endodermal cells on their outer surface; most important, those
clumps which do have endodermal cells at the time of detachment never
have a complete outer endodermal cell layer. Yet, within 24–28 h of
detachment the outer cells have acquired the characteristics of endodermal cells. The most likely explanation for this observation is that
those clumps which did form endoderm while still attached to the
culture dish did so only on their free surface (Fig. 7). The completion

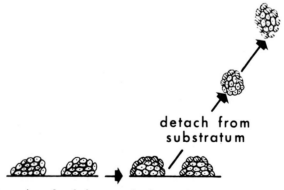

detach from
substratum

Fig. 7. The formation of endoderm on the free surface of embryonal carcinoma cell
clumps.

of the outer endodermal cell layer requires the exposure of the previously attached embryonal carcinoma cells to the stimulus of being on the free surface of the clump. Thus the mechanism by which endoderm forms in cultures of teratocarcinoma stem cells is probably the same as in the embryo, and the differentiation of the cells is in some way stimulated by their position on the outer surface of the clump.

The subsequent differentiation which occurs within the embryoid bodies and which parallels normal embryonic development may also be the consequence of some sort of positional information. It has been suggested that the proamniotic cavity of the normal embryo is formed as a consequence of programmed cell death (Bonnevie, 1950; reviewed by Lockshin and Beaulaton, 1975). That is, the cells which are at the centre of the inner cell mass are programmed to die, so that a cavity may be formed. The first cavity which appears in the cystic embryoid bodies described above, and which is analogous to the proamniotic cavity of the normal mouse embryo, probably forms by cell death. It should be noted, however, that there is considerably more necrotic cell debris in the cystic embryoid bodies formed by embryonal carcinoma cells *in vitro* than in embryos.

EMBRYONAL CARCINOMA CELLS AS NORMAL EMBRYO CELLS

In conclusion, I would like to present a summary of the similarities between teratocarcinoma stem cells and normal early embryo cells, which includes the pattern of differentiation which has been described in detail above. These data have been reviewed recently by Damjanov and Solter (1974a) and by Martin (1975).

(1) Pluripotency: Embryonal carcinoma cells, like the cells of the early embryo, are pluripotent. That is, they are capable of proliferating in the undifferentiated state and also of giving rise to differentiated cell types which represent derivatives of all three "primary germ layers".

(2) Ultrastructural, biochemical and cell surface properties: Embryonal carcinoma cells and the pluripotent cells of the early embryo are similar on the ultrastructural level; both cell types contain high levels of the enzyme alkaline phosphatase, and they have at least one specific cell surface antigen in common.

(3) Formation of embryoid bodies *in vitro* by clonal embryonal carcinoma cell lines: Some embryonal carcinoma cell lines differentiate via the formation of embryoid bodies. The fact that endoderm is the first differentiated cell type formed by embryonal carcinoma cells *in vitro* suggests that, on a functional level, mouse embryonal carcinoma cells resemble the cells of the inner cell mass of the 4-day mouse embryo.

One explanation for the close similarity between teratocarcinoma stem cells and normal early embryo cells is that the former have been derived from the latter by some process of genetic or epigenetic "transformation". Alternatively, it has been suggested that embryonal carcinoma cells are normal early embryo cells which behave abnormally (i.e. they form tumours) because they are not in their normal environment. Two observations support this latter hypothesis.

(1) The high efficiency of tumour formation by ectopic embryo implantation. D. Solter has found that almost 100% of all ectopically implanted embryos will give rise to tumours (personal communication), although the ratio of benign to malignant teratoma tumours is dependent on the strain of embryo and host being used (Damjanov and Solter, 1974b).

(2) Ability of embryonal carcinoma cells to participate in the formation of a normal mouse. In brief, it has been found that when embryonal carcinoma cells are injected into a mouse blastocyst, and the operated embryo placed in a pseudopregnant foster mother, normal chimeric (allophenic) mice are born which contain tissues derived from the host blastocyst and the injected embryonal carcinoma cells (Brinster, 1974, 1975; Mintz et al., 1975; Papaioannou et al., 1975; Illmensee and Mintz, 1976). The experiments which most conclusively demonstrate the ability of the embryonal carcinoma cells to behave as normal embryonic cells are those which have shown that injected embryonal carcinoma cells can form functional sperm in such chimeric mice (Mintz and Illmensee, 1975).

The data described above can be summarized in the following diagram (Fig. 8), which illustrates the relationship between embryonal carcinoma cells and normal embryo cells, as it is presently understood. The isolation of pluripotent cells from normal embryos has yet to be accomplished (Sherman, 1975b), and is therefore included as a dotted line in the figure. All available evidence suggests that when such cells are isolated, they will behave like embryonal carcinoma cells isolated from teratocarcinomas.

ACKNOWLEDGEMENTS

I am grateful to Drs S. Rapisardi and L. Wiley for their invaluable help in preparing the histological sections described here, and for numerous discussions. I would also like to thank Mr D. Akers for his excellent photographic reproductions, and Dr G. S. Martin for his helpful criticism of this manuscript. My support during the period when the original work described here was accomplished came from the US

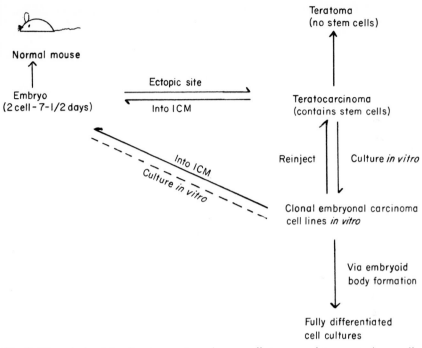

Fig. 8. The relationship of embryonal carcinoma cells to normal mouse embryo cells.

Public Health Service, National Cancer Institute, fellowship No. F22 CA04012-02.

REFERENCES

Artzt, K. and Bennett, D. A genetically caused embryonal ectodermal tumor in the mouse. *J. nat. Cancer Inst.* **48**, 141–158 (1972).

Bonnevie, K. New facts on mesoderm formation and proamnion derivatives in the normal mouse embryo. *J. Morph.* **86**, 495–545 (1950).

Brinster, R. L. The effect of cells transferred into the mouse blastocyst on subsequent development. *J, exp. Med.* **140**, 1049–1056 (1974).

Brinster, R. L. Can teratocarcinoma cells colonize the mouse embryo? in M. Sherman and D. Solter (ed.), Teratomas and Differentiation, pp. 51–58, Academic Press, New York (1975).

Damjanov, I. and Solter, D. Experimental teratoma. *Curr. Top. Path.* **59**, 69–130 (1974a).

Damjanov, I. and Solter, D. Host-related factors determine the outgrowth of teratocarcinomas from mouse egg-cylinders. *Z. Krebsforsch.* **81**, 63–69 (1974b).

Evans, M. J. and Martin, G. R. The differentiation of clonal teratocarcinoma cell cultures in vitro, in M. Sherman and D. Solter (ed.), Teratomas and Differentiation, pp. 237–250, Academic Press, New York (1975).

Gearhart, J. D. and Mintz, B. Contact-mediated myogenesis and increased acetylcholinesterase activity in primary cultures of mouse teratocarcinoma cells. *Proc. nat. Acad. Sci.* (*Wash.*) **71**, 1734–1738 (1974).

Gearhart, J. D. and Mintz, B. Creatine kinase, myokinase, and acetylcholinesterase activities in muscle-forming primary cultures of mouse teratocarcinoma cells. *Cell* **6**, 61–66 (1975).

Herbert, M. C. and Graham, C. F. Cell determination and biochemical differentiation of the early mammalian embryo. *Curr. Top. develop. Biol.* **8**, 151–178 (1974).

Hsu, Y.-C. and Baskar, J. Differentiation *in vitro* of normal mouse embryos and mouse embryonal carcinoma. *J. nat Cancer Inst.* **53**, 177–185 (1974).

Illmensee, K. and Mintz, B. Totipotency and normal differentiation of single teratocarcinoma cells cloned by injection into blastocysts. *Proc. nat. Acad. Sci.* (*Wash.*) **73**, 549–553 (1976).

Kahan, B. W. and Ephrussi, B. Developmental potentialities of clonal *in vitro* cultures of mouse testicular teratomas. *J. nat. Cancer Inst.* **44**, 1015–1036 (1970).

Kleinsmith, L. J. and Pierce, G. B., Jr. Multipotentiality of single embryonal carcinoma cells. *Cancer Res.* **24**, 1544-1552 (1964).

Lehman, J. M., Speers, W. C., Swartzendruber, D. E. and Pierce, G. B. Neoplastic differentiation: characteristics of cell lines derived from a murine teratocarcinoma. *J. cell. Physiol.* **84**, 13–28 (1974).

Levine, A. J., Torosian, M., Sarokhan, A. J. and Teresky, A. K. Biochemical criteria for the *in vitro* differentiation of embryoid bodies produced by a transplantable teratoma of mice. The production of acetylcholinesterase and creatine phosphokinase by teratoma cells. *J. cell. Physiol.* **84**, 311–318 (1974).

Lockshin, R. A. and Beaulaton, J. Programmed cell death. *Life Sci.* **15**, 1549–1565 (1975).

Martin, G. R. Teratocarcinomas as a model system for the study of embryogenesis and neoplasia: Review. *Cell* **5**, 229–243 (1975).

Martin, G. R. and Evans, M. J. The differentiation of clonal lines of teratocarcinoma cells: formation of embryoid bodies *in vitro*. *Proc. nat. Acad. Sci.* (*Wash.*) **72**, 1441–1445 (1975a).

Martin, G. R. and Evans, M. J. The formation of embryoid bodies *in vitro* by homogeneous embryonal carcinoma cell cultures derived from isolated single cells, in M. Sherman and D. Solter (ed.), Teratomas and Differentiation, pp. 169–187, Academic Press, New York (1975b).

Martin, G. R. and Evans, M. J. Multiple differentiation of clonal teratocarcinoma stem cells following embryoid body formation *in vitro*. *Cell* **6**, 467–474 (1975c).

Mintz, B. and Illmensee, K. Normal genetically mosaic mice produced from malignant teratocarcinoma cells. *Proc. nat. Acad. Sci* (*Wash.*) **72**, 3585–3589 (1975).

Mintz, B., Illmensee, K. and Gearhart, J. D. Developmental and experimental potentialities of mouse teratocarcinoma cells from embryoid body cores, in M. Sherman and D. Solter (ed.), Teratomas and Differentiation, pp. 59–82, Academic Press, New York (1975).

Nicholas, J.-F., Dubois, P., Jakob, H., Gaillard, J. and Jacob, F. Teratocarcinome de la souris: Differenciation en culture d'une lignée de cellules primitive à potentialités multiples. *Ann. Microbiol.* (*Inst. Pasteur*) **126A**, 3–22 (1975).

Papaioannou, V. E., McBurney, M. W., Gardner, R. L. and Evans, M. J. Fate of teratocarcinoma cells injected into early mouse embryos. *Nature* (*Lond.*) **258**, 70–73 (1975).

Pierce, G. B., Jr., Dixon, F. J. and Verney, E. L. Teratocarcinogenic and tissue-forming potentialities of the cell types comprising neoplastic embryoid bodies. *Lab. Invest.* **9**, 583–602 (1960).

Rosenthal, M. D., Wishnow, R. M. and Sato, G. H. *In vitro* growth and differentiation of clonal populations of multipotential mouse cells derived from a transplantable testicular teratocarcinoma. *J. nat. Cancer Inst.* **44**, 1001–1014 (1970).

Rossant, J. Investigation of the determinative state of the mouse inner cell mass. II. The fate of isolated cell masses transferred to the oviduct. *J. Embryol. exp. Morph.* **33**, 991–1001 (1975).

Sherman, M. I. Differentiation of teratoma cell line PCC4 : aza 1 *in vitro*, in M. Sherman and D. Solter (ed.), Teratomas and Differentiation, pp. 189–205, Academic Press, New York (1975a).

Sherman, M. I. The culture of cells derived from mouse blastocysts. Review. *Cell* **5**, 343–349 (1975b).

Solter, D. and Knowles, B. The immunosurgery of the mouse blastocyst. *Proc. nat. Acad. Sci. (Wash.)* **72**, 5099–5102 (1975).

Solter, D., Skreb, N. and Damjanov, I. Extrauterine growth of mouse egg cylinders results in malignant teratoma. *Nature (Lond.)* **227**, 503–504 (1970).

Stevens, L. C. Embryology of testicular teratomas in strain 129 mice. *J. nat. Cancer Inst.* **23**, 1249–1295 (1959).

Stevens, L. C. Embryonic potency of embryoid bodies derived from a transplantable testicular teratoma of the mouse. *Develop. Biol.* **2**, 285–297 (1960).

Stevens, L. C. Testicular teratomas in foetal mice. *J. nat Cancer Inst.* **28**, 247–268 (1962).

Stevens, L. C. Experimental production of testicular teratomas in mice. *Proc. nat. Acad. Sci. (Wash.)* **52**, 654–661 (1964).

Stevens, L. C. The biology of teratomas. *Advanc. Morphogen.* **6**, 1–31 (1967).

Stevens, L. C. The development of teratomas from intra-testicular grafts of tubal mouse eggs. *J. Embryol. exp. Morph.* **20**, 329–341 (1968).

Stevens, L. C. The development of transplantable teratocarinomas from intratesticular grafts of pre- and postimplantation mouse embryos. *Develop. Biol.* **21**, 364–382 (1970).

Stevens, L. C. A new inbred subline of mice (129/tμ Sv) with a high incidence of spontaneous congenital testicular teratomas. *J. nat. Cancer Inst.* **50**, 235–242 (1973).

Stevens, L. C. and Varnum, D. The development of teratomas from parthenogenetically activated ovarian mouse eggs. *Develop. Biol.* **37**, 369–380 (1974).

Tarkowski, A. K. and Wroblewska, J. Development of blastomeres of mouse eggs isolated at the 4 and 8 cell stage. *J. Embryol. exp. Morph.* **18**, 155–180 (1967).

Teresky, A. K., Marsden, M., Kuff, E. L. and Levine, A. J. Morphological criteria for the *in vitro* differentiation of embryoid bodies produced by a transplantable teratoma of mice. *J. cell. Physiol.* **84**, 319–332 (1974).

Wiley, L. M. and Pederson, R. A. Morphology of mouse egg cylinder development *in vitro*: a light and electron microscope study. *J. exp. Zool.* (in press).

II
Positional Information and Morphogenetic Signals

Positional Information and Morphogenetic Signals: An Introduction

L. WOLPERT

Department of Biology as Applied to Medicine, The Middlesex Hospital Medical School, London W1P 6DB, England

In this introduction some general points are made and attention is drawn to some recent work not covered in this volume. Therefore the important new work in relation to compartments (see Lawrence, p. 89) and the development of neural connections (see Cooke, p. 111 and Hunt, p. 97) will not be dealt with.

The basic idea of positional information is that pattern formation may result from cells first having their positions specified, and then interpreting this according to their genome and developmental history (Wolpert, 1969, 1971). This proposed two-step process for pattern formation is closely associated with the principle of non-equivalence (Lewis and Wolpert, 1976). This states that cells of the same differentiation class many have intrinsically different internal states and they will as a rule be non-equivalent if they give rise to structures differing in shape, pattern or function. This means that cells are characterized not merely by their histological type but more specifically by their position in the body. This idea is quite well known in relation to the nervous system under the heading of specificity. However, a strong case can be made for other systems, such as the vertebrate limb, where hind limb mesenchyme is non-equivalent to fore-limb mesenchyme, and where cartilage in one rudiment is non-equivalent in anotther. For example, the cartilage in the wrist grows much less than that in adjacent regions (Lewis, 1975; Summerbell, 1976). The principle of non-equivalence makes it clear that there are many more cell states characterized by positional values than there are characterized by well-defined terminal cytodifferentiation. This is nicely illustrated by the cockroach leg where intercalary regeneration takes place when cells of different

8o L. WOLPERT

positional value are placed together such that there is a discontinuity in the coordinate system (Bohn, 1970; French and Bullière, 1975). The coordinate system is a polar one and seems to be the same in each segment. It thus provides excellent evidence for the distinction between positional value and overt cytodifferentiation: cells with the same circular value have that parameter in common irrespective of the structures they give rise to.

That the behaviour of a cell in the development in pattern is largely specified by its position within the developing field is a very old one going back to Driesch in the 1890s. There are numerous experiments both classical and recent in the embryolial literature to justify this view (see reviews by Wolpert, 1971; Lawrence, 1973; and Cooke, 1975). A recent experiment using clonal analysis illustrates the point. Garcia-Bellido et al. (1976) have carried out a clonal analysis of wing development in *Drosophila* using the *Minute* mutant. Cells which are heterozygous for this mutant grow more slowly than the wild type, but nevertheless a normal fly develops. By appropriate genetic techniques a clone of wild type cells can be formed in a *Minute* background. This clone then grows much faster than the surrounding cells and can give rise to almost half the cells of the wing blade. In the normal wing the clone from a similar cell would have been very much smaller. We thus have two situations in which the progeny of a single cell forms in the one a small, and in the other a large, part of the wing, yet in both cases the pattern of cellular differentiation such as vein formation is quite normal. This result forces one to the view that the behaviour of the progeny of the cell could not be specified by the unequal distribution of cytoplasmic determinants in the initial cell or any mechanism based on a programmed cell lineage (see, for example, Holliday and Pugh, 1975). The behaviour of the cells must be specified by their relative position in the wing disc by some sort of cellular interaction. Even where the early development of the egg is based on the qualitative or quantitative differences in the egg the specification is very coarse-grained compared to later development, the important exception being the specification of germ cells by pole plasm (Illmensee, 1976). Thus while the general anterior/posterior pattern may be specified very early, single cells from the blastoderm can give rise to a wide variety of cell types and tissues. For example, Wieschaus and Gehring (1976) have shown in *Drosophila* that even after 10 hours of development—that is some 7 hours after blastoderm formation—clones can include cells that form parts of both the antennae and eye. Clearly cell interactions are involved.

POSITIONAL VALUE AND CELL AUTONOMY

Patterns based on positional information need not involve any cellular interactions other than those involved in specifying the positional value, and regulation and regeneration would primarily involve respecification of these values. The cells could interpret the positional value autonomously and be unaffected by the behaviour of adjacent cells. This is a rather extreme and perhaps simplistic view, and is contrary to that of many workers who believe inductive processes, extracellular matrices, and the immediate environment of the cell play a dominant role in specifying its development. How autonomous is the interpretation of positional information? The studies with genetic mosaics in *Drosophila* are probably the most impressive in this respect as most pattern mutants are autonomous with respect to position. The classical examples include the mutant *aristapaedia* in which the antennae form leg parts: in genetic mosaics, leg and antennal structures are intermingled according to position and genome (Postlethwait and Schneiderman, 1973). Recent studies such as those of Morata and Garcia-Bellido (1976) on the *bithorax* system show that it behaves similarly. *Bithorax* converts the anterior haltere compartment of the metathorax into an anterior wing compartment. Clones of *bithorax* in anterior haltere form wing structures appropriate to their position.

For the chick limb the overwhelming impression is that interpretation of positional information is autonomous and does not rely on the presence of adjacent structures. We may consider the chick limb in terms of a three-dimensional coordinate system and the extreme view would be that the patterns of cartilage, muscle and tendons are independent of one another and it is as if the limb is sculpted by numbers. Our previous grafting studies suggested a high degree of autonomy of the cartilaginous elements (Wolpert *et al.*, 1975). For example, grafting an early distal tip to an older stump resulted in limbs in tandem. Recently we have been investigating whether our model for the cartilaginous elements is also applicable to the more complex pattern of muscles and tendons (Shellswell and Wolpert, 1977). It was thus of great interest to know how autonomous their development is. Does, for example, the development of a muscle require the presence of the tendon or cartilaginous site of insertion? Does a tendon require its muscle for development? Or is the development of each autonomous with respect to its position? Our experiments thus far suggest that the latter is the case. For example, grafts of early distal regions of the limb develop hands with appropriate patterns of tendon, even though the muscles to which they attach are absent. Truncated limbs have more or less normal muscles even though their tendons are absent. Limbs reduplicated by grafts of an additional zone of polarizing activity (ZPA) have the pattern of muscles appropriate to the position even though cartilaginous elements are missing, and it thus seems that the same signals are responsible for the specification of both muscle tendon and cartilage. It is worth emphasizing that the positional signals may be very simple and it is the cell's response that is crucial (Wolpert and Lewis, 1975). Thus, it was of great satisfaction to us when we demonstrated that the ZPA of a mouse would induce additional chick digits in a chick wing bud (Tickle et al., 1976).

We have also investigated whether the development of the elbow joint is autonomous or is due to interaction of the radius and ulna with the humerus. Holder (1976), by removing distal regions at early stages, has shown that the shape of the distal end of the humerus which characterizes the articulation is independent of the presence of radius and ulna.

POSITIONAL SIGNALLING

Positional information, implying as it does a coordinate system, requires a boundary or reference region, with respect to which position is measured, a scalar which gives the distance from the boundary, and a vector to give the polarity or the direction in which the position is measured. It now seems clear that when positional fields are being specified the distances involved are small, about 1 mm or 100 cells— and the time is long, in the order of hours (Wolpert, 1971). As Crick (1970) has shown, a diffusion gradient fits these requirements well, and we have obtained evidence which is consistent with the inhibitory signal in hydra being a diffusible morphogen (Wolpert et al., 1972). However, the direct evidence for a diffusible morphogen is still weak, and while it is very attractive to think of gap junctions being the channel for such signals, this too remains no more than an interesting speculation. This problem is well illustrated by the presence of gap junctions allowing the passage of small ions at the segmental boundar in insects where a discontinuity in the gradient has been inferred (Warner and Lawrence, 1973; Lawrence and Green, 1975).

A particularly interesting model for setting up the gradients has been proposed by Gierer and Meinhardt (1972). This model, based as it is on defined molecular kinetics, does not rely on localized sources as in the more formal models of, for example, Wolpert et al. (1974). Their model is essentially based on the interaction of an autocatalytic substance (activator) with a short-range diffusion and a more rapidly diffusing antagonist the inhibitor. This can result in a localized high concentration of the activator, which is the site of production of the inhibitor, which forms a concentration gradient which can act as a positional signal (Meinhardt, 1976). Thus, while the source is somewhat distributed, it is similar to the model we have proposed for hydra and, like it, requires two gradients, and reduction in inhibitor concentration can result in the formation of a new activator peak, which is analogous to a new source. This model has recently been used quantitatively to account for a number of experiments on the specification of the basic body plan in insects such as *Euscelis* and *Smittia* (Meinhardt, 1977). Strong formal similarities can also be drawn between early sea urchin development and early insect development: e.g. the micromeres are analogous to the posterior cytoplasm of *Euscelis* and can bring about polarity reversal, and I have no doubt that with appropriate initial starting conditions, the same model with its two gradients could account for the sea urchin. That the same sort of model may be applicable to hydra regeneration, and early insect and sea urchin development, encourages one to believe that similar fundamental mechanisms are involved. We have also argued that the signal from the zone of polarizing activity in chick limb development is very similar to such systems, and the antero-posterior axis of the chick wing appears to be specified by a graded signal from this region (Tickle *et al.*, 1975). In this system there are two points worth noting, (1) placing an additional ZPA immediately adjacent to the normal one in the limb has no effect and suggests that if we are dealing with a diffusion gradient, the source is maintained at a constant concentration, (2) whenever the signal is attenuated by, for example, chemical treatment or distance, it is structures at the lowest level of the gradient that are formed.

Regulation may be viewed in terms of specifying new positional values when a field is perturbed. Epimorphic regulation may be distinguished from morphallactic regulation (Wolpert, 1971). In epimorphosis the positional values in the original field remain constant over most of the field and new values are generated in regions of localized growth. In morphallaxis, on the other hand, positional values change over most of the field, growth is not involved, and new boundary regions which are capable of positional signalling are formed: hydra and sea urchin and most early embryos fall into this class. An important new advance, both theoretical and experimental, has been made by French *et al.* (1976) with respect to epimorphic regulation. They have proposed a model which accounts formally and in a simple and unified form for pattern regulation in epimorphic fields in both vertebrates and invertebrates. The basis of the model is that positional information is specified in polar co-ordinates. For the cockroach leg, for example, the outer circle represents the proximal boundary of the limb field while the field centre is at the distal tip of the limb. For the imaginal disc of *Drosophila* the centre is usually the presumptive distal tip of the appendage. Two rules are proposed for behaviour of such systems, (1) when normally non-adjacent positional value in either the circular or radial sequence is confronted in a graft combination, growth occurs at the graft junction until all the intermediate positional values have been intercalated. For the circular values intercalation is by the shortest route, (2) the cells will generate a complete set of distal radial values when a complete set of circular positional values are exposed at an amputation surface or are generated by intercalation. This model provides important support for the idea of positional information and for the idea that positional value is a distinct entity. It also encourages one, once again, to believe in universal mechanisms. The model also raises difficult problems as to how the polar coordinate system is set up and what sort of interactions are involved. Does a Cartesian system become a polar one at a later stage? What sort of mechanism can provide polar coordinates?

It is important to remember that positional information may be specified in ways other than by a diffusible positional signal. The phase shift model of Goodwin and Cohen (1969) makes use of time differences in propagation of two waves, and the progress zone model of Summerbell *et al.* (1973) makes use of the time—possibly the number of cell divisions—cells spend in a particular region. Intercalary regeneration might involve interaction between a positional signal and a positional value causing localized cell division.

INTERPRETATION OF POSITIONAL INFORMATION

In spite of the different ways positional information might be specified, it seems reasonable to assume that ultimately the positional value of the cells will be governed by the concentration of one or more chemical compounds. The interpretation of this chemical concentration by specifying a particular cell's state is the key event in pattern formation, for the overt pattern in no way reflects the positional information; positional information could always be the same, and it is the interpretation that gives rise to the pattern. We have considered this problem in terms of thresholds; for concentrations just above the threshold the cells will adopt one state, just below it, another (Lewis *et al.*, 1977). Thresholds are also related to cell determination and memory since differences in a positional field must persist long after the positional signal in the form of a concentration gradient has disappeared. Models based on cooperative binding of control molecules to an allosteric enzyme do not seem to be satisfactory, both because the change is not sharp enough with plausible values for both the concentration differences between adjacent cells and the Hill coefficient, and because the system provides no memory. A kinetic model, on the other hand, which incorporates positive feedback, has many of the desired properties. A very small change in the concentration of the signal substance can bring about the change in the system from one stable steady state, in which a gene may be off, to another, in which it may be on. Moreover, the system has memory, and once the concentration has been raised above a threshold, removal of the signal will leave those genes still turned on. Thus, the system will now only respond to increased concentrations of the signal. Biological evidence for such a phenomenon is available from studies on the chick limb's response to a grafted polarizing region (Tickle *et al.*, 1975) and from the early development of the insect egg (Meinhardt, 1977). We have also shown that with such a model a concentration gradient across a field of 100 cells could easily define as many as 30 distinct cell states in a reliable sequence. The basic factor limiting the precision is the random variability of the individual cells.

TOWARDS MOLECULAR MECHANISMS?

There has been good progress in understanding how pattern formation and regulation can be understood in terms of positional information, but it has to be recognized that this is at the level of phenomenology. We have rules and models which can explain many observed phenomena. But is this understanding? A widely held view is that until the mechanisms described are placed on a firm molecular basis—e.g. the molecular nature of the gradients or positional signals—no real progress will have been made. Not surprisingly the author finds such a view unacceptable. At the very least it denies the existence of different levels of organization at which one may meaningfully investigate biological processes. Of course we would like to know the molecular basis, but at present there is no obvious way whereby we can find it out, primarily because of the lack of a biochemical assay. This is a problem for all those who wish to reduce cellular responses to molecular terms, especially where it involves intracellular reactions leading to change in cell state or behaviour. Many of the difficulties can be seen in relation to the control and initiation of cell division where the systems are orders of magnitude easier to handle and nevertheless the biochemical basis of the control of cell division remains unknown. Perhaps we should be less apologetic and point out that genetics was and is useful at levels other than that of DNA; that unless we have the right phenomenology we do not know what we are trying to explain; and that we have a long way to go and need hard work, inspiration and luck.

ACKNOWLEDGEMENT

This work is supported by the Medical Research Council.

REFERENCES

Bohn, H. Interkalare Regeneration and segmentale Gradienten bei ein Extremitaten von *Leucophaea*—Larven (Blattoria). I. Femur und Tibia. *Wilhelm Roux' Arch. Entwickl.—Mech. Org.* **165**, 303–341 (1970).

Cooke, J. The emergence and regulation of spatial organization in early animal development. *Ann. Rev. Biophys. Bioeng.* **4**, 185–217 (1975).

Crick, F. H. C. Diffusion in embryogenesis. *Nature (Lond.)* **225**, 420–422 (1970).

French, V., Bryant, P. J. and Bryant, S. V. Pattern regulation in epimorphic fields. *Science* **193**, 969–981 (1976).

French, V. and Bullière, D. Etude de la détermination de la position des cellules épidemiques: ordonnancement des cellules autour d'un appendice de Blatte; démonstration du concept de génératrice. *C.R. Acad. Sci. (Paris)* **280**, 195–298 (1975).

Garcia-Bellido, A., Ripoll, P. and Morata, G. Developmental compartmentalization in the dorsal mesothoracic disc of *Drosophila. Develop. Biol.* **48**, 132–147 (1976).

Gierer, A. and Meinhardt, H. A theory of biological pattern formation. *Kybernetik* **12**, 30-39 (1972).

Goodwin, B. C. and Cohen, M. H. A phase shift model for the spatial and temporal organization of developing systems. *J. theor. Biol.* **25**, 49–107 (1969).

Holliday, R., and Pugh, J. E. DNA modification mechanisms and gene activity during development. *Science* **187**, 226–232 (1975).

Illmensee, K. Nuclear and cytoplasmic transplantation in *Drosophila*, in P. A. Lawrence (ed.), Insect Development, pp. 76–96, Blackwell's Scientific Publications, Oxford (1976).

Lawrence, P. A. The development of spatial patterns in the integument of insects, in S. J. Counce and C. H. Waddington (ed.), Developmental Systems: Insects, pp. 157–211, Academic Press, London (1973).

Lawrence, P. A. and Green, S. M. The anatomy of a compartment border. *J. Cell Biol.* **65**, 373–382 (1975).

Lewis, J. H. Fate maps and the pattern of cell division. *J. Embryol. exp. Morph.* **33**, 419–434 (1975).

Lewis, J., Slack, J. M. W. and Wolpert, L. Thresholds in development. *J. theor. Biol.* **65**, 579–596 (1977).

Lewis, J. H. and Wolpert, L. The principle of non-equivalence in development. *J. theor. Biol.* **62**, 479–490 (1976).

Meinhardt, H. A model of pattern formation in insect embryogenesis. *J. Cell Sci.* **25**, 117–139 (1976).

Postlethwait, J. H. and Schneiderman, H. A. Developmental genetics of *Drosophila* imaginal discs. *Ann. Rev. Genet.* **7**, 381–433 (1974).

Shellswell, G. and Wolpert, L. The pattern of muscle and tendon development in the chick wing, in M. Balls, D. A. Ede and J. R. Hinchliffe (ed.), Vertebrate Limb and Somite Morphogenesis, Cambridge Univ. Press, in press (1977).

Summerbell, D. A descriptive study of the rate of elongation and differentiation of the skeleton of the developing chick wing. *J. Embryol. exp. Morph.* **35**, 241–266 (1976).

Summerbell, D., Lewis, J. H. and Wolpert, L. Positional information in chick limb morphogenesis. *Nature (Lond.)* **244**, 492–496 (1973).

Tickle, C., Shellswell, G., Crawley, A. and Wolpert, L. Positional signalling by mouse limb polarizing region in chick wing bud. *Nature (Lond.)* **259**, 396–397 (1976).

Tickle, C., Summerbell, D. and Wolpert, L. Positional signalling and specification of digits in chick limb morphogenesis. *Nature (Lond.)* **254**, 199–202 (1975).

Warner, A. E. and Lawrence, P. A. Electrical coupling across developmental boundaries in insect epidermis. *Nature (Lond.)* **245**, 47–48 (1973).

Wieschaus, E. and Gehring, W. Clonal analysis of primordial disc cells in the early embryo of *Drosophila melanogaster*. *Develop. Biol.* **50**, 249–263 (1976).

Wolpert, L. Positional information and pattern formation. *Curr. Top. develop. Biol* **6**, 183–224 (1971).

Wolpert, L., Clarke, M. R. B. and Hornbruch, A. Positional signalling along hydra. *Nature New Biol. (Lond.)* **239**, 101–103 (1972).

Wolpert, L., Hornbruch, A. and Clarke, M. R. B. Positional information and positional signalling along hydra. *Amer. Zool.* **14**, 647–663 (1974).

Wolpert, L. and Lewis, J. H. Towards a theory of development. *Fed. Proc.* **34**, 14–20 (1975).

Wolpert, L., Lewis, J. H. and Summerbell, D. Morphogenesis of the vertebrate limb, in Cell Patterning, Ciba Foundation Symposium 29, pp. 95–119, Associated Scientific Publishers, Amsterdam (1975).

Compartments in the Development of *Drosophila*: a Progress Report

PETER A. LAWRENCE

MRC Laboratory of Molecular Biology,
Hills Road, Cambridge CB2 2QH, England

The discovery of compartments in the wing of *Drosophila* was first reported by Garcia-Bellido *et al.* in 1973. In this chapter an attempt is made to indicate why the author regards that paper as an important landmark in developmental biology, and to present a very brief review of the more recent work that it has stimulated.

The technique exploited by the Madrid school was to analyse cell lineage of the normally developing *Drosophila*. This was achieved by marking individual cells with a mutant label (such as *multiple wing hairs*, which changes the cuticular secretion of each cell homozygous for the mutant, Fig. 1). Each marked cell gives rise to a clone of cells which is seen as a patch on the adult. In addition, Morata and Ripoll (1975) have devised a method which makes the marked cell grow much more rapidly than its unmarked neighbours: the same recombination event which combines the mutant alleles of the marker into one cell also recombines *out* a deleterious allele (called a *Minute*). The resulting cell is consequently both marked and fast-growing. We call this method the *Minute* technique. Using this method Garcia-Bellido *et al.* (1973, 1976) found that the cells which generate the wing disc of *Drosophila* are subdivided into two populations from a very early age, each forming a precise region of the wing and the thorax. These regions are called compartments. Even a rapidly growing cell was unable to produce progeny that could cross the line separating the two compartments (Fig. 2).

Studies of cell lineage in another insect, the Lygaeid bug *Oncopeltus*, had shown that the abdominal segments also developed from a small group of founder cells, precise borderlines separating one segment from

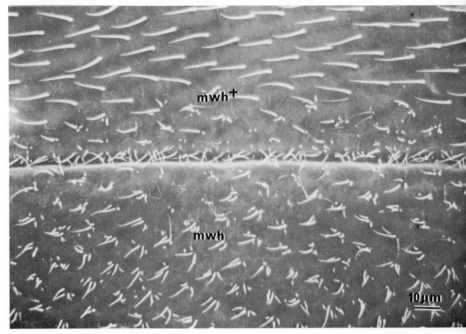

Fig. 1. Scanning electron micrograph of a wing of *Drosophila* heterozygous for *multiple wing hairs* (mwh⁺) showing part of a clone homozygous for the mutant (mwh).

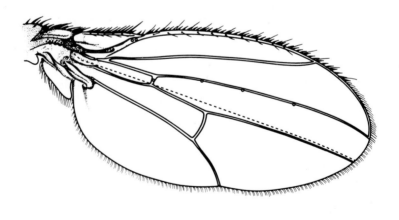

Fig. 2. Drawing of a wing of *Drosophila*, to show the position of the antero-posterior compartment border (dashed line). The border runs in the same place in both dorsal and ventral surfaces of the wing (from Crick and Lawrence, 1975).

another (Lawrence, 1973). Therefore each segment could also be regarded as a compartment. Here it was possible to watch the growth of clones and to show that their failure to cross the segmental boundaries was not due to the segmental epithelia being remote during growth and coming close together in the adult. Indeed, an ultrastructural study of the compartment border showed no special features which could explain either its straightness or the failure of the clones to cross it (Lawrence and Green, 1975). Comparison of clone shapes and sizes in both *Drosophila* and *Oncopeltus* showed that each compartment begins existence as a group of cells (a "polyclone"), each cell having a variable role in different individual insects, even though the prospective fate of the founder group as a whole is precisely defined (see Crick and Lawrence, 1975, for review and references). In this view the adult insect is analogous to a jig-saw puzzle, each piece being a compartment. The pattern runs continuously from one piece to another, but the border-lines of each piece are identically placed in every individual.

One surprising feature of the antero-posterior compartment boundary of the *Drosophila* wing (Fig. 2) is that it does not coincide with any particular structure; without an analysis of cell lineage it could not have been detected. Surprises often lead to new ways of looking at old information, as happened in this case. Garcia-Bellido *et al.* (1973) realized that the line separating the two wing compartments circumscribed the region affected by certain mutations of *Drosophila* which transform one part of the body into another (such mutations are called homoeotic). For example, *bithorax* transforms the anterior part of the haltere into the anterior part of the wing, *postbithorax* the posterior part of the haltere into the posterior part of the wing (Lewis, 1963) and *engrailed* the posterior part of the wing into the anterior part (Garcia-Bellido and Santamaria, 1972). The regions being altered by the mutations appeared to correspond to the compartments. These observations led Garcia-Bellido (Garcia-Bellido *et al.*, 1973, 1976; Garcia-Bellido, 1975) to suggest that the wild type function of homoeotic genes is to control the development of compartments, the genes combining additively to generate a simple binary code. He called such genes "selector genes" because they select a particular developmental program in the cells where they are active.

These exciting ideas have led to a flurry of experiments in the last three years. Steiner (1976) showed that the leg was also compartmentalized with a precisely positioned line separating the anterior from the posterior half. It was a surprise to find that the cells which generate the two wing compartments are already subdivided into two polyclones very early indeed: at or very soon after the blastoderm stage (Steiner,

1976; Lawrence and Morata, 1977) when the embryo is a simple
monolayer of cells and the segment pattern is not visible. Cells marked
at about this time can contribute to both wing and leg (Wieschaus and
Gehring, 1976), although they will not cross from anterior to posterior
in either (Steiner, 1976; Lawrence and Morata, 1977). Thus the cells
which make all the cuticular structures of the mesothoracic cuticle are
divided into only two populations. It is difficult to estimate the precise
number of cells in the mesothoracic polyclones at blastoderm, partly
because we do not know what other structures derive from them. For
example, it is possible that the larval epidermis and the mesothoracic
muscles may be constructed by the same polyclones. However, the
number cannot be large because after the one or two divisions that
occur during postblastodermal development in the embryo (Wieschaus
and Gehring, 1976) the number of cells in the wing disc which generate
cuticle has been estimated at about 20 cells (Lawrence and Morata,
1977).

It was important to see whether the realm of effect of a homoeotic
gene precisely coincided with the compartment, *engrailed*, a homoeotic
mutant which transforms the posterior part of the wing and leg compartments so that they resemble the anterior ones (Garcia-Bellido and
Santamaria, 1972) was the mutant chosen. According to the selector
gene hypothesis the *engrailed*[+] product should be required exclusively in
the posterior polyclone, while the gene should be inactive in the anterior
polyclone. Using the *Minute* technique it was found that mutant
engrailed cells, which were growing in a wild type wing, could fill the
anterior compartment of the wing without affecting its structure, and
defined the antero-posterior border. By contrast any *engrailed* cells in
the posterior compartment showed the mutant phenotype. This experiment showed that all the posterior cells were affected by the mutant
and none of the anterior cells (Morata and Lawrence, 1975; Lawrence
and Morata, 1976).

Furthermore, posterior *engrailed* cells crossed the compartment
border, and could extend some way into anterior territory. This result
suggested that the *engrailed* gene is responsible for "labelling" the cells
of the posterior compartment, so that they will not intermingle with
anterior cells during development (Morata and Lawrence, 1975; Lawrence and Morata, 1976). This effect of selector genes on cell affinities
has been underlined by Morata and Garcia-Bellido (1976) who made
marked clones of cells mutant for *bithorax* in the anterior haltere. These
clones, now lacking the *bithorax*[+] product, sorted out *in situ* and frequently
formed completely separated vesicles of wing structures inside the haltere (Fig. 3). This observation is interesting because it shows that sorting

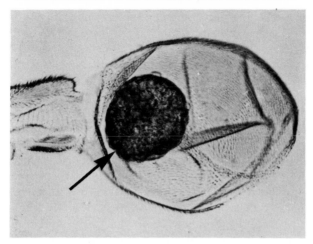

Fig. 3. A haltere of a fly heterozygous for *bithorax*, containing a clone of wing cells homozygous for *bithorax*. These cells have sorted out *in situ* to form an isolated vesicle (arrow) (from Morata and Garcia-Bellido, 1976).

out does not only occur in the test tube following dissociation and reaggregation, but can occur in an intact epithelium. Clearly, cell affinities are important during normal development; they ensure that differently determined polyclones do not intermingle during growth.

Little progress has been made in understanding how polyclones become subdivided, the evidence suggesting that the line is drawn geographically (Crick and Lawrence, 1975). A new mutant, *wingless*, which may affect this process, has recently been described (Sharma and Chopra, 1976) and studied in some detail (Morata and Lawrence, 1977). In mutant flies, in the majority of the wing and haltere segments, the appendages are missing and replaced by an additional set of thoracic structures—the remaining wing and haltere discs develop normally. This all-or-none phenotype, which is quite unlike other homoeotic mutants, might be caused by an error when the wing and haltere discs are subdivided into appendages and thoracic structures. It is noteworthy that the anterior and posterior polyclones (which were separated a long time previously) are always affected coordinately, so that when *wingless* affects a disc the entire wing is transformed into its corresponding anterior and posterior thoracic parts. This suggests that the subdivision is a single positional event in the entire disc and does not occur independently in the anterior and posterior polyclones (Morata and Lawrence, 1977).

Analysis of cell lineage has shown that *in situ* once a cell becomes part of a polyclone it always contributes only to the compartment or compartments that the polyclone generates. The cell is "determined", determination being to a particular area, rather than to a particular cell type (Crick and Lawrence, 1975). Haynie and Bryant (1976) have shown that this state of determination can be broken if the discs are removed, cut and allowed to regenerate in an adult host. Here cells which *in situ* would have given rise only to either dorsal or ventral compartments can now generate parts of both. The stability of determination is not a fixed parameter; it varies with the experimental conditions: e.g. Haynie and Bryant (1976) showed that, under normal conditions, thoracic portions of the wing disc will only generate thorax, but when mixed with other wing cells they can then generate wing structures.

Compartmentalization of the nervous system is another interesting area, although I know of no further progress since the observations of Hasenfuss (1973) and Lawrence (1975) that peripheral nerve axons do not cross segmental boundaries in several insect species. Whether the peripheral nervous system is compartmentalized elsewhere is unknown, and we know nothing about the central nervous system in this regard.

The possibility that compartments are a general feature of animal development remains an open and interesting question. The early binary subdivision of the mammalian embryo into trophoblast and inner cell mass is reminiscent of compartmentalization, in that some changes in cell affinity appear to accompany it and the number of cells is small. There is evidence that at least one further binary subdivision occurs in the two groups of cells, i.e. into mural and polar trophoblast on the one hand, and ectoderm and endoderm on the other (Gardner and Rossant, 1976). The discovery of a defined line respected by branching nerve axons in the amphibian limb is also suggestive of a compartment boundary (Diamond *et al.*, 1976). But much work is needed before we know whether compartments, with their discrete genetic control elements, are quirks of *Drosophila* development or common to animals in general.

ACKNOWLEDGEMENT

I am grateful to Dr Gines Morata for help with the manuscript, and for many of the ideas expressed in it.

REFERENCES

Crick, F. H. C. and Lawrence, P. A. Compartments and polyclones in insect development. *Science* **189**, 340–347 (1975).

Diamond, J., Cooper, E., Turner, O. and Macintyre, L. Trophic regulation of nerve sprouting. *Science* **193**, 371–377 (1976).

Gardner, R. L. and Rossant, J. Determination during embryogenesis, in Embryogenesis in Mammals, Ciba Foundation Symposium 40, pp. 5–18, Elsevier, Amsterdam (1976).

Garcia-Bellido, A. Genetic control of wing disc development in *Drosophila*, in Cell Patterning, Ciba Foundation Symposium 29, pp. 161–182, Associated Scientific Publishers, Amsterdam (1975).

Garcia-Bellido, A., Ripoll, P. and Morata, G. Developmental compartmentalization of the wing disc of *Drosophila*. *Nature New Biol. (Lond.)* **245**, 251–253 (1973).

Garcia-Bellido, A., Ripoll, P. and Morata, G. Developmental segregations in the dorsal mesothoracic disc of *Drosophila*. *Develop. Biol.* **48**, 132–147 (1976).

Garcia-Bellido, A. and Santamaria, P. Developmental analysis of the wing disc in the mutant *engrailed* of Drosophila melanogaster. *Genetics* **72**, 87–107 (1972).

Hasenfuss, I. Vergleichend-morphologische Untersuchung der sensorische Innervierung der Rumpfwand der Larven von *Rhyacophila nubila* Zett und *Galleria mellonella* L. (Lepidoptera). *Zool. Jb. Anat.* **90**, 1–54, 175–253 (1973).

Haynie, J. L. and Bryant, P. J. Intercalary regeneration in imaginal wing disc of *Drosophila melanogaster*. *Nature (Lond.)* **259**, 659–662 (1976).

Lawrence, P. A. A clonal analysis of segment development in *Oncopeltus* (Hemiptera). *J. Embryol. exp. Morph.* **30**, 681–699 (1973).

Lawrence, P. A. The structure and properties of a compartment border: the intersegmental boundary in *Oncopeltus*, in Cell Patterning, Ciba Foundation Symposium 29, pp. 3–23, Associated Scientific Publishers, Amsterdam (1975).

Lawrence, P. A. and Green, S. M. The anatomy of a compartment border. *J. Cell Biol.* **65**, 373–382 (1975).

Lawrence, P. A. and Morata, G. Compartments in the wing of *Drosophila:* a study of the *engrailed* gene. *Develop. Biol.* **50**, 321–337 (1976a).

Lawrence, P. A. and Morata, G. The early development of mesothoracic compartments in *Drosophila*. An analysis of cell lineage, fate mapping and an assessment of methods. *Develop. Biol.* **56**, 40–51 (1977).

Lewis, E. B. Genes and developmental pathways. *Amer. Zool.* **3**, 33–45 (1963).

Morata, G. and Garcia-Bellido. A. Developmental analysis of some mutants of the *bithorax* system of *Drosophila*. *Wilhelm Roux' Årch. Entwickl.-Mech. Org.* **179**, 125–143 (1976).

Morata, G. and Lawrence, P. A. Control of compartment development by the *engrailed* gene of *Drosophila*. *Nature (Lond.)* **255**, 614–617 (1975).

Morata, G. and Lawrence, P. A. The development of *wingless*, a homeotic mutation of *Drosophila*. *Develop. Biol.* **56**, 227–240 (1977).

Morata, G. and Ripoll, P. Minutes: Mutants of *Drosophila* autonomously affecting cell division rate. *Develop. Biol.* **42**, 211–221 (1975).

Sharma, R. P. and Chopra, V. L. Effect of the *wingless* (*wg*) mutation on wing and haltere development in *Drosophila melanogaster*. *Develop. Biol.* **48**, 461–465 (1976).

Steiner, E. Establishment of compartments in the developing leg imaginal discs of *Drosophila melanogaster*. *Wilhelm Roux' Arch. Entwickl.-Mech. Org.* **180**, 9–30 (1976).

Wieschaus, E. and Gehring, W. Clonal analysis of primordial disc cells in the early embryo of *Drosophila melanogaster*. *Develop. Biol.* **50**, 249–263 (1976).

Positional Signalling and Nerve Cell Specificity

R. KEVIN HUNT

Jenkins Biophysical Laboratories, Johns Hopkins University,
Charles and 34th Streets, Baltimore, Md. 21218, USA

The optic fibers differ from one another in quality according to the particular locus of the retina in which the ganglion cells are located. The retina apparently undergoes a polarized, field-like differentiation during development, which brings about local specification of the ganglion cells.... The functional relations established by the optic fibers in the brain centers are patterned in a systematic way on the basis of this refined specificity.

Sperry (1950)

For the latter day neuroembryologist, nurtured on the central dogma and the dynamic chemistry of the cell surface, it is perhaps impossible to imagine it any other way. Certainly, it is difficult to bring together the sorts of mystical "nerve energies" and "functional communions" that used to be thought to bring young neurons into synaptic connection. Yet, that Sperry's words ring obvious to the modern ear reflects, more than anything else, how completely the Theory of Neuronal Specificity has come to dominate contemporary thinking about the development of nerve circuits.

To be sure, the emergence of the Theory has in part followed the emergence of the neuron in contemporary cell biology: the wealth of anatomic, physiologic and biochemical data cataloguing all that is characteristic, recognizable, invariant and (hence) specific about individual neurons and neuronal cell types (Hunt and Jacobson, 1974a). But the Theory is also and foremost an explanatory model, in broad outline, for the self-assembly of interneuronal connections on the basis of unique differentiated properties, cell recognition and differential cytoaffinity among developing neurons. And on this level, the Theory has been supported by the amassing of evidence, especially in the retinotectal system of *Xenopus* and other anurans, which systematically excludes all of the obvious alternatives.

The 70 000 retinal ganglion cells of *Xenopus* make central connections in a number of visual centres; those in the principal and best-studied visual centre, the midbrain optic tectum, are remarkably precise and invariant (from animal to animal) in the three dimensions of tectal space. The topography of connections mirrors, against the tectal surface, the topographic relationships among the corresponding ganglion cell bodies in the retina; and the connections are distributed, according to the physiologic subclass of the ganglion cells ("sustained" responders, "event" detectors, and "dimming" detectors), to different depths beneath the tectal surface. Studies on regeneration of the adult optic nerve in *Xenopus* have shown a re-establishment of patterned connections (including both topography and depth distribution); moreover, rotated retinae regenerate rotated maps, only the correct subpopulation of optic fibres may regenerate into a partially extirpated tectum, fibres can regenerate a normal map via abnormal routes, and visual function appears to play no role in directing optic axons to particular tectal sites (review Gaze, 1970; Jacobson, 1970; Hunt and Jacobson, 1974a; Hunt, 1976a). Developmental studies have related the synaptic behaviour of the ganglion cells to critical periods of determination and differentiation (Jacobson, 1968a; Jacobson and Hunt, 1973). Finally, while the histogenetic processes which could organize retinotectal connections have been shown to be sufficiently ordered to exert "helper effects" in map assembly, they are *not* sufficiently ordered to generate point-to-point connectivity in the absence of true *locus specificity* (Hunt and Jacobson, 1972b). Cell birthdates have been delineated in both structures, but they can be dissociated from map assembly mechanisms (Hunt, 1975b; Chung and Cooke, 1975); cell deaths occur in both retina and tectum, but are insufficient in number to explain map formation by random connectivity and selective survival; lastly, the timing of fibre arrivals into the tectum, though normally organized into waves of successively more peripheral ganglion cells, can be disrupted without serious perturbation of the connectivity pattern (Hunt and Jacobson, 1972a; Hunt, 1975b).

DEVELOPMENTAL PROGRAMMING IN THE RETINA

If this and other evidence compel the inference that retinal ganglion cells must use unique informational properties to target their axons to the correct sites in the retinotectal map, the molecular basis of these

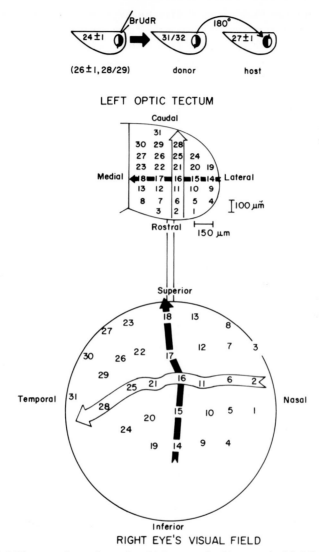

Fig. 1. (*Top*) Diagram of experiment in which an eye-bud is treated with 5-Bromodeoxyuridine at stage 24 ± 1, a procedure that suppresses cytogenesis of all retinal neurons, and then transplanted at stage 31/32, with 180°-rotation, into a normal stage 27 host orbit. (*Bottom*) Visual projection to the optic tectum from an eye treated as above: each numbered position in the visual field shows the optimal position of the stimulus that evoked potentials recorded at the electrode position given by the same number on the tectum. The retinotectal map in this animal is normal, indicating that the eye-bud had retained the capacity to respond to extraocular signals past the usual time to stage 31/32.

locus specificities remains as elusive today as it was when Sperry first invoked these properties 30 years ago. To be sure, the lack of a molecular assay is a severe limitation to the study of the embryonic origins of these properties. Yet, an operational treatment of the properties is possible, using the map itself as an assay of (some parameters of) the set of locus specificites in the retina. Thus, it has been fruitful to perturb early retinal rudiments with surgery, drugs or tissue-culture methods, and observe, from the maps they ultimately assemble with a normal tectum, the effects on positional differentiation of the ganglion cells.

Our earliest experiments were concerned with the problem of how the retina *orients* its pattern of differentiations with respect to the embryo as a whole. Following upon clues from the early studies of Stone (1960) and Jacobson (1968a), we performed a simple input-output analysis in which an essentially binary output (normal or rotated orientation of the final retinotectal map) was interpreted against many different inputs (orientational histories of experimental eyes). For example, early eye-buds at optic vesicle stages 22–28 were grafted in 180°-rotated orientation into a stage 27 host orbit, either (1) directly or (2) after 3 to 5 days of intervening tissue culture; both groups developed upside-down eyeballs, but the orientation of the retinotectal map was *normal* in the first experiment and 180°-rotated in the second. We inferred that orientational information for the specificity pattern was present in stage 22–28 anlage (and stable enough to survive *in vitro*) but *erasable* under the influence of new orientational information from the extraocular microenvironment. In contrast, stage 31 or older eyes always gave rotated maps after both experiments, indicating that the orientational information had now been permanently locked in, or "specified", by stage 31 (Hunt and Jacobson, 1972b, 1973; Jacobson and Hunt, 1973; Hunt, 1975a). By "stacking" additional experimental procedures at the front end of such experiments (e.g. prior drug treatment, or transfer to an older or younger host embryo), we were able to show that the specification events were triggered from within the eye-bud itself and were coupled to cytogenesis of the first ganglion cells (cf. Fig. 1; Hunt and Jacobson, 1974b; Hunt et al., 1977).

Fig. 2. Visual projections to the tectum in five (a–e) *Xenopus* froglets. These animals were final carriers of eyes, in normal orientation, which had spent three days in *rotated* orientation on a *Rana pipiens* temporary host. The same sequence of electrode positions was probed in all five animals and only one tectum is shown (*top, left*). Though a few animals gave normal (a) or disorganized (b) maps, most maps were 180°-rotated (c, d). A few "hybrid" maps (e) developed, in which the AP-component derived from the *Rana* host, and the DV-component was normal and originated from either the donor or carrier *Xenopus*.

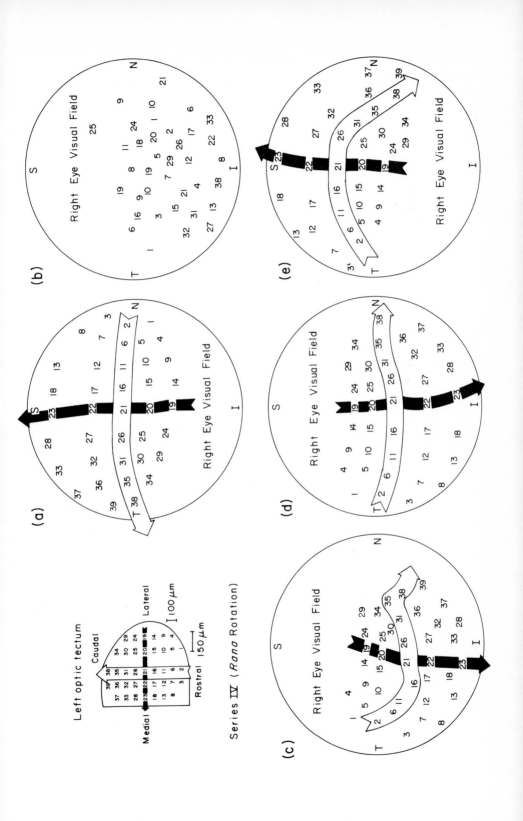

(a)

N
S
Right Eye Visual Field
T
I

(b)

S
N
Right Eye Visual Field
T
I

(c)

S
N
Right Eye Visual Field
T
I

(d)

S
N
Right Eye Visual Field
T
I

(e)

S
N
Right Eye Visual Field
T
I

Left optic tectum

Caudal
Lateral
Medial
Rostral

100 μm
150 μm

Series IV (*Rana* Rotation)

In more ambitious experimental designs, serial rotations were carried out to study orientational signalling between the embryonic tissues and early eye-buds (Hunt and Jacobson, 1972a; Hunt, 1975a; Hunt and Piatt, 1977). For example, a stage 22–28 eye was grafted in 180°-rotated orientation into a temporary host and, 3 days later when the eye was well past stage 31, *rotated* back to normal orientation into the orbit of a final carrier embryo; in such experiments, the eyeball grows up in anatomically normal orientation but, for many kinds of test hosts, assembles a 180°-rotated map across the carrier tectum. Such a map indicates that ganglion cell specificities were patterned according to orientational information from the test host, and provides a sensitive assay for the presence of effective orientational signals in the test host. Thus, when a foreign species such as *Rana pipiens* (Fig. 2) or *Ambystoma maculatum* (Hunt and Piatt, 1977) served as a test host with this result, the orientational signals were shown to cross genus and species barriers; the use of flank sites in *Xenopus* hosts or of *Xenopus* hosts of much earlier or later stage established the ubiquity of such signals in space and time in *Xenopus* embryogenesis; shortening the period of time spent on the temporary host showed that signalling is very rapid and completed in 2 to 6 hours (Hunt and Jacobson, 1974).

This last experiment also showed that orientational information has two components, one related to the orientation of the eye-bud with respect to the anteroposterior (AP) axis of the embryo's body and one related to the dorsoventral (DV) axis, and that the signalling events between an embryo and a rotated early eye-bud proceed much more rapidly for the AP component than the DV component. Thus, serially rotated eyes (especially those rotated for brief intervals) may assemble "hybrid" maps which show the rotated orientation in the AP axis (i.e. are backwards across the rostrocaudal axis of the tectum) and the normal orientation in the DV axis (see Fig. 2e). The independence of these two axial components of orientation is also evident in the fact that they appear to be locked in and specified at slightly different times between stages 28 and 31 in *Xenopus* (Jacobson, 1968a; Hunt and Jacobson, 1972b), although such a temporal disparity has not been detected in other species (cf. Stone, 1960; Hunt and Piatt, 1974).

What emerges from these studies is a view of the retina in which a developmental program for ganglion cell patterning, a program with two orthogonal axial components that will be used for positional reference by all 70 000 ganglion cells the retina will ultimately contain, is gradually evolved during the early phases of ocular morphogenesis. The optic vesicle contains information about its AP and DV orientation on the embryo and yet retains responsiveness to extraocular AP and

DV signals, signals which are ubiquitous over much of the embryo and over many stages of development. These same signals might be used for organizing a number of organ rudiments in the embryo (Wolpert, 1971) which could explain their ubiquity and their conservation during evolution of the amphibia; the protracted period of responsiveness of *Xenopus* eye-buds to these cues might serve as an error-correction mechanism to compensate for geometrical distortions in the eye-bud during its outgrowth and invagination (Hunt, 1975a); but such speculations require further experimental test. Nevertheless, precisely when the anlage needs a definitive plan for positional differentiation of its ganglion cells, as the first ganglion cells commence withdrawal from DNA synthesis at stages 28–31, the two axial components of the retinal program are locked in. The temporal constraint imposed by the appearance of the first ganglion cells appears to have been met by triggering the specification events from within the eye and coupling the trigger directly to the cytogenesis of the first cells (Hunt et al., 1977).

COMBINATORIAL SPECIFIERS ON RETINAL GANGLION CELLS

If programming events at stages 28–31 establish the axial plan for retinal locus specificities—a decisive sequence of events on the second day of embryonic life—the execution of the plan is a protracted and poorly understood process extending through larval and early post-metamorphic stages. The 70 000 ganglion cells which comprise the full adult complement are added to the retina gradually at its peripheral margin from the monotonic division of a ring of neurogenic stem cells at the ciliary margin of the eye (Jacobson, 1968b; Straznicky and Gaze, 1971). What is known about the way in which individual ganglion cells interface with the axial plan is fragmentary and has been reviewed elsewhere (Hunt, 1975a, 1976b). Positional signalling is known to continue *within* growing retinae, despite their insensitivity to signals from outside the eye; the ring of precursor cells may plan a role in modulating positional signals, though they are individually dispensable to the execution of the plan; and there is evidence for important interactions between the nascent neuroblasts at the extreme retinal periphery with the subjacent older neurons generated earlier. Nevertheless, the details of the process by which individual ganglion cells actually acquire information about their retinal positions remain a mystery, obscured to a large extent by our ignorance of the results of this process: that is, of the way specificity is encoded across the final ganglion cell population. Recently, however, it has become possible to exploit our knowledge of orientational signalling to produce "scrambled" retinotectal maps and so address the coding problem directly.

The magnitude of the coding problem is perhaps best conveyed by the precision with which the adult map has been documented electrophysiologically—not only in normal animals but in frogs subjected to the experimental manipulations from which the inference of specificity derives. The surface topography of the map has been documented at ten ganglion cell diameters resolution, identifying some 700 unique "subspaces" in the adult retina; and each of these sends its three subclasses of ganglion cells to precise depths beneath the tectal surface. The system is required to make 2100 discriminations in the course of map assembly in order to generate the biological result of a ten cell diameter map. If each ganglion cell contains only one informational codeword, or *specifier*, on its fibre tip, then a system of 2100 such specifiers is required by the system. It is perhaps somewhat premature at this stage to speculate on the limits of the genome, in generating cell diversity, and of cells, in recognizing it (Hunt, 1976b), based on precedents from genetics and cell physiology. Nevertheless, a combinatorial code based on several specifiers per cell could generate 2100 discriminations with much less informational strain. In fact, the retina of *Xenopus* uses such a combinatorial code and may require as few as sixty-three specifiers from three different classes to assemble the entire map.

Using what was known about the susceptibilit of early eye-buds to orientational signals from the embryo (see above), we rotated eye-buds two to six times at short intervals between stages 24 and 33. In a substantial number of cases the signal-response apparatus in the anlage was "jammed" and the eye developed without a definitive axial plan. The resulting retinotectal map in the final carrier embryo was completely scrambled with respect to the surface topography of the tectum (Figs 3 and 4). To make absolutely certain that the abnormal visual projection was not due to secondary optical or retinal defects, we recorded from many individual optic fibres at their terminals and from many individual intertectal relay neurons. The optic fibres had normal response properties and normal-sized receptive fields, while the intertectal cells had huge and patchy receptive fields. Thus, the individual ganglion cells appeared normal, yet each tectal locus received terminals from ganglion cells randomly distributed across the retina (Fig. 4).

In a second recording session, frogs with completely scrambled surface maps were studied for the depth distribution of optic fibre units in the tectum. Remarkably, the three physiologic classes of ganglion cells segregated to different depths beneath the tectal surface with a pre-

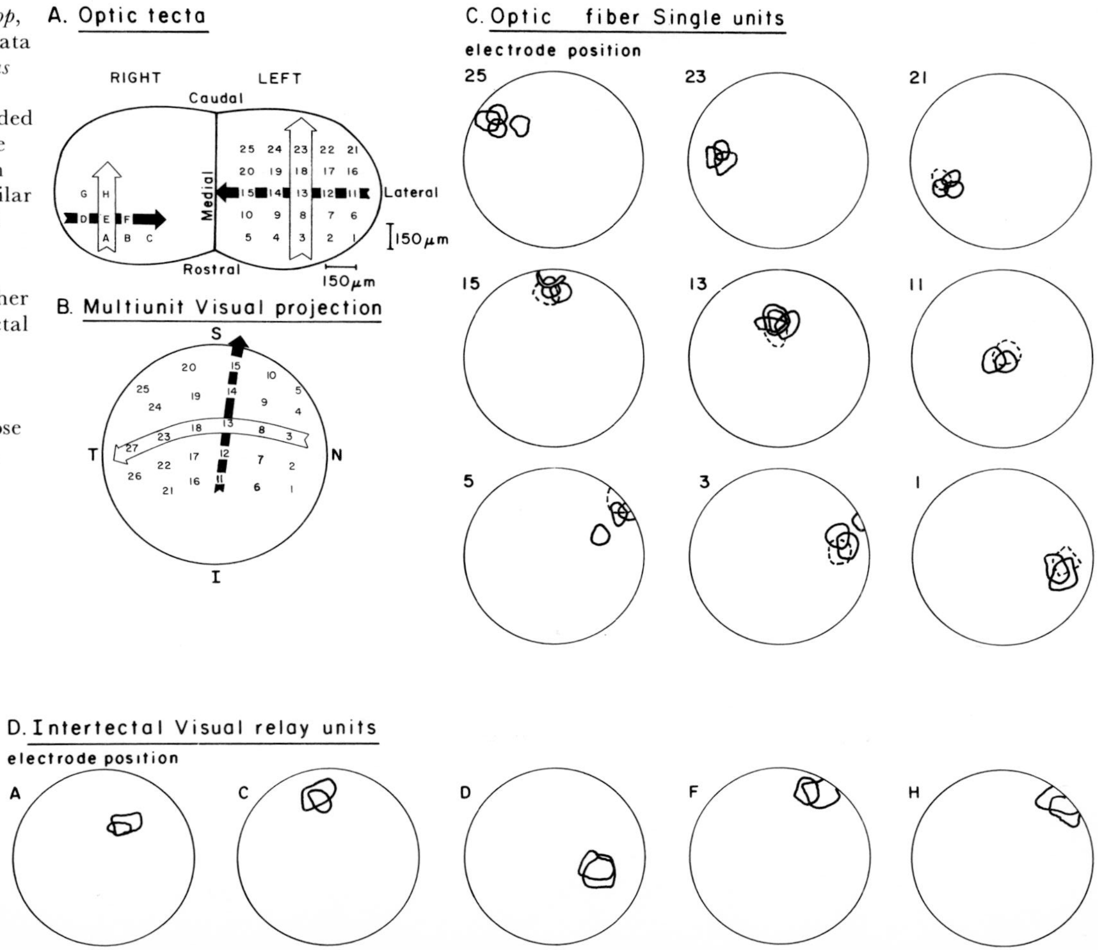

Fig. 3. Visual map (*top*, *left*) and single-unit data from a normal *Xenopus* froglet. Optic fibre single units (C) recorded at any given electrode position in the tectum were driven from similar positions in the visual field, indicating that their ganglion cell bodies lay close together in the retina. Intertectal visual relay units (D) showed similar clustering, indicating that ganglion cells close together in the retina converged on single tectal output cells.

A. Optic tecta

C. Optic fiber Single units
electrode position

B. Multiunit Visual projection

D. Intertectal Visual relay units
electrode position

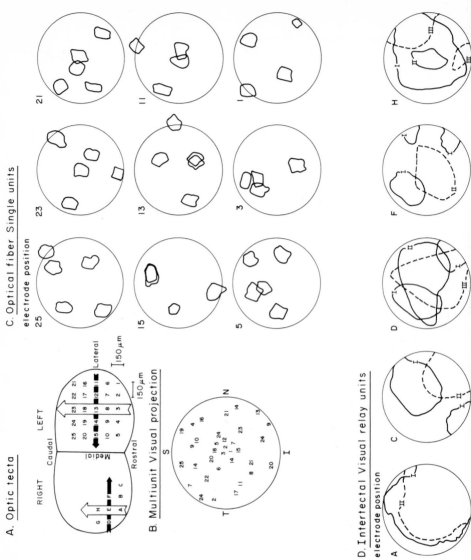

Fig. 4. Scrambled visual map (*top, left*) produced as described in the text, showing random targeting of single optic fibre units (C) and large and scattered range of input to single intertectal units (D).

A. Optic tecta

B. Multiunit Visual projection

C. Optical fiber Single units
electrode position

D. Intertectal Visual relay units
electrode position

cision indistinguishable from control frogs (Fig. 5). The results provided an important internal control that we were not recording from fibres of passage in our efforts to document scrambling of the surface topography of the map (see Hunt and Jacobson, 1974a). More importantly, however, they demonstrate the existence of a distinct specifier, related to ganglion cell class, which (independent of surface topography) discriminates map depth in the tectum.

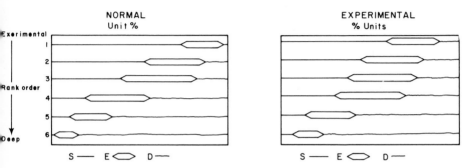

Fig. 5. Depth distribution of the three classes of ganglion (sustained cells, S; event cells, E; dimming cells, D) in a normal (*left*) and a scrambled map experimental (*right*) froglet. On line 1, all units which were the first unit encountered in a penetration normal to the tectal surface are considered, and the percentage of each class is shown. On line 2, all units recorded second in a penetration are considered, and so forth. Depth distribution in the frog whose retinal map was scrambled with respect to surface topography of the tectum, as in Fig. 4, is essentially normal.

With variations on the embryologic manoeuvres, such as repeated left-to-right transplantation of eye-buds, we have been able to produce other varieties of partially scrambled maps which are 1) accurately ordered with respect to the AP axis of retina and tectum but chaotic with respect to the DV axis of the retina and mediolateral axis of the tectum, or 2) accurately ordered with respect to the DM/ML axes and scrambled in the AP axes (Hunt, 1976a). Single-unit studies revealed that the optic fibres showed normal responses and receptive fields of normal size and organization; yet, instead of mapping point-to-point, the ganglion cells had mapped into the tectum as a family of nearly vertical or nearly horizontal lines across the retina. The depth distribution was again normal and, as expected, individual intertectal cells showed long "banana-shaped" receptive fields matching the ganglion cell lines in orientation. We inferred that at least two additional specifiers exist on the tip of each optic fibre: one related to the cell's AP position in the retina that independently discriminates rostrocaudal (AP) map level in the tectum; and a second related to the cell's DV position in the retina that independently discriminates mediolateral map level in the tectum.

Since the adult retina contains about 300 ganglion cells across its equator, and the map is documented at ten cell diameters, the 2100 discriminations required to assemble the map in *Xenopus* may require as few as thirty AP specifiers, thirty DV specifiers and three class specifiers distributed in a Cartesian pattern across the retinal ganglion cell population. These results offer many implications, but two are particularly worth noting. The first is that developmental information may flow and evolve along channels which are largely independent and non-overlapping: from AP orientational signals to AP orientational information, to specification of a definitive AP reference asix at stages 28–31, to AP positional information and finally an AP specifier on the fibre tip (Hunt, 1977). The second is that such an XYZ code could effect extraordinary economy in neurogenetic processes beyond the retina: simply by varying the "class specifier", and defining class compatibility as a necessary condition for synaptogenesis, the organism could use the same set of thirty by thirty topographical specifiers to assemble all the topographic projections in the central nervous system.

It is hoped that future experiments will test these and other hypotheses, and that these initial sorties into the specifity code will prove useful beyond our immediate understanding in the evaluation of molecular differences across the retina and of positional signalling during retinal growth. In particular, since the information for map assembly is Cartesian at the outset and Cartesian at the end-point, it may well prove to be Cartesian throughout. This may provide a starting point for sorting out the complicated and nearly uninterpretable results on growing post-stage 31 eyes, in which surgical rearrangement and recombination of retinal fragments have produced bizarre departures from the original axial plan, on the basis of radial transformation of Cartesian information during annular growth of the eye.

SUMMARY

This chapter has presented a brief review of the evidence for unique informational properties on retinal ganglion cells, the origin of this locus specificity in the early orientational signals of the embryo and programming event in the optic cup, and the evolution of AP and DV orientational information into a ternary code of specifiers that guide the fibre to its map target in the tectum. Returning to Sperry (1950), his Theory of Neuronal Specificity may now seem obvious or even superfluous amid the data on the mechanics of patterning and the diversification of neurons, but a sense of history only illuminates its power. Indeed, perhaps the best evidence for the magnitude of his private heresy is the fact that Sperry predicted (1950), before the existence of the map was even known, the AP and DV specification of the retina as well as the existence of molecular specifiers on the fibre tip, whose identity will remain a major focus of experimental attention for years to come.

ACKNOWLEDGEMENT

This work was supported by research grants from the US National Institutes of Health NS-12606) and the US National Science Foundation (BMS-75-18998).

REFERENCES

Chung, S.-H. and Cooke, J. Polarity of structure and of ordered nerve connections in the developing amphibian brain. *Nature (Lond.)* **258**, 126–132 (1975).

Gaze, R. M. The Formation of Nerve Connections, Academic Press, New York (1970).

Hunt, R. K. Developmental programming for retinotectal patterns in cell patterning. *Ciba Foundation Symposium* **29**, 131–159 (1975a).

Hunt, R. K. The cell cycle, cell lineage and neuronal specificity. *Res. Probl. Cell Differ.* **7**, 43–62 (1975b).

Hunt, R. K. Informational properties of growing optic axons. *Biophys. J.* **16** (1976a).

Hunt, R. K. Position-dependent differentiation of neurons, in D. McMahon and C. F. Fox (ed.), Developmental Biology, W. A. Benjamin, Menlo Pk., California (1976b).

Hunt, R. K. Combinatorial specifiers on retinal ganglion cells for retinotectal map assembly in *Xenopus*. *Nature (Lond.)* In press (1977).

Hunt, R. K. and Jacobson, M. Development and stability of positional information in *Xenopus* retinal ganglion cells. *Proc. nat. Acad. Sci. (Wash.)* **69**, 780–783 (1972a).

Hunt, R. K. and Jacobson, M. Specification of positional information in retinal ganglion cells of *Xenopus*: Stability of the specified state. *Proc. nat. Acad. Sci. (Wash.)* **69**, 2860–2864 (1972b).

Hunt, R. K. and Jacobson, M. Assays for the unspecified state. *Proc. nat. Acad. Sci. (Wash.)* **70**, 503–507 (1973).

Hunt, R. K. and Jacobson, M. Neuronal specificity revisited. *Curr. Top. develop. Biol.* **8**, 203–259 (1974a).

Hunt, R. K. and Jacobson, M. Intraocular control of the time of specification. *Proc. nat. Acad. Sci (Wash.)* **71**, 3616–3620 (1974b).

Hunt, R. K. and Jacobson, M. Rapid realignment of retinal axes in embryonic *Xenopus* eyes. *J. Physiol. (Lond.)* **241**, 90–91P (1974c).

Hunt, R. K. and Piatt, J. Axial specification in salamander embryonic eye. *Anat. Rec.* **178**, 515 (1974).

Hunt, R. K. and Piatt, J. Cross-species axial signalling with reversal of retinal polarity in embryonic *Xenopus* eyes. *Develop. Biol.* In press (1977).

Hunt, R. K., Bergey, G. K. and Holtzer, H. 5-Bromodeoxyuridine: Localization of a developmental program in *Xenopus* optic cup. *Develop. Biol.* In press (1977).

Jacobson, M. Development of neuronal specificity in retinal ganglion cells of *Xenopus*. *Develop. Biol.* **17**, 202–278 (1968a).

Jacobson, M. Cessation of DNA synthesis in retinal ganglion cells correlated with the time of specification of their central connections. *Develop. Biol.* **17**, 219–232 (1968b).

Jacobson, M. Developmental Neurobiology, Holt, New York (1970).

Jacobson, M. Histogenesis of retina in the clawed frog with implications for the pattern of development of retinotectal connections. *Brain Res.* **103**, 541–546 (1976).

Jacobson, M. and Hunt, R. K. The origins of nerve-cell specificity. *Sci. Amer.* **228 (2)**, 26–35 (1973).

Sperry, R. W. Neuronal specificity, in P. Weiss (ed.), Genetic Neurology, University of Chicago Press, Chicago (1950).

Stone, L. S. Polarization of the retina and development of vision. *J. exp. Zool.* **145**, 85–93 (1960).

Straznicky, K. and Gaze, R. M. Growth of the retina in *Xenopus laevis*: An autoradiographic study. *J. Embryol. exp. Morphol.* **26**, 69–79 (1971).

Wolpert, L. Positional information and pattern formation. *Curr. Top. develop. Biol.* **6**, 183–224 (1971).

Organizing Principles for Anatomical Patterns and for Selective Nerve Connections in the Developing Amphibian Brain

JONATHAN COOKE

Developmental Biology, The National Institute for Medical Research, Mill Hill, London NW7 1AA, England

INTRODUCTION

Two principles of organization seem to be involved in generating the spatial pattern of a functioning central nervous system in vertebrates. First, there is the anatomical pattern involving separate brain parts, each with its distinctive location and histological structure, and then there are the systems of neural projection, whereby arrays of cells within those parts send precisely patterned sets of connections to arrays in other parts. I hope here to convey the essentials of some experimental work, on the developing brain of the amphibian *Xenopus*, which suggests that at least analogous processes may be at work in controlling both the polarity of the brain's anatomical pattern, and that of a neural projection system within it.

During early development of the neural plate as a whole, various territories become delimited, occupying characteristic spatial positions and proportions, wherein determination occurs for development of each anatomical region of the central nervous system. Since the plate and early neural tube are, topologically, a sheet of cells, this problem is essentially one of the interpretation of pattern-forming information along two dimensions (medio-lateral and antero-posterior), to give rise to a subsequent programme of determined, sequential cell divisions on the part of each group of neuroblasts. On this basis is probably produced the series of differently specialized neuron types that make up the radial histological structure and circuitry characterizing each brain region (Sidman, 1974; Rakic, 1974), e.g. cerebellum, cerebrum, thalamic nuclei, etc.

The gross anatomical patterning of the brain can fairly well be accounted for in principle in this way, but there is no strong evidence that the early neural plate is ever itself the site of a single field of positional information (Wolpert, 1969). From studies of primary embryonic induction, and the transplantation of mesodermal inducing tissue between embryos (Spemann, 1938; Cooke, 1972), it rather seems that the pattern of differently determined regions within the plate is specified indirectly from the underlying cell sheet, the inducing mesoderm. Positional information within the latter, which does have true regulative capacity after surgical disturbances, is expressed (1) through local development of particular mesodermal structures and (2) through a patterned modulation of the inducing activity so that the various local neural structures are specified in overlying neurectoderm.

Earlier work (Jacobson, 1964; Corner, 1963) has shown that irreversible specification of the pattern of brain parts, discussed above, occurs during the rolling up of the plate into a neural tube. This is still apparently undifferentiated, but surgical extirpation or rotation of quite small areas within it leads to development of brains with defined parts either missing or developed in reverse spatial sequence.

What of the second principle of organization, that of selective nerve connection? Many surgical studies have been performed on the eye-cup, the peripheral component of the model amphibian retino-tectal projection system (Gaze, 1970). In this system, visual space is represented through the strictly polarized pattern of retinal cell connections made across the primary visual area of the brain, the tectum of the midbrain. It is known that the eyecup becomes polarized with respect to its tendencies for connection to the future tectum at an early embryonic stage. Later on, when the first neurons have differentiated at its centre and are sending axons which will within 24 h (in *Xenopus*) colonize the tectum, this polarization in the eyecup becomes irreversibly determined. In these experiments, the eyecup is treated as a unit in transplantation and rotation operations.

The normal topography of this system is shown schematically in Fig. 1, from which I hope that the retino-tectally uninitiated can derive the essential information, free of too much neuroanatomical jargon. Small areas of the retina's radial structure connect, via the ganglion cell axons, only to small areas of the tectum radial structure, to create a map there with a systematic, polar organization along the dorso-ventral as well as the antero-posterior visual axes. Only the latter

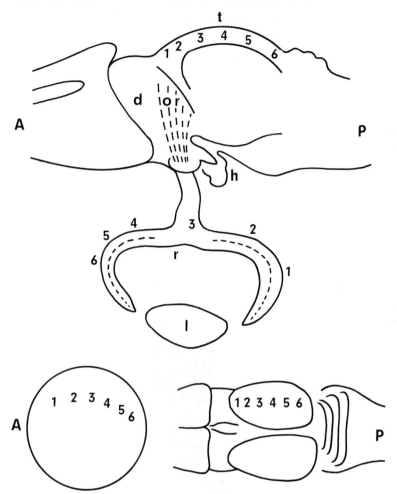

Fig. 1. The normal retino-tectal projection. Electrode positions, on a row along the optic tectum, are seen in schematic longitudinal section. Corresponding visual field positions, stimulation at which evokes retinal ganglion-cell axon activity at each tectal location, form an antero-posterior row, representing in inverted form an equivalent zone in the retina (see in horizontal section). The "map" in terms of visual field positions from a typical tectal row is shown below. t, tectum; d, diencephalon; o.r., optic radiation (optic nerve after entry into brain, ascending the side walls at the diencephalic/midbrain junction); r, retina; l, lens; h, hypophysis; A, anterior; P, posterior.

axis will be considered here. In normal animals anterior visual stimuli, stimulating of course posterior retinal cells via the lens, cause recordable physiological activity only at anterior positions on the tectum, and systematic tracking with stimuli towards the posterior visual field (anterior retina) results in a corresponding shift of the location of recordable activity towards the posterior tectum.

The nature of the patterned information across the population of retinal precursor cells that results in setting up such an ordered array of central connections is unknown, but its overall polarity can be reversed within eyecups by transplanting them, before their critical stage of development mentioned above, in reversed anatomical polarity (back to front) into their own or another orbit. Such an operation results in a normal map, and this must involve most of the cells in re-assessment of their properties as the precursors of cells whose axons will compete with those of their neighbours in establishing connections at the tectum. After the critical embryonic stage of eyecup development, such re-assessment (repolarization at the whole pattern level) cannot occur, and the operation necessarily results in an abnormal map of visual space at the tectum, ordered but with reversed polarity.

I purposefully introduce the idea of competitive properties just here. On various grounds (Gaze and Keating, 1972; Gaze et al., 1974; Prestige and Willshaw, 1975) the prevailing view is that the mechanism of setting up such systems of selective connections involves competition rather than the matching up of exact sets of labels between members of one set of cells and their targets in the other. An appropriate concept seems that of a graded competitive property, and hence the need for a polarized system of pattern-forming information developmentally, across the cell array of the retina. So far as the tectum is concerned, a similarly polar property is required during development, but not all theories would require an actual graded signal across it. An "arrow" property, directing the wanderings of invading axons, might be sufficient (Hope et al., 1976), or even a single reference point at one edge.

For the eyecup (retina), the experiments tell us that at the critical developmental stage, the capacity to preserve the graded array of cell properties, and its polarity during further growth, is rather suddenly fixed within the tissue. By combining physiological mapping studies of the developed retino-tectal system with the type of operation mentioned earlier for studying anatomical patterns and their fixation in the early brain rudiment, one can ask the following interesting questions. Is the developmental polarity for selective nerve connections within the tectum or its precursor cells also "fixed" at some critical time in embryonic life, before the connecting axons arrive? If it is, then surgical reversal of tracts of brain tissue, after this time, will cause reversed mapping to occur (relative to the animal's nose and tail) from a normal eyecup. Further, what is the timing relationship, if any, between events that irreversibly polarize brain tissue with respect to the first principle of organization, the spatial positioning of brain regions as such, and those that polarize tissue *within* such regions with respect to the principle of organized neural projections? With respect to this second principle, the differences amongst cells within each of the arrays that are to be matched up are much subtler than those between adjacent anatomical regions of the brain, to say the least. They are inaccessible to physiology, microscopy or biochemistry, and may only now be yielding to the most fine-grained, quantitative immuno-logical techniques available (Edelman, 1976). But the intercellular signalling responsible for organizing the two types of spatial specificity in developing brain, which we might call the anatomical and the functional, may be similar or even identical.

BRAIN OPERATIONS AND THE RETINO-TECTAL MAP

Dr Shin-Ho Chung and the author at Mill Hill, have been able to begin investigating some of these questions, and I should like to try and transmit the essential results to date. Figure 2 shows the type of brain operation that can be performed on *Xenopus* at any time from stage 21, when the future brain is anatomically a simple, newly rolled-up tube, to stage 37 when only a few tracts of rotated tissue heal into the brain in an integrated way. Subsequently to this, the differentiating tissues have proved too fragile and sticky for us to operate successfully, although the tectal rudiment is not invaded by the optic fibres until, say, stage 43 some 15 h later. Anatomical results of the operation all confirm that, before stage 21, the neural plate/tube has become a mosaic of determined regions for the basic brain parts. Such results are of several categories. We are often able to derive electrophysiological maps of the connections, from the untouched eye, to tectal tissue in the transformed brains at the time of metamorphosis, when such maps in normal tadpoles are well developed.

Figures 3 and 4 exemplify the commonest class of abnormal brains, and the maps that they exhibit. Instead of one discrete tectum on each side, two variably asymmetrical pairs of tecta in series are seen. For various reasons we are confident that the grooves between such pairs of tecta represent the position of the posterior cut in the operation, whereby one tract of the tissue determined as midbrain roof (tectum) has been rotated and has differentiated to form a unit of tectal structure of its own, followed by another unit derived from the remaining, unrotated embryonic midbrain tissue.

In normal development there is a well-defined time-sequence across the tectum, for the final expression of the cell programme to give the

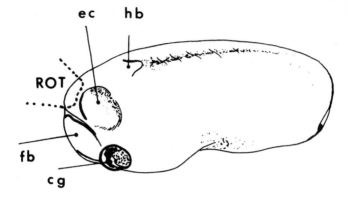

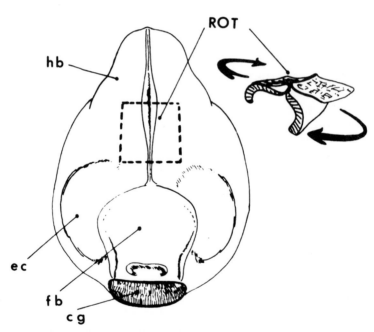

Fig. 2. Operations producing abnormal anatomical sequences of brain structure. During 20 s stages, the brain rudiment looks a relatively undifferentiated tube, but rotation of dorsal sectors of this tube through 180° results in reversed sequences of structure in the brains that then develop. The early system is thus a mosaic one with respect to basic brain anatomy. ROT, rotated piece; ec, eyecup; hb, outline of hindbrain; fb, outline of forebrain; cg, cement gland.

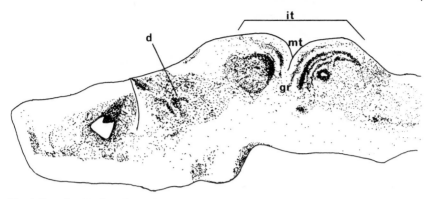

Fig. 3. Longitudinal section of brain showing doubled tectal structure. d, diencephalon; mt, mature tectal structure; it, immature tectal structure; gr, groove between tecta.

definitive radial histological structure (see Introduction, p. 111) at each level. In longitudinal section, such differentiation is seen earliest at the front edge only, and proceeds posteriorly during larval and postmetamorphic life, with cells at each level ceasing division and beginning the sequence of histogenesis. In the abnormal brains being considered, that tissue which we belive to have been rotated often appears to have differentiated in reverse, tailward-to-headward sequence. That is, it has kept to its original presumptive schedule of developmental maturation despite rotation. This might be because such a schedule is an intrinsic part of the polarity for anatomical pattern, already fixed at operation stages. There is an alternative explanation, however. We have electrophysiological and histological evidence that, in such brains, the optic nerve input as a whole invades the tectum from the intermediate groove, or even from the rear of a completely reversed tectum, rather than as normally from the forebrain/midbrain boundary at the front. Local guidance factors for this are presumably already imbued in certain cells at the time of operation, that are often included in the rotation. It could be that invasion of the optic axons, which is progressive across the tectum during development of the projection or "map", also controls the progressive onset of postmitotic cell states and histogenesis with time. The timing of cell development, in this case, would be determined more immediately by cell–cell interactions than is assumed on the previous hypothesis. Autoradiographic studies in conjunction with these operations are proceeding, and should help us distinguish between the hypotheses of intrinsic programming and of response to optic innervation to explain the time gradient of tectal maturation.

The most striking information to come out of such altered brain

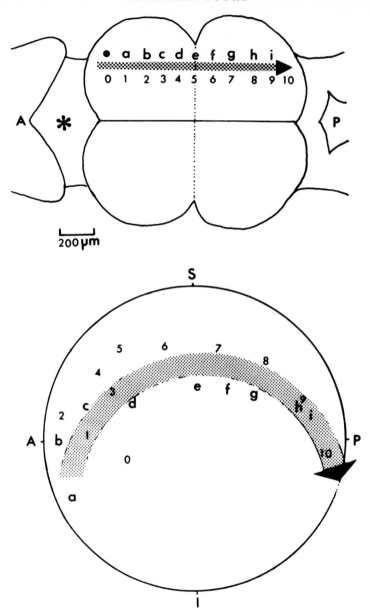

Fig. 4. Normally polarized, single map of visual space occupying both tecta of a brain like that of Fig. 3. Innervation was from the intertectal groove, corresponding with middle visual field (see text).

structures as these concerns the functional polarity of the tectal tissue for map formation by the retinal axon array. This is normal, relative to the whole animal's head–tail axis. Thus a single, whole, normally polarized representation of the visual field or retina is spread over the total available tectal tissue regardless of the presence of (1) intertectal grooves, (2) abnormal position of invasion of the optic fibres of (3) the sequence of histogenesis. This means that information for polarity, whatever its nature, can be re-organized within transposed or rotated tracts of tissue so that it coordinates with other tissue to form a coherent set of targets for the incoming array of optic fibres. The most posteriorly originating of these fibres form systematically more stable terminations at the front end of all tectal tissue, as in the normal brain.

This result has been seen when rotation was as late as stage 37. It means that until this stage at least, many hours after stable polarization of the eyecup, tissue outside the tectal rudiment generates some form of signal to which the whole array of tectal cells responds, causing re-organization of polarity for functional connections within it.

Such maps also suggest that there may be no time of irreversible polarization within the tectum, prior to actual invasion by optic input, but that if there is such a time it is very much later than that for polarized determination of the brain's anatomical pattern of parts. I should say that another line of experimental work, on *regenerating* maps after optic nerve section in adult goldfish and frogs, is suggesting that once innervation and original setting up of the map has occurred there *is* then a fixed polarity within tectal tissue for reconnections with the eye. But in investigating the original development of maps, and then their regeneration in adult life, we may be studying two partially different sets of phenomena.

If cells outside the tectal area are continuously emitting a signal during development, that can organize polarity within tectal tissue, where are these cells and what is the nature of the polarizing information? It is recognized that polarizing information for the more familiar anatomical patterns in regulative embryos has so far eluded investigation at cellular and biochemical levels. The polarity organizing the brain's anatomical pattern, which ultimately derives from the mesoderm beneath, is a classical example. We don't expect, therefore, a rapid answer to the second part of the question, but can note that the actual information used for achieving first anatomical and then functional patterning within the brain need not be different. Only the times of critical cellular response, by determination, are definitely different.

A further class of results for early brain operations (Figs 5 and 6) appears to be providing a guide to the location of origin of the func-

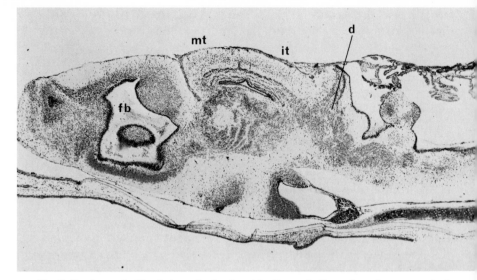

Fig. 5. Section as in Fig. 3, but of a brain with diencephalic region absent anteriorly and developing behind the tectum. d, diencephalon; mt, more mature tectal structure; it, less mature tectal structure; fb, telencephalon (cerebrum).

tional polarizing information. We find that whenever, following operations, a significant unit of diencephalic (thalamic) structure develops behind some or all of the tectal tissue, then the map does show reversals of specificity within that tract of tissue. According to the various abnormal maps we obtain, the rule that is followed is that axons deriving from the rear of the normal retina (stimulated by anterior visual field) find preferential connections amongst tectal cells nearest to diencephalic tissue, and the map is organized accordingly.

It is difficult reliably to adjust the operation to produce, systematically, brains either with normally situated diencephalon immediately anterior to tectum, or else with such structures interpolated elsewhere. We take this to mean that the group of cells determined to form diencephalon is very small in the 20 s stages brain rudiment, and is just sometimes included in the rotated piece to be transferred posteriorly. Thus various combinations are possible. If diencephalon develops between two tracts of tectum in the antero-posterior axis, then the map across the anterior tectum shows reversed polarity while that across the posterior one is normal. Sometimes two sets of diencephalic structures appear, one in normal place and one behind an intervening tectum. We assume that in such cases part of the group of cells was left *in situ* and part transferred posteriorly. The complex maps we then get

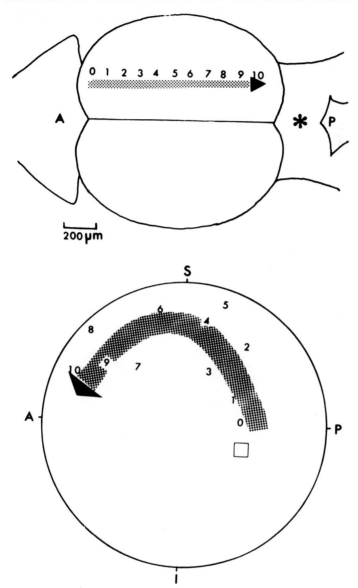

Fig. 6. Reversed map of visual space, relative to the animal's nose and tail, in a brain similar to that of Fig. 5.

indicate that each set of cells competes with the other in organizing the intervening projection. Thus some posterior and some anterior retinal fibres project to each end of the tectum, or else anterior retina projects only to the middle of the tectum, giving a bipolar representation of the visual field physiologically.

Part or all of the optic nerve input usually invades tectal tissue from the rear in cases where diencephalic structure develops there. We should expect this, since the previous results show us that factors guiding the optic tract are intrinsic to cells just posterior to the normal diencephalic region. The whole set of results thus reveals that map polarity and polarity of innervation are separable features, the former controlled only by the location of cells outside the tectum, the latter probably by cells at its very anterior edge.

THE ORGANIZER AND THE RETINO-TECTAL MAP

There is an interesting link between the foregoing story, and the earlier control of the pattern of anatomical differentiation in the gastrula mesoderm (and hence, the overlying neurectoderm). The primitive diencephalic area overlies, and is specifically induced by, just that part of the embryonic mesoderm whose initial position at the start of gastrulation was in the beginning dorsal lip. In grafting experiments, only implants of dorsal lip that include this tissue characteristically organize a *whole* new body pattern of differentiation in conjunction with host embryos' tissues. Later lips, or those where some of the inner, earliest migrating cells are left behind in the donor, usually produce patterns that are incomplete anteriorly. On this basis, this region of the presumptive mesoderm is the organizer or boundary region (Wolpert, 1969) that controls positional information for the whole body pattern.

Now the polar information system within brain tissue, that allows selective nerve connections, may have the form of a gradient in an unknown cellular variable. Alternatively, it could be only of an "arrow" nature—a polar cellular property existing at each point—and thus different from true positional information. With this proviso, a group of cells acts as the organizer for patterns of nerve connection within adjacent tissue, in a way quite analogous with the earlier effects of the very group of mesoderm cells that induced it to be what it has become. Are we observing merely analogous principles at work, or is there a real similarity, perhaps identity, in physiological activity within the two regions at successive times?

We have started a series of experiments at Mill Hill to investigate these questions. Pieces of organizer mesoderm (i.e. originally early dorsal lip deep cells) from late gastrulae and neurulae are used as implants in the post-orbital head region of larvae whose eyecups are just before the stage of polarity specification (stage 28). In another series, diencephalic implants from larvae of the same age as the host are used instead. Control animals have comparable tissue implants from other parts of gastrula mesoderm or larval brain, and in all cases the eye contralateral to the graft also serves to give a control map.

Implants frequently self-differentiate within the host head structure, causing a local area of morphological disturbance, but because of the advanced age of the hosts we should not expect, and do not see, any organization of new differentiation patterns within their own tissues due to the grafts. All responsiveness of the primary embryonic field to new pattern-forming information has by this stage been lost. Eyecups remain mechanically and morphologically undisturbed throughout larval life, as shown by symmetrical size, shape and ventral fissure position on the two sides of the head. The eye contralateral to the operated side gives a normal map to its tectum at time of metamorphosis. On the operated side, however, the eye has been exposed, during stages of polarity fixation, to any competing influences emanating from the implant, which is situated at a very disparate position from that occupied by the host's own original organizer cells (near the nose, under the forebrain).

Our results to date show that tissues derived from cells having known organizer status (i.e. only the anterior gastrula mesoderm or larval diencephalon) exert a disturbing effect on the map formed from the adjacent eye to its tectum. This effect is not understandable as polarity reversal in part of the eyecup, which would be the simplest expected phenomenon, but as a more complex "de-focussing" of the localizations of termini from each retinal region at the tectum. Thus although the maps are ordered, a standard electrode position on the tectum picks up terminal activity from a wider retinal area, extending more anteriorly in the retina, that would be the case in the equivalent projection from a normal eye. The effect is most pronounced in the retinal quadrant opposite the implant site. This phenomenon is striking, but difficult to display by the means conventionally used to represent retino-tectal projections hitherto in the literature, and will be reported fully elsewhere. Control implants have no such effects, although similar degrees of mechanical growth disturbance are involved. Tantalum foil and nuclepore membrane barriers are also inactive.

Earlier work (Hunt and Jacobson, 1973) suggests that the eyecup takes its functional polarity, for tectal connectivity, originally from the body gradient of the larva, which is a reflection of the primary embryonic field due to the organizer (gastrula dorsal lip). It would now seem that intercellular communication within the larval body allows this polarizing information to be disturbed, by ectopic grafts of tissues that possess physiological properties of organizers at their normal sites. It remains to be seen whether the molecular and cellular nature of such information will first be elucidated for one of the classical pattern-forming "fields" of embryology, or for a system of selective nerve connections or neural projection.

REFERENCES

Cooke, J. Properties of the primary organization field in the embryo of *Xenopus laevis*. II. Positional information for axial organisation in embryos with two head organisers. *J. Embryol. exp. Morph.* **28**, 27–46 (1972).

Corner, M. Development of the brain in *Xenopus laevis* after removal of parts of the neural plate. *J. exp. Zool.* **153**, 301–311 (1963).

Edelman, G. Surface modulation in cell recognition and cell growth. *Science* **192**, 218–226 (1976).

Gaze, R. M. The Formation of Nerve Connections, Academic Press, London, New York (1970).

Gaze, R. M. and Keating, M. J. The visual system and "neuronal specificity". *Nature (Lond.)* **237**, 375–378 (1972).

Gaze, R. M., Keating, M. J. and Chung, S.-H. The evolution of the retino-tectal map during development of *Xenopus*. *Proc. roy. Soc.* **B**, **185**, 301–330 (1974).

Hope, R. A., Hammond, B. J. and Gaze, R. M. The arrow model: Retinotectal specificity and map formation in the goldfish visual system. *Proc. roy. Soc.* **B**. **194**, 447–466 (1976).

Hunt, R. K. and Jacobson, M. Specification of positional information in retinal ganglion cells of *Xenopus*: Assays for analysis of the unspecified state. *Proc. nat. Acad. Sci. (Wash.)* **70**, 507–511 (1973).

Jacobson, C.-O. Motor nuclei, nerve roots and fibre patterns in the medulla oblongata after reversal experiments on the neural plate of Axolotl larvae. *Zöol. Bidrag. Uppsala* **36**, 73–160 (1964).

Prestige, M. C. and Willshaw, D. J. On a role for competition in the formation of patterned neural connections. *Proc. roy. Soc.* **B**, **187**, 449–459 (1975).

Rakic, P. Neurons in Rhesus monkey cortex: Systematic relation between time of origin and eventual disposition. *Science* **183**, 425–427 (1974).

Sidman, R. L. Contact interactions among developing mammalian brain cells, in A. A. Moscona (ed.), The Cell Surface in Development, John Wiley and Sons, Inc., New York (1974).

Spemann, H. Embryonic Development and Induction (1938), Rep. Hafner, New York (1967).

Wolpert, L. Positional information and the spatial pattern of cellular differentiation. *J. theor. Biol.* **25**, 1–48 (1969).

Regulation of Proximo-distal Pattern Formation in the Developing Limb

Madeleine Kieny

Laboratoire de Zoologie et Biologie animale, Université Scientifique et Médicale de Grenoble, 38041 Grenoble, France

INTRODUCTION

It has been known for many years that the chick embryo limb bud, like many other organs endowed with developmental or regenerative properties, is capable of extensive regulation along its proximo-distal axis. Thus when it is deprived of part of its material or supplemented with excess material between 3 and 4 days of incubation, it nonetheless is able to give rise to a harmonious three-segmented limb. For instance, when a whole stage 19 wing bud, composed of its three presumptive segments—stylopod (s), zeugopod (z) and autopod (a)—is grafted on a stage 22 wing stump whose presumptive autopod has been severed, the composite excedentary wing bud $sz : sza$ develops into a regulated three-segmented wing, whose zeugopod, however, frequently contains one or two supernumerary skeletal elements (Kieny, 1964a). Likewise, when the presumptive autopod of a stage 22 wing bud is grafted on a stylopodial stage 22 wing stump, the deficient $s : a$ wing bud, whose presumptive zeugopod was removed, is able to form a three-segmented wing with a completely regulated intercalary zeugopod containing its two normal skeletal elements (Kieny, 1964a). The same holds for the leg bud (Hampé, 1959).

It is clear that in these two examples of excess or deficiency regulation, some of the cells of the experimental limb bud are reprogrammed according to their new position in the proximo-distal morphogenetic field. Indeed when the presumptive segments are isolated and cultured on the chick CAM, they give rise to nothing else than the structures corresponding to their initial prospective values. However, when they

are brought together in abnormal sequences, like the excedentary
sz : *sza* or deficient *s* : *a* combinations just mentioned, somehow the confronted cells at the plane of contact exchange morphogenetic informations leading to a redetermination of part of the associated tissues. It
may be worth reminding here that control grafts where homologous
levels of the proximo-distal axis are brought in contact do not undergo
any regulative processes. The question then arises by what mechanism
the regulation occurs. In particular, what are the tissues that are reprogrammed: how do the five prospective segments *sz* : *sza* or the two prospective segments *s* : *a* collaborate to develop into a harmonious *sza*
structure?

An answer to this question has already been provided by heterotopic
recombinations of wing parts grafted on a leg stump and conversely
of leg parts grafted on a wing stump (Kieny, 1964b, c). In the first
type of recombinants, the surplus limb buds *SZ* : *sza* (capital letters
stand for leg parts, lower-case letters for wing parts) gave rise to a regulated three-segmented limb whose stylopod was of mixed origin and
displayed heteromorphic skeletal element of femuro-humeral morphology and whose zeugopod and autopod were typically of wing nature;
the deficient limb buds of this type *S* : *a* likewise formed a three-segmented limb whose stylopod was a typical thigh, and whose zeugopod
and autopod had the skeletal and cutaneous characters of a forearm
and a hand. Conversely, in the second type of recombinants, the surplus
limb buds *sz* : *SZA* developed into a regulated three-segmented limb
whose stylopod contained an unusual heteromorphic skeletal element
with neither humerus nor femur characters, while the zeugopod and
the autopod were typical lower leg and foot; the deficient limb buds
of this type *s* : *A* also formed a three-segmented limb whose stylopod
was a typical arm, and whose zeugopod and autopod were undoubtedly
recognizable as lower leg and foot.

From these heterotopic recombination experiments it was concluded that (1) in the surplus limb buds, it was the three proximal prospective segments (stylopod, zeugopod and stylopod) that underwent
reprogrammation in order to give rise to a single stylopodial segment;
(2) in the deficient limb buds, it was the distal autopodial prospective
segment that gave rise to at least part if not all of the intermediate zeugopod; and consequently that, in both instances, cells and tissues of
a distal part of the field had changed their developmental fate for the
construction of more proximal ones.

This interpretation, however, was questioned by Wolpert and his
group (Summerbell *et al.*, 1973; Wolpert *et al.*, 1975), on the grounds
that, according to their "progress zone" theory, cells belonging to a

certain level of the proximo-distal field could only change their prospective values towards the more distal zones of the field, and would thus be unable to give rise to structures more proximal than those corresponding to their initial position in the field.

In order to clarify the situation, and also to gain a more precise knowledge of the fate of the recombined prospective segments, new experiments were performed (Kieny and Pautou, 1976) taking advantage of the heterospecific chick/quail associations advocated by Le Douarin and Barq (1969). Due to the specific nucleolar marker of the quail nuclei, cells of the two species can be easily recognized and accurately located after Feulgen staining of xenoplastic limbs (see Le Douarin, this volume, p. 172).

In the following two sections the essential results of the heterospecific excedentary and deficient recombinations will be presented. They are illustrated by the description of some representative cases of both series (Figs 1, 14, 20).

EXCEDENTARY RECOMBINATIONS

The graft, consisting of a whole stage 19–20 quail wing or leg bud, was transplanted on a stage 22 chick wing stump composed of the prospective segments of stylopod and zeugopod. The recombinant heterospecific bud thus had the following constitution: chick *sz* : quail *sza* or chick *sz* : quail *SZA*. Two kinds of controls were run along each recombination in order to ascertain the precise level of the cuts: (1) the severed presumptive autopod of the host wing was grafted on the chick CAM, and the cut was termed successful when it gave rise to nothing else than hand parts: (2) the quail donor was allowed to develop until it could be ascertained that it did not form anything else than girdle parts at the site of excision. In order to ensure a good and rapid take of the graft it was found essential to use donors that were younger than the hosts and to perform the cuts in a single stroke of the Pascheff-Wolff scissors, thus yielding smooth and clean section surfaces. The grafts were held in place by three silver hooklets.

In most of the cases (14 out of 21 retained recombinants), the recombinant buds gave rise to harmoniously three-segmented wings resp. "wing-legs". The skeletal constitution was generally less well regulated than the external shape, although some wings and some wing-legs displayed a perfect skeletal regulation and contained no excess skeletal elements.

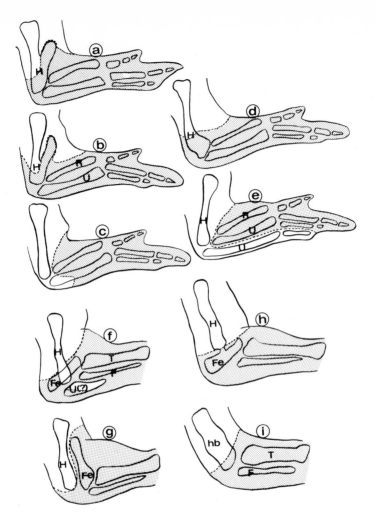

Fig. 1. Schematic illustration on nine cases of surplus regulation that were analysed histologically. a–e, wings originating from the graft of quail wing buds on chick wing stumps (*sz : sza* combinations); f–i, appendages originating from the graft of quail leg buds on chick wing stumps (*sz : SZA* combinations); *stippled*, mesodermal tissues composed of quail cells; *white*, mesodermal tissues composed of chick cells; F, fibula; Fe, femur; H, humerus; hb, heteromorphic stylopodial bone; R, radius; T, tibiotarsus; U, ulna.

Figs 2–4. Doubled humerus (H) shown at three different levels of the stylopod.

Fig. 5. Detail of the distal epiphysis of the humerus shown in Fig. 2.

Fig. 6. Detail of the distribution of chick and quail cells in the humerus shown in Fig. 3.

Fig. 7. Higher magnification of the cartilaginous cells shown in Fig. 6.

Fig. 8. Detail of the proximal part of the quail humeral portion surrounded by quail mesodermal cells.

Fig. 9. Higher magnification of the bispecific skin (quail dermis/chick epidermis) shown in Fig. 8.

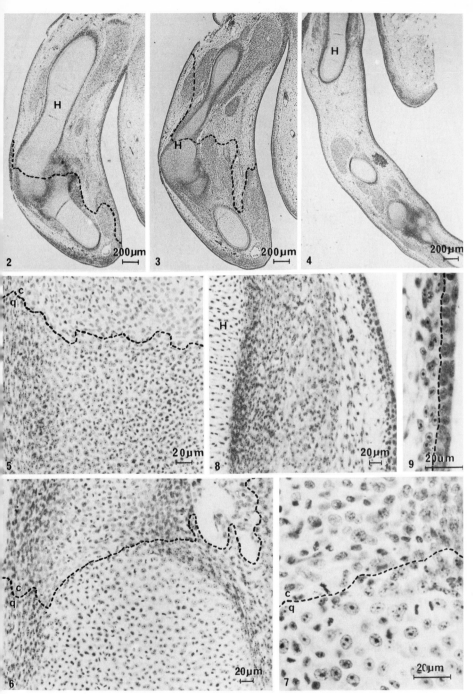

Figs 2–9. Regulation in a surplus recombinant (*sz:sza*) constituted by a whole quail wing bud grafted on a chick wing stump. This case is shown diagrammatically in Fig. 1b. The interrupted lines represent the limit between chick (c) and quail (q) cells.

Wing:wing recombinants

In the case of the perfectly regulated wings, the limit between proximal chick mesodermal tissues and distal quail mesodermal tissues ran either through the distal third of the humerus, so that the distal epiphysis of the humerus and the entire zeugopod and autopod were of quail origin, or between humerus and radius and the proximal fourth of the ulna, so that the major part of the zeugopod—with the exception of the proximal epiphysis of the ulna—was again of quail origin.

When surplus skeletal elements had formed, they were found either inside the stylopod as a partly doubled humerus with a fused distal epiphysis (Figs 2–9) or inside the zeugopod as an excedentary ulna.

In the former configuration, the preaxial branch and the distal epiphysis of the doubled humerus were of quail origin. In the latter one, the zeugopod contained two ulnae, of which one was of chick origin and lay postaxially alongside the other one which was of quail origin. The chick ulna was frequently prolonged by a chick-originated extra metacarpal.

Wing:leg recombinants

In the perfectly regulated wing-legs, the stylopodial skeletal element was heteromorphic and bore resemblance to neither humerus nor femur. In this respect it looked very much like the heteromorphic bones that had been obtained and described previously in homospecific heterotopic chick:chick recombinations (Kieny, 1964c). This heteromorph was composed of a larger proximal part of chick origin and a smaller distal part (the distal epiphysis) of quail origin. In the wing-legs containing surplus elements, a short partially developed quail-originated femur was found at the level of the elbow either oriented at right angles or parallel with respect to the chick-originated humerus. It was sometimes fused to the humerus, which occasionally was prolonged inside the zeugopod by an ulnar (?) rudiment.

These heterospecific recombination experiments fully confirm the previously obtained results of the homospecific heterotopic grafts. Excedentary regulation can take place harmoniously in xenoplastic chick-:quail limb buds. As a rule it leads to the formation of a limb externally shaped into three correctly angulated segments, which may, however, contain non-integrated excedentary skeletal elements. Two main configurations are observed: in one, the entire zeugopod (and also of course the autopod) originates from the graft, in which case the limit between graft and host mesodermal tissues lies within the stylopod and runs more or less at right angles to the proximo-distal axis of the limb, with a propensity, however, for the graft tissues to extend proximally on the preaxial side and for the host tissues to extend distally on the postaxial side. In the other configuration, which may be interpreted as an exaggeration of the preceding one, a more or less extended portion of the postaxial zeugopod (and sometimes also autopod) originates from the host, while the graft builds up the remainder of the zeugopod, i.e. its preaxial portion, and the autopod. Finally the fate of the five prospective segments of the excedentary limb buds is now known with sufficient accuracy to state the following correspondence:

host stylopod	→	entire or proximal postaxial part of stylopod
host zeugopod	→	not expressed or partly expressed as more or less developed postaxial elements in zeugopod
graft stylopod	→	not expressed or partly expressed as preaxial or distal elements in stylopod
graft zeugopod	→	entire or distal preaxial major part of zeugopod
graft autopod	→	entire or preaxial major part of autopod

Since there is not an abnormally high rate of cell death in the limb-bud recombinants, it can be concluded that no appreciable amount of cells is eliminated and that most of the cells participate in morphogenesis. It is thus clear that, in this experimental situation, the developmental fate of some prospective segments is shifted in either direction, proximally or distally. For example, the host prospective zeugopod may become completely integrated into the final stylopod and thus experiences a proximal shift of its initial prospective value. Conversely, the graft prospective stylopod may become completely integrated into the final zeugopod and thus undergoes a distal shift of its initial prospective value. Evidently, cells at the abutting section surfaces and their neighbours along relatively long distances become naive again and then reprogrammed according to their new position in the proximo-distal field.

DEFICIENT RECOMBINATIONS

It was known from previous experiments that deficient $s : a$ limb buds, from which the intermediate zeugopodial presumptive segment had been removed, usually give rise to regulated three-segmented limbs. Here again, however, as in the preceding excedentary recombinations, the skeletal regulation is frequently imperfect: for instance, in the case of leg : leg recombinants $(S:A)$, the fibula is generally shorter than normal or absent; in the case of wing : wing recombinants $(s:a)$, the zeugopodial elements are often fused to the humerus, so that the elbow joint does not form, or the radius is missing (Figs 10–13). Despite these defects, it is evident that an intercalary regulated zeugopod is formed in most of the cases, provided the grafted presumptive autopod is obtained from an embryo not older than stage 23.

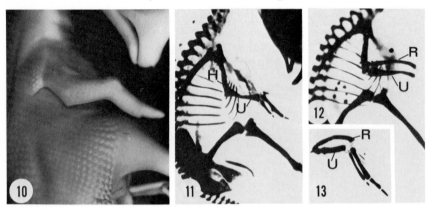

Figs 10–13. Case of homospecific deficiency regulation $(s:a)$ in the chick wing with its controls. H, humerus; R, radius; U, ulna.

Fig. 10. External morphology of the three-segmented wing obtained from a stage 22 wing autopod grafted on a stage 21 wing stylopod.

Fig. 11. Skeletal equipment of the regulated wing: the radius is missing.

Fig. 12. *Autopod control:* after sectioning of the presumptive autopod, the remainder of the wing bud developed a complete set of stylopodial and zeugopodial elements.

Fig. 13. *Stump control:* the sectioned distal portion of the operated wing was cultured on the chick CAM, where it formed forearm and hand.

In order to answer the question of the origin (stump or graft, or both, and to what extent) of the regulated intercalary segment, heterospecific quail : chick and chick : quail recombinations were performed.

When a stage 22 quail presumptive autopod was grafted on the presumptive stylopod of a stage 19–21 chick host, no intercalary regulation or only very limited regulation was obtained. The host's limb bud tissues gave rise to a stylopod (sometimes prolonged by a fused proximal portion of an ulna (?)) which was jointed to the graft-originated autopod. This was interpreted as resulting from the dyschrony between the developmental rates of chick and quail embryos, the tissues of the later probably differentiating faster than those of the chick (quail embryos hatch after 16 days of incubation as compared to the 20 days of the chick). The grafted quail autopod, then, probably became too quickly involved into the morphogenesis of the autopod to be able to participate in that of the missing zeugopod.

However that may be, the reverse combination of a stage 22 chick presumptive autopod with a stage 20–21 quail presumptive stylopod regularly resulted in the formation of a regulated three-segmented limb. Because, in our hands, quail host embryos were found to have a poor survival rate after operation, the following procedure was designed. Stage 19–20 chick hosts were prepared whose wing bud was severed at the presumptive level of the scapulo-humeral joint. The removed wing bud was cultured on the chick CAM as a control for the accuracy of the cut. On that chick base, consisting of the prospective shoulder girdle, a presumptive stage 20–21 quail stylopod was transplanted, oriented in concordance with the host's three axis of polarity, and held in place by three silver hooklets. On the distal section surface of this presumptive quail stylopod a stage 22 chick presumptive autopod was then grafted, again in orthopolar orientation, and secured by two long silver hooklets that ran through the quail stylopod and were thus anchored into the host's presumptive girdle tissues. Three additional controls were run parallel to each recombinant. The quail donor was allowed to develop to a stage when it could be ascertained that it had not formed anything more than the girdle elements at the site of excision. The distal portion of the quail stylopod donor limb bud was cultured on the chick CAM where it could be verified that the explant did not form less than a complete set of zeugopodial and autopodial elements. Finally the chick autopod donor was likewise allowed to develop to a stage when it could be ascertained that its operated limb bud had formed a complete set of stylopodial and zeugopodial elements. For final analysis, only those recombinants were retained whose four controls met the prescribed standards.

In all, 37 cases were thus retained. Thirty of them had formed a regulated intercalary zeugopod, and 16 were used for histological analysis. In 2 cases, the zeugopod was contributed exclusively by the presumptive autopod; the limit between chick and quail tissues was found to run transversely across the limb at the level of the elbow. In one case, it originated exclusively from the presumptive stylopodial base; the quail : chick limit was located at the level of the wrist. In the remaining 13 cases, the intercalary segment was of mixed origin, the presumptive stylopod contributing predominantly to the preaxial part, and the presumptive autopod to the postaxial part of the zeugopod. Figures 14–

23 give a choice of serial sections illustrating the respective participation of presumptive autopod and stylopod to the construction of the zeugopod. The differential pre/postaxial distribution of quail and chick cells frequently resulted in the radius being predominantly built up by quail cells and the ulna by chick cells (Figs 19, 22, 23). The borderline was often seen to run along the postaxial edge of the radius, so that cartilage cells were of quail origin and inner perichondrial cells of chick origin. In other cases, or in other sections of the same cases, the musculature between radius and ulna was of mixed bispecific constitution (Fig. 20).

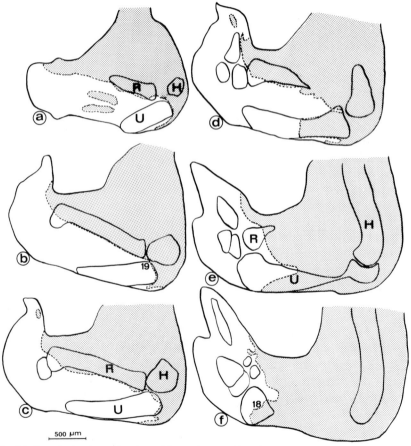

500 μm

Fig. 14. Regulation in a wing issued from the combination (*s* : *a*) of a chick presumptive hand grafted on a quail presumptive arm. Histological distribution of quail (*stippled*) and chick (*white*) mesodermal cells over a thickness of 450 μm (60 × 7.5 μm sections) from (a) to (f). The numbered areas are photographically illustrated in the following corresponding figure numbers. H, humerus; R, radius; U, ulna.

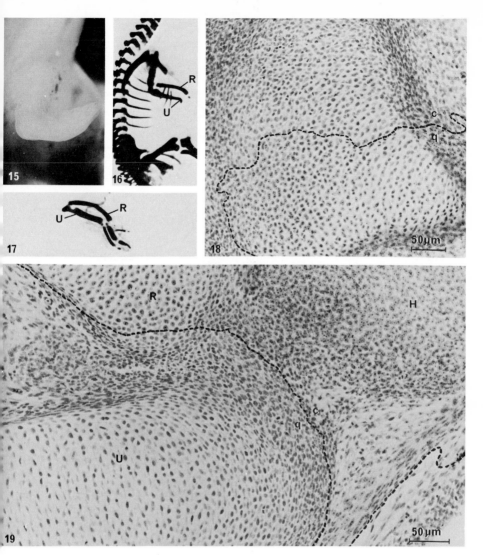

Figs 15–19. Photographic illustration of the regulated wing described in Fig. 14. The interrupted lines represent the limits between chick (*c*) and quail (*q*) cells. H, humerus; R, radius; U, ulna.

Fig. 15. External morphology of the regulated three-segmented wing 5 days after grafting.

Fig. 16. *Chick autopod donor control:* a complete arm and forearm formed after removal of the presumptive hand.

Fig. 17. *Quail stump donor control:* the distal portion of the quail wing was grafted on the chick CAM and formed a complete set of forearm and hand elements.

Fig. 18. Bispecific constitution of the distal epiphysis of the ulna.

Fig. 19. Distribution of chick and quail cells at the elbow level.

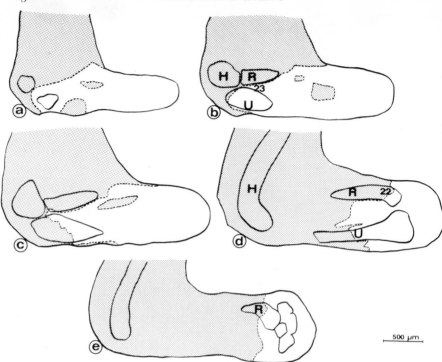

Fig. 20. Other case of regulation in a wing issued from the combination (s:a) of a chick presumptive hand grafted on a quail presumptive arm. Histological distribution of quail (*stippled*) and chick (*white*) mesodermal cells over a thickness of 345 μm (46 × 7.5 μm sections) from (a) to (e). The numbered areas are photographically illustrated in the corresponding following figure numbers.

In no cases were chick (presumptive autopodial) cells found inside the stylopod, nor did quail cells extend distally further than the zeugo-auto-podial joint, except in one case (Fig. 14) when a small islet of quail cells was found in the preaxial metacarpal zone. Finally orthotopic (wing : wing s : a) recombinants and heterotopic (wing : leg s : A) recombinants led to similar results, with the important particularity that heterotopic s : A buds gave rise to a three-segmented limb whose stylopod was an arm and zeugopod and autopod were a lower leg and a foot, a result which is in line with the findings of the previous homo-specific heterotopic recombinations.

In a last series of experiments of the same kind as the ones just de-scribed, the quail presumptive wing stylopod was inverted along its proximo-distal axis before it was grafted onto the chick host stump and topped by the chick presumptive wing or leg autopod. These xenoplas-

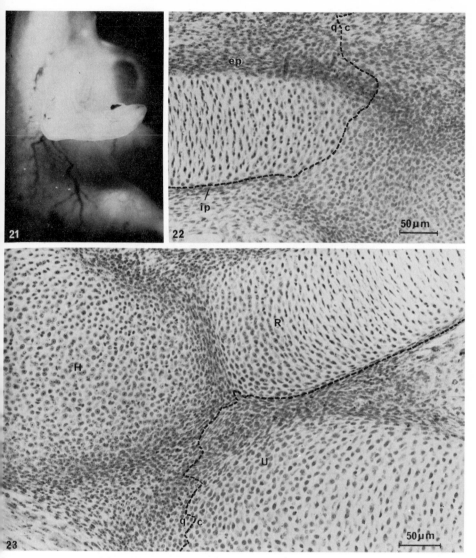

Figs 21–23. Photographic illustration of the case of deficiency regulation described in Fig. 20. The interrupted lines represent the limits between chick (c) and quail (q) cells. ep, external perichondrium; H, humerus; R, radius; U, ulna.

Fig. 21. External morphology of the three-segmented wing 5 days after grafting.

Fig. 22. Bispecific constitution of the distal portion of the radius. The epiphysis and the internal perichondrium (ip) are composed of chick cells.

Fig. 23. Distribution of chick and quail cells at the elbow level.

tic recombinants did not regulate. Each of the presumptive segments gave rise to the structures corresponding to their normal developmental fate: the quail tissues formed a stylopod with an inverted humerus oriented at right angles to the chick-derived hand or foot (Figs 24, 25).

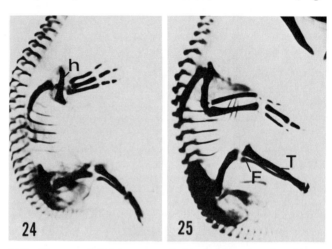

Figs 24 and 25. Absence of deficiency regulation in the case of grafting a quail leg autopod on a chick stylopod whose proximo-distal axis has been reversed.

Fig. 24. Experimental upper appendage with the reversed humerus (h).

Fig. 25. Autopod donor control: after excision of the presumptive foot the remainder of the leg bud formed a thigh and a complete lower leg with tibiotarsus (T) and fibula (F).

In short, the whole set of deficient recombinations shows that intercalary regulation can take place between associated tissues of different specific origin. In the majority of the cases the regulated zeugopod has a bispecific origin, which demonstrates that both presumptive stylopod and autopod participate in its formation. Furthermore the regulated zeugopod can arise exclusively from either presumptive autopodial or stylopodial tissues, the variation being probably due to the precise relative state of differentiation of the abutted tissues and to other yet unknown factors. It is clear then that presumptive autopodial cells can become respecified to a more proximal morphogenetic fate, as well as presumptive stylopodial cells can be reprogrammed towards the expression of more distal structures. Finally the absence of intercalary regulation in the case of a proximo-distally inverted presumptive stylopod can be interpreted in the following way, if one remembers that cell differentiation proceeds in the limb bud in proximodistal sequence. Presumably the cells at the proximal surface section of the presumptive stylopod are already too far along their differentiative pathway to both undergo respecification and transmit meaningful messages to the adjacent presumptive autopodial cells, so that they neither become involved in zeugopodial regulation nor elicit regulative reprogramming of the autopodial cells.

CONCLUSION

Limb bud tissues are endowed with extensive regulative properties. Excedentary buds with five presumptive segments and deficient buds with only two give rise, in the vast majority of cases, to harmoniously shaped three-segmented limbs. In this process, cells in the vicinity of— and apparently also cells further away from—the initial plane of contact between tissues belonging to different prospective levels of the proximo-distal axis become respecified according to the morphogenetic value of their immediate neighbours, i.e. according to their position in the re-establishing proximo-distal field. The shift in their developmental fate may occur, depending on experimental conditions, in either proximal or distal direction. Any theory, like Wolpert's "progress zone" in which the positional value of cells can only move towards more distal structures as development proceeds, seems to be in contradiction with our results.

ACKNOWLEDGEMENT

This work was supported in part by DGRST and CNRS.

REFERENCES

Hampé, A. Contribution à l'étude du développement et de la régulation des déficiences et des excédents dans la patte de l'embryon de Poulet. *Arch. Anat. micr. Morph. exp.* **48**, 345–478 (1959).

Kieny, M. Régulation des excédents et des déficiences du bourgeon d'aile de l'embryon de Poulet. *Arch. Anat. micr. Morph. exp.* **53**, 29–44 (1964a).

Kieny, M. Etude du mécanisme de la régulation dans le développement du bourgeon de membre de l'embryon de Poulet. I. Régulation des excédents. *Develop. Biol.* **9**, 197–229 (1964b).

Kieny, M. Etude du mécanisme de la régulation dans le développment du bourgeon de membre de l'embryon de Poulet. II. Régulation des déficiences dans les chimères "aile-patte" et "patte-aile". *J. Embryol. exp. Morph.* **12**, 357–371 (1964c).

Kieny, M. and Pautou, M. P. Régulation des excédents dans le développement du bourgeon de membre de l'embryon d'oiseau. Analyse expérimentale de combinaisons xénoplastiques caill/poulet. *Wilhelm Roux' Arch. Entwickl.-Mech. Org.* **179**, 327–338 (1976).

Le Douarin, N. and Barq, G. Sur l'utilisation des cellules de la caille japonaise comme "marqueurs biologiques" en embryologie expérimentale. *C.R. Acad. Sci. (Paris)* Sér. D, **269**, 1543–1546 (1969).

Summerbell, D., Lewis, J. H. and Wolpert, L. Positional information in chick limb morphogenesis. *Nature (Lond.)* **244**, 492–496 (1973).

Wolpert, L., Lewis, J. and Summerbell, D. Morphogenesis of the vertebrate limb, in Cell Patterning, Ciba Foundation Symposium **29**, pp. 135–161, Associated Scientific Publishers, Amsterdam (1975).

III
Morphogenetic Tissue
Interactions

Embryonic Induction:
A Historical Note

Sulo Toivonen

Department of Zoology, University of Helsinki,
SF-00100 Helsinki 10, Finland

Early in this century an important discovery was made in Spemann's laboratory. At the gastrula stage in amphibian embryos the blastopore lip, when invaginated to form the archenteron roof, guides subsequent differentiation. Until 1932 the nature of this action was an enigma. During that year Spemann's group showed that even if the blastopore lip was killed by one means or another, it still induced differentiations of the competent ectoderm, and that various adult tissues, when brought in contact with competent ectoderm, also acted as inducers. It thus became evident that inductive action was chemical in nature.

In the following years, the chemical nature of inductive action was corroborated by various means. Many theories were proposed to explain the induction process and especially the regional determination of the developing embryo.

Of these theories, the first one of importance was the concept of Dalcq and Pasteels, later known as the quantitative theory of primary induction. According to that theory, regionalization was due to a single agent acting in different quantities. In 1938, Chuang and I were the first to report that different adult tissues seemed to possess qualitatively different inductive capacities, some tissues inducing head structures alone, others trunk and tail formations. We both interpreted these results as evidence for the existence of chemically different principles with different specific actions. Thus, in contrast to the quantitative theory, we assumed that there must be more than one inductive principle. This was later called the qualitative theory of primary induction, and was clearly in conflict with the earlier theory.

In 1949 the discrepancy between the two theories was resolved in favour of the qualitative theory. Its supporters were able to demonstrate a chemical difference between the two principles, later to be called the neuralizing and mesodermalizing principles. But it still remained to be shown how the two principles together determine the regionality of the developing embryo.

This riddle was solved when neuralizing and mesodermalizing inducers were applied simultaneously to the same gastrula or to an ectodermal vesicle. In this way, with my co-worker L. Saxén, I was able to induce formations belonging to all regions of the normal embryo and in the normal sequence from head to tail. From these results, we worked out a scheme, in which the two inducers form gradients on the dorsal side of the embryo: in the forebrain region the neuralizing principle acts alone, but in the hindbrain region its action is combined with a weak mesodermalizing action, and in the trunk and tail the mesodermalizing principle is strongest.

We could later corroborate this scheme by various modified experiments, and to my mind it is still in general valid. Our later results established that, in the first step, the neuralizing principle alone causes the differentiation of the neural plate, and in the next step the mesodermalizing principle brings about regionalization of the embryo, for instance the spatial differentiation of the CNS. The former action doesn't need any cell contact, the latter seems to need it.

Although the basic mechanism of primary induction seems now to be understood, the chemical nature of the inductive triggers still remains to be solved, despite the many attempts to do so during the last fifty years.

Morphogenetic Tissue Interactions: An Introduction

Lauri Saxén

*Third Department of Pathology, University of Helsinki,
SF-00290 Helsinki 29, Finland*

The significance of communication between embryonic cells of different origin and different developmental history has been known for more than 50 years. Their necessity for normal, organized development has been conclusively shown in isolation and transplantation experiments, and the importance of disrupted communication in abnormal development has been demonstrated in embryos of various mutant strains (reviews: Grobstein, 1967; Fleischmajer and Billingham (ed.), 1968; Wessells, 1973; Saxén *et al.*, (1976). Yet the molecular basis of these morphogenetic or "inductive" tissue interactions is poorly understood and the progress in the field has been slow.

BIOLOGY OF MORPHOGENETIC TISSUE INTERACTIONS

Two features of the biological nature of the morphogenetic tissue interactions should be emphasized: Cytodifferentiation and morphogenesis being progressive events during embryogenesis, it is to be expected that the interactions guiding this process are *sequential*. Another characteristic is the varying "*specificity*" of these processes as shown in experiments with heterotypic tissue combinations; in many interactive events the "induction" can be mimicked by various heterotypic tissues or tissue extracts, whereas in some instances only the normal counterpart can release the proper trigger. In addition to the examples provided by the various chapters in this section, a few others can be mentioned.

The sequential nature of morphogenetic tissue interactions

Neuralization of the gastrula ectoderm and the regionalization of the neural plate are successive events regulated by two waves of interactions between the target cells and the archenteron roof. Experimentally this can be demonstrated as follows: If the ectodermal cells are treated with heterogeneous inductors of certain type, they will exclusively develop towards structures of the forebrain region, but if they are subsequently exposed to the axial mesoderm, they can be converted into hindbrain or spinal cord structures (Saxen *et al.*, 1964). The same conclusion is reached by experiments in which cells of the forebrain region of the neural plate were combined with cells from the caudal axial mesoderm. A series of CNS structures developed, the regional type being a function of the amount of mesodermal cells combined with the neuralized ones (Toivonen and Saxén, 1968).

A similar sequential interactive process is guiding the development of the liver: Presumptive liver endoderm is programmed in chick embryos at the 5-somite stage through its intimate association with the hepatocardiac mesoderm (Le Douarin, 1964). The hepatocytes determined at this stage are still developmentally interdependent with the adjacent tissues, now the hepatic mesenchyme. In an experimental situation, this homotypic mesenchyme can be replaced by any type of mesenchyme derived from the lateral plate, and the morphogenesis, proliferation and functional maturation of the liver cells can be achieved (Le Douarin, 1976).

Directive versus permissive influences

As already suggested by the above examples, interactions guiding determination, proliferation, morphogenesis and organogenesis may vary in their "specificity". I would suggest that we distinguish between two basic types: *directive* and *permissive*. Directive influences refer to those interactive events in which the target cells possess more than one developmental option, and in which the inductive effect leads to selection of one of them. Permissive interactions refer to those in which one tissue acts upon another already determined towards its final fate, but which still requires an exogeneous stimulus for the expression of its phenotype. These two basic types of tissue interactions can be thought to regulate the progressive differentiation of the embryonic cells: with gradual restriction of the number of developmental options the interactive events shift from predominantly directive to predominantly permissive ones (Saxén *et al.*, 1976). This quite schematic view finds support from various experimental studies.

There are many examples of directive influences in this volume: The decision to become a cell of the inner cell mass rather than an outer cell at an early embryonic stage is determined by the position at an 8- to 16-cell stage (Hillman *et al.*, 1972; Graham and Kelly, this volume), the competent amphibian ectoderm can be experimentally converted to various structures normally derived from the mesoderm or entoderm (review: Saxén and Toivonen, 1962; Tiedemann, 1971; Toivonen, this volume) and the dental mesenchyme can induce a non-dentogeneous epithelium to become an enamel organ with its specific protein metabolism (Kollar and Baird, 1970; Thesleff, this volume). Embryonic epidermis is a very good example of a cell population with many developmental options to be selected by the influence of dermal mesenchyme. Heterospecific combination experiments have conclusively demonstrated that while the epidermis of various species show their own "class-specific" repertoire of developmental options restricting the type of the appendicular structures, the size and distribution of these are directed by a dermal influence (Dhouailly, 1975; Sengel and Dhouailly, this volume). Permissive influences have been demonstrated in many interactive events. Tubule formation of the mesenchymal cells in the metanephric blastema can be triggered by a variety of heterotypic tissues ranging from embryonic nervous tissue to certain neoplastic cells (Grobstein, 1955a; Auerbach, 1972). Yet none of these inducers will initiate tubulogenesis in non-kidney mesenchyme (Saxén, 1970). At the 30-somite stage mouse embryo, the pancreatic epithelium seems to have lost most of its developmental options, but would not differentiate into acini without exogenous support normally provided by the mesenchymal stroma. This normal counterpart can, however, be experimentally replaced by almost any heterotypic mesenchyme, by certain tissue extracts and by purified protein preparations (Rutter *et al.*, 1964; Pictet *et al.*, 1975; Pictet and Rutter, this volume).

MECHANISM OF MORPHOGENETIC TISSUE INTERACTION

A complete understanding of the molecular basis of interactive events would involve chemical characterization of the signal substances, their transmission mechanism, and, finally, their mode of action on the target cell resulting in an expression of a new phenotype. None of the morphogenetic tissue interactions has yet been brought down to this level.

Signal substances

Reasonably well-characterized molecules with a morphogenetic effect as tested on embryonic cells are few, and their number might be restricted to three: the vegetalizing factor acting upon competent amphibian ectoderm (Tiedemann, 1973), certain proteoglycans releasing the chondrogenic bias of embryonic somites (Kosher *et al.*, 1973, Lash, this volume), and the mesenchymal factor involved in pancreatic development (Pictet *et al.*, 1975; Pictet and Rutter, this volume). In addition to these rather large molecules, divalent cations studied by Barth and Barth (1964, 1967) should be mentioned. Cations like Li^{++}, Mg^{++}, Ca^{++} and K^+ have shown to exert a determinative action on the cells of Amphibian gastrula ectoderm, and these authors suggest that embryonic induction is initiated by changes in the intracellular cation composition (Barth and Barth, 1968). However, in no interactive system has an active compound been isolated and characterized from the *normal inductor tissue*.

Transmission of inductive signals

As long as the chemistry of the (hypothetic) signal substances remains unknown, their localization and transmission in tissues are a pertinent problem. Various hypotheses, including free diffusion, interaction of matrix molecules and the importance of actual cell-to-cell contacts, have been presented over the last 50 years (reviews: Grobstein, 1955b: Saxén, 1972, Saxén *et al.*, 1976). Here, a somewhat modified classification of transmission mechanisms will be presented (Table I).

TABLE I

Transmission of inductive signals

A. *Long-range*	(50 000 nm)
1. Diffusion	
2. Matrix interaction	
B. *Short-range*	(5 nm)
1. Exchange of small molecules	
2. Surface-associated compounds	

The suggestion of long-range interaction carried by diffusible signal substances or compounds in the extracellular matrix has been based on experiments employing membrane filters. Judging from the thickness of such filters believed to prevent actual cell contacts, a transmission distance of the order of 25 000 to 75 000 nm is presumed. The alternative hypothesis on the role of actual cell contacts has recently been elaborated by our group and is based on findings showing a close correlation between the establishment of intimate cell contacts and an inductive interaction. In these situations, the cell membranes were separated by a narrow gap only, being of the order of 5 to 10 nm (Wartiovaara *et al.*, 1974; Saxén, 1975; Lehtonen, 1976; Nordling *et al.*, this volume).

The hypothesis of long-range transmission of signal substances is supported by studies on primary embryonic induction: both transfilter experiments and direct morphological observations suggest that neuralization of the ectoderm might not require close apposition of the archenteron roof mesoderm (review: Toivonen *et al.*, 1976; Tarin, this volume). The work of Lash and his group (*op. cit.*) demonstrating a morphogenetic effect of extracellular matrix compounds in one permissive induction lends support to another type of long-range transmission, namely that termed "matrix. interaction" by Grobstein (1956).

Close contacts between interacting heterotypic cells have been reported in many instances: ureter epithelium/metanephric mesenchyme (Lehtonen, 1975), bronchial epithelium/pulmonary mesenchyme (Bluemink *et al.*, 1976), enamel epithelium/odontoblastic mesenchyme (Slavkin and Bringas, 1976; Slavkin *et al.*, this volume) (review: Saxén *et al.*, 1976). These findings together with the experimental evidence of transfilter studies in the tubule induction system (review: Saxén, 1975; Lehtonen, 1976; Nordling *et al.*, this volume) suggest that the hypothesis of contact-mediated interactions should be seriously reconsidered. Whether this is transmitted by an exchange of small molecules, by complementary surface molecules or by some other mechanisms is still a matter of speculation.

An "intermediate" communication mechanism is suggested by the findings of Meier and Hay (1976): in the interaction guiding corneal differentiation, cell-substrate type contact was apparent and the hypothesis finds further support from the results by Thesleff (this volume) dealing with the requirements for odontoblast differentiation.

CONCLUDING REMARKS

To conclude, morphogenetic tissue interactions seem to vary in their biological "specificity" and both directive and permissive types have been demonstrated. Keeping in mind this biological diversity and the scattered observations on the chemistry of signal substances and their transmission characteristics, it becomes evident that we should not search for too many common denominators nor should we at this stage of knowledge generalize our findings or build unifying concepts on the mysterious problem termed long ago "embryonic induction".

REFERENCES

Auerbach, R. The use of tumors in the analysis of inductive tissue interactions. *Develop. Biol.* **28**, 304–309 (1972).

Barth, L. G. and Barth, L. J. Sequential induction of the presumptive epidermis of the *Rana pipiens* gastrula. *Biol. Bull.* **127**, 413–427 (1964).

Barth, L. G. and Barth, L. J. The uptake of Na-22 during induction of presumptive epidermis cells of the *Rana pipiens* gastrula. *Biol. Bull.* **133**, 495–501 (1967).

Barth, L. G. and Barth, L. J. The role of sodium chloride in the process of induction by lithium chloride in cells of the *Rana pipiens* gastrula. *J. Embryol. exp. Morph.* **19**, 387–396 (1968).

Bluemink, J. G., Van Maurik, P. and Lawson, K. A. Intimate cell contacts at the epithelial/mesenchymal interface in embryonic mouse lung. *J. Ultrastruct. Res.* **55**, 257–270 (1976).

Dhouailly, D. Formation of cutaneous appendages in dermo-epidermal recombinations between reptiles, birds and mammals. *Wilhelm Roux' Arch. Entwickl.-Mech. Org.* **177**, 323–340 (1975).

Fleischmajer, R. and Billingham, R. E. (ed.), Epithelial-Mesenchymal Interactions, pp. 1–326, Williams & Wilkins Co., Baltimore, Md (1968).

Grobstein, C. Inductive interaction in the development of the mouse metanephros. *J. exp. Zool.* **130**, 319–340 (1955a).

Grobstein, C. Tissue interaction in the morphogenesis of mouse embryonic rudiments in vitro, in D. Rudnick (ed.), Aspects of Synthesis and Order in Growth, pp. 233–256, Princeton Univ. Press, Princeton, N.J. (1955b).

Grobstein, C. Mechanisms of organogenetic tissue interaction. *Nat. Cancer Inst. Monogr.* **26**, 279–299 (1967).

Hillman, N., Sherman, M. I. and Graham, C. The effect of spatial arrangement on cell determination during mouse development. *J. Embryol. exp. Morph.* **28**, 263–278 (1972).

Kollar, E. J. and Baird, G. R. Tissue interaction in embryonic mouse tooth germs. II. The inductive role of the dental papilla. *J. Embryol. exp. Morph.* **24**, 173–186 (1970).

Kosher, R. A., Lash, J. W. and Minor, R. R. Environmental enhancement of *in vitro* chondrogenesis. IV. Stimulation of somite chondrogenesis by exogenous chondromucoprotein. *Develop. Biol.* **35**, 210–220 (1973).

Le Douarin, N. Induction de l'endoderme préhépatique par le mésoderme de l'aire cardiaque chez l'embryon de poulet. *J. Embryol. exp. Morph.* **12**, 651–664 (1964).

Le Douarin, N. An experimental analysis of liver development. *Med. Biol.* **53**, 427–455 (1975).

Lehtonen, E. Epithelio-mesenchymal interface during mouse kidney tubule induction *in vivo*. *J. Embryol. exp. Morph.* **34**, 695–705 (1975).

Lehtonen, E. Transmission of signals in embryonic induction. *Med. Biol.* **54**, 108–128 (1976).

Meier, S. and Hay, E. D. Stimulation of corneal differentiation by interaction between cell surface and extracellular matrix. I. Morphometric analysis of transfilter "induction". *J. Cell Biol.* **66**, 275–291 (1975).

Pictet, R. L., Filosa, S., Phelps, P. and Rutter, W. J. Control of DNA synthesis in the embryonic pancreas: Interaction of the mesenchymal factor and cyclic AMP, in H. C. Slavkin and R. C. Greulich (ed.), Extracellular Matrix Influences on Gene Expression, pp. 531–540, Academic Press, New York (1975).

Rutter, W. J., Wessels, N. K. and Grobstein, C. Control of specific synthesis in the developing pancreas. *Nat. Cancer Inst. Monogr.* **13**, 51–65 (1964).

Saxén, L. Failure to demonstrate tubule induction in a heterologous mesenchyme. *Develop. Biol.* **23**, 511–523 (1970).

Saxén, L. Interactive mechanisms in morphogenesis, in D. Tarin (ed.), Tissue Interactions in Carcinogenesis, pp. 49–80, Academic Press, London (1972).

Saxén, L. Transmission and spread of kidney tubule induction, in H. C. Slavkin and R. C. Greulich (ed.), Extracellular Matrix Influences on Gene Expression, pp. 523–529, Academic Press, New York (1975).

Saxén, L., Karkinen-Jääskeläinen, M., Lehtonen, E., Nordling, S. and Wartiovaara, J. Inductive tissue interactions, in G. Poste and G. L. Nicolson (ed.), Cell Surface Interactions in Embryogenesis, pp. 331–407, North-Holland Division of ASP Biological and Medical Press, Amsterdam (1976).

Saxén, L. and Toivonen, S. Primary Embryonic Induction, pp. 1–271, Logos Press, Academic Press, London (1962).

Saxén, L., Toivonen, S. and Vainio, T. Initial stimulus and subsequent interactions in embryonic induction. *J. Embryol. exp. Morph.* **12**, 333–338 (1964).

Slavkin, H. C. and Bringas, P., Jr. Epithelial-mesenchyme interactions during odontogenesis. IV. Morphological evidence for direct heterotypic cell–cell contacts. *Develop. Biol.* **50**, 428–442 (1976).

Tiedemann, H. Pretranslational control in embryonic differentiation, in E. K. F. Bautz (ed.), Regulation of Transcription and Translation in Eukaryotes, pp. 59–80, Springer-Verlag, Berlin (1973).

Toivonen, S. and Saxén, L. Morphogenetic interaction of presumptive neural and mesodermal cells mixed in different ratios. *Science* **159**, 539–540 (1968).

Toivonen, S., Tarin, D. and Saxén, L. Transmission of morphogenetic signals from amphibian mesoderm to ectoderm in primary induction. Review. *Differentiation* **5**, 49–55 (1976).

Wartiovaara, J., Nordling, S., Lehtonen, E. and Saxén, L. Transfilter induction of kidney tubules: Correlation with cytoplasmic penetration into Nucleopore filters. *J. Embryol. exp. Morph.* **31**, 667–682 (1974).

Wessells, N. K. Tissue interactions in development, in Addison-Wesley Module, Vol. 9, pp. 1–43, Addison-Wesley Publ., Reading, Mass (1973).

Tissue Interactions in Amniote Skin Development

PHILIPPE SENGEL and DANIELLE DHOUAILLY

Laboratoire de Zoologie et Biologie animale,
Université scientifique et médicale de Grenoble, 38041 Grenoble, France

INTRODUCTION

It is now well established that skin and cutaneous adnexa arise during embryonic development from precisely timed dermo-epidermal interactions. This knowledge has been gained mostly through the use of various types of tissue recombination experiments performed with embryonic skin. The most frequently used species for these experiments have been the chick and the duck, so that more data are now available on the mechanisms of feather development than on any other amniote appendage (Sengel, 1971, 1976a, b). In recent years, however, experiments have also been conducted on reptilian and mammalian embryos (Dhouailly, 1973, 1975). A striking similarity was thus revealed between the mechanisms of feather development and those of reptilian scales and mammalian hairs.

While it is still entirely unknown how cells and tissues exchange informations during skin morphogenesis, much has recently been learned about the content of the exchanged messages through the use of *heterotypic* dermo-epidermal recombinants. Three main species, lizard, chick and mouse, representatives of the three classes of amniotes, were used for the experiments. Accessorily, two other species were also employed, the duck and the quail. Pieces of skin were obtained either from glabrous body sites or from regions bearing various cutaneous adnexa differing in their morphology and distribution pattern. Combinations of skin tissues from different body regions lead to *heterotopic* recombinants, while the association of dermis and epidermis from two different species constitutes a *heterospecific* recombination. Many kinds of heterotopic and/or heterospecific recombinations have been performed. In order to yield meaningful results, they were analysed in comparison to the performances of the corresponding control *homotypic* recombinants, where dermis and epidermis originate from the same piece of skin or from pieces of skin with equivalent morphogenetic fate.

Heterotypic and homotypic control recombinants were prepared in two different ways differing in the stage at which the prospective cutaneous tissues or the cutaneous tissues were brought in contact. Either prospective skin tissues were juxtaposed at a very early stage in development, long before the actual differentiation of skin, by combining limb bud mesoderm and ectoderm; the resultant limb bud would thus be covered by recombinant skin; this type of recombination was performed on bird embryos only. Or pieces of skin were taken from various body regions at later stages, just before or by the time the differentiation of cutaneous appendages becomes visible, and split into dermis and epidermis by the usual treatment with calcium- and magnesium-free saline solution of pancreatic enzymes. In all recombinants—either limb or skin—the associated tissues were oriented in their normal anatomical relationship.

Recombinant limb buds were grown as grafts on a host embryo's wing stump for about 2 weeks. Recombinant skins were cultured on the chorioallantoic membrane of chick or duck embryos for at least 3 days and usually for as long as 8 or 10 days, after which their morphogenetic performance could be evaluated by macroscopical and histological observation.

In what follows, we are going to examine successively the morphogenetic properties of reptilian, avian and mammalian dermis in various homo- and heterotypic recombinations.

PROPERTIES OF REPTILIAN DERMIS

Lizard skin was obtained from the dorsum of 15- to 20-day (post-laying) embryos, i.e. prior to the visible onset of scale formation, and from the ventrum of 15- to 25-day embryos, i.e. after the beginning of scale rudiment differentiation. Size and shape of dorsal and ventral scales differ conspicuously. Ventral scales are large, wider than long, overlapping, and arranged in three longitudinal rows on each side of the midventral line. Dorsal scales are small, polygonal, non overlapping, and arranged in a hexagonal pattern.

Let us first examine the homospecific recombinants. When dorsal dermis was recombined to dorsal epidermis, the explants formed a large number of small tubercular dorsal-type scales. Likewise, when ventral dermis was reassociated with ventral epidermis, large ventral-type scales were formed. These homotopic control recombinants thus simply show that region-specific characters of lizard skin are expressed normally in chorioallantoic grafts. Apparently the relative high temperature of the avian host does not noticeably interfere with scale morphogenesis (Dhouailly, 1975).

The heterotopic homospecific recombinants led to clear and constant results. The association of ventral dermis with dorsal epidermis formed large, overlapping ventral-type scales that were characteristically arranged in three longitudinal rows (pieces of ventral skin were dissected symmetrically on each side of the midventral line). However it was noted that a few of these large scales, in some of the explants, were subdivided into smaller scutes by abnormally placed interscutellar furrows. This resulted in a few scales being less wide than normally.

The reverse combination of dorsal dermis with ventral epidermis gave rise to small, tubercular dorsal-type scales, arranged in a hexagonal fashion. These scales were not or but little overlapping. Again, a few of them, however, were larger than might have been expected from dorsal scales (Dhouailly, 1975).

It is thus clear, from these heterotopic recombination experiments, that in the lizard embryo the dermis is responsible for the regional specification of the scales. Shape, size and arrangement of the scales are dependent on the regional origin of the dermis. However, a weak, but unmistakable, influence of the epidermis appears to interfere with the dermal message. Indeed, in both situations, the size of some few scales seemed to be influenced by the epidermis. Whether there is really a conflict between a "strong" dermal induction and a "weaker" epidermal induction is not known. The results might be due simply to the fact that the epidermis has already received from the underlying dermis, by the time it is isolated, a slight morphogenetic influence, which it partially expresses when brought in contact with the heterotopic dermis. Despite this possible early "pre-determination", it is obvious that the epidermis remains quite malleable and competent to respond to the dermal region-specificity.

The heterospecific recombinants of lizard dermis and chick or mouse epidermis did not perform particularly well. When dorsal lizard dermis was combined with dorsal 7-day chick epidermis, only a few (an average of three per explant) atypical ingrowing buds were formed. These may be interpreted as an attempt of the epidermis to build a barb ridge in the absense of any feather development. The recombinants of ventral lizard dermis and dorsal chick epidermis did not give rise to any epidermal structures, except the moulding of the epidermis on the dermal scale elevations. When dorsal or ventral lizard dermis was associated with dorsal 13.5- or 14.5-day mouse epidermis, a few (an average of five per explant) typical stage 2 (after Hardy, 1951) hair buds were formed. Despite prolonged cultivation, these buds did not develop beyond stage 2. They were termed *arrested* hair buds. In comparison, the control homospecific chick/chick and mouse/mouse recombinants developed a large number of feather filaments and stage 5 or 6 hair follicles respectively (Dhouailly, 1973).

It thus appears that, in both heterospecific situations, lizard dermis is not able to elicit much morphogenesis in foreign epidermis. Although a small number of hair buds were formed in the mouse/lizard recombinants, their distribution pattern bore certainly no resemblance to either dorsal or ventral lizard scale pattern. Consequently, intertissue communication between lizard dermis and avian or mammalian epidermis is apparently rather limited. However limited, it is nonetheless existant, since neither chick nor mouse epidermis alone or recombined to homospecific dermis from glabrous areas is able by itself to form any epidermal derivative (Sengel *et al.*, 1969; Dhouailly, 1973).

PROPERTIES OF AVIAN DERMIS

In birds, not all skin regions bear feathers or scales. Some areas are bare and form no cutaneous appendages. What are the morphogenetic properties of the dermis from these glabrous regions, as compared to those of dermis from prospective feather tracts, where feathers are arranged in the typical hexagonal pattern of bird plumage? To answer that question, a number of heterotopic recombinations were performed involving dorsal feather-forming epidermis and non-appendage-forming dermis, such as from the comb (Sengel, 1958; Lawrence, 1971), or the midventral apterium of the chick embryo (Sengel *et al.*, 1969). Such recombinants did not form any appendages, while the homotopic control recombinants of dorsal epidermis and dorsal dermis formed normal feathers arranged in a hexagonal pattern. The reverse recombinants of feather-forming dermis and epidermis from these glabrous areas, or others such as the extra-embryonic membranes (Kato and Hayashi, 1963; Mizuno, 1972), or the cornea (Coulombre and Coulombre, 1971), developed well-formed feather filaments. These results demonstrate that the feather-forming ability resides in the dermis from prospective feather tracts and that the dermis from glabrous regions is deprived of any such morphogenetic activity. Conversely epidermis from glabrous regions is able to respond to the dermal feather-forming influence by differentiating normal feathers.

The dermis exerts a similar determining influence in the choice between different regional types of cutaneous appendages, such as dorsal feathers and shank scales (Sengel, 1958; Rawles, 1963). For instance, when 7-day dorsal dermis is recombined with 11– to 13-day tarsometatarsal epidermis, feathers are obtained. Conversely, when 13-day tarsometatarsal dermis is associated with 7-day dorsal epidermis, scales are formed. However, the latter often bear at their distal rim ill-formed abortive feather filaments, as if morphogenesis had started out by the construction of feather buds and had later deviated toward the building of scales. Thus, although it is clear that the dermis is responsible for the region specificity of the cutaneous appendages, feather morphogenesis appears somehow to predominate over scale morphogenesis. This predominance is even more marked if the tarsometatarsal dermis is obtained from younger 11- or 12-day embryos. In these conditions, scale formation may be completely or partly depressed in favour of feather development.

This supremacy of feathers over scales is further demonstrated by recombinations of 3-day wing bud ectoderm with leg bud mesoderm. The heterotopic limb bud, when grafted on a host's wing or leg stump, develops into a leg (Kieny and Pautou, 1967; Pautou, 1968), the foot and tarsometatarsus of which, however, are covered with feathers and scales (or, more specifically, with feathered scales), and sometimes with feathers only, the scales being completely suppressed (Sengel and Pautou, 1969).

That the regional morphology of feathers is also determined by the dermal component of skin was demonstrated by transplantation of a block of mesoderm from the prospective thigh region of a leg bud to the prospective upper arm region of the wing bud (Saunders and Gasseling, 1957). The implanted leg bud mesoderm induced the differentiation of typical thigh feathers into the humeral apterium of the wing.

Thus, in birds, the information for region-specific skin differentiations is contained within the dermis. Epidermis appears to be indifferent as regards region-specificity and is able to respond to the specific instructions of the dermis, with the restriction that avian epidermis appears to possess an intrinsic capacity to form feathers, unless told otherwise.

Let us now turn to the heterospecific recombinations, and examine first the results of those that deal with intraclass (bird/bird) heterospecificity. Experiments were performed with chick and duck embryos, either as dorsal skin recombinants cultured on the chorioallantoic membrane, or as limb bud recombinants grafted on a host's wing stump

(Dhouailly, 1967, 1970). The results were essentially the same and may be summarized as follows.

Recombinants form well-developed feathers that grow to the neoptile or even teleoptile stage, when their specific characters can be analysed with ease and accuracy. Dermis is responsible for all morphological features of the feathers, except for the number of barbule cells per barbule and the shape of barbule cells, both of which are strictly epidermis-dependent. The amino-acid composition of the keratins of those feathers is also determined by the specific origin of the epidermis (Sengel *et al.*, 1975). Thus the heterospecific chick/duck and duck/chick recombinants give rise to chimeric feathers whose general architecture (presence or absence of rachis, number of barbs, length of calamus, etc.) is dictated by the specific origin of the dermis, but whose barbules and keratins bear the specific characters of the epidermis. It must be added that the dermis also determines the number and distribution pattern of the feathers in the pterylae.

Now to come to interclass heterospecific recombinations involving chick dermis. When 7-day dorsal chick dermis was associated with either dorsal or ventral lizard epidermis, scale buds formed. These had the morphological appearance of early symmetrical scale rudiments on the back of the 25-day post-laying lizard embryo. However, the keratization of their epidermis was more advanced than that of normal scale buds of the same size and shape. In many cases, the distribution pattern of these scale buds could be recognized as being akin to the hexagonal feather pattern. These scales did not develop further, however, and were consequently termed "arrested scales" (Dhouailly, 1975). After an equal length of culture, homospecific lizard/lizard control recombinants had formed typical asymmetrical scales with well-differentiated inner and outer faces. Nine- to 12-day tarsometatarsal chick dermis in combination with dorsal lizard epidermis induced the formation of typical asymmetrical scales, the shape, size and distribution pattern of which bore the characters of the anterior face of the chick shank (Dhouailly, 1975).

When dorsal or tarsometatarsal chick dermis was associated with 11.5- or 12.5-day dorsal mouse epidermis, the recombinants formed arrested stage 2 hair buds. With dorsal chick dermis, these were arranged in a more or less distorted hexagonal pattern; with tarsometatarsal chick dermis, it was striking to see that each scale field, i.e. the area corresponding to the construction of one scale, was occupied by one hair bud (Dhouailly, 1973).

In short, in all these interclass heterospecific recombinations, the nature of the appendages produced is strictly epidermis-dependent:

lizard epidermis forms scales, mouse epidermis hairs. However, the size, shape and distribution pattern of these appendages is dictated by the dermis. In all cases, the chimeric appendages stop to develop at a relatively early stage, except for the recombinant of tarsometatarsal chick dermis and lizard epidermis, where shank scales are formed with an avian shape and pattern, but with a reptilian keratization.

PROPERTIES OF MAMMALIAN DERMIS

In the mouse, as in the chick, not all skin regions are covered with appendages. Some areas, such as the foot sole, remain bare. There are also regional differences in the types of hair produced by different areas of skin. Particularly, the body as a whole is covered by small pelage hairs, which are arranged in typical trio groups (the triad pattern). Each trio group is composed of one central primary follicle (which appears first in development) and two lateral primary follicles (which form later). A number of smaller secondary follicles are added still later. In addition to pelage hairs, the upper lip (the mystacial zone) bears large sensory hairs, the vibrissae. The latter—also called whiskers in this area—are arranged, starting from one dorso-ventral row of four largest pre-ocular vibrissae, in a typical rectangular pattern extending in five oculo-nasal rows of follicles of decreasing sizes.

In the mouse embryo, as in the chick embryo, region-specificity is dermis-dependent. Indeed when foot sole plantar dermis is associated with dorsal epidermis, no hairs are formed. Conversely, recombinants of dorsal dermis with plantar epidermis give rise to hair follicles (Kollar, 1970).

The problem of the regional determination of vibrissa versus pelage hair has recently been cleared up by new experiments involving dorsal skin tissues from 12.5- and 14.5-day mouse embryos and upper lip skin tissues from 11-, 11.5-, 12- and 12.5-day mouse embryos (Dhouailly, 1976).

The control homotopic recombinants of 12-day upper lip epidermis and 12-day upper lip dermis gave rise to an average of 12 hair follicles per explant of which 3 to 5 had a mean diameter of 71 μm, which is characteristic of whisker follicles and larger than any pelage hair will ever grow (the largest pelage hair follicles measure some 55 μm in diameter). In addition, the follicles in this recombinant presented the typical vibrissal constriction above the hair bulb. The other controls, of course, where dorsal 12.5-day epidermis was recombined with 12.5-day dorsal dermis, did not form any large whisker-type follicles, but a large number (a mean of 82 per explant) of pelage hair follicles arranged in the typical triad pattern.

The heterotopic recombinants of 12.5-day upper lip dermis and 12.5-day dorsal epidermis developed a mean number of 13 large hair follicles per explant, of which 2 to 8 had a diameter equal to or larger than 59 μm (they had a mean diameter of 77μm), and displayed the characteristic "hour-glass" shape due to the suprabulbar constriction. None of the follicles were arranged in trio groups, but rather in a more or less distorted rectangular vibrissal pattern (Fig. 3).

The reverse heterotopic recombination of 14.5-day dorsal dermis and upper lip epidermis was performed with mystacial epidermis from embryos of graded ages, namely 11, 11.5, 12 and 12.5 days of gestation. Regardless of the age of the epidermis, the recombinants either formed a single population of pelage follicles (Figs 1 and 2) or gave rise to two populations of follicles. In the latter case, small follicles, not exceeding 58 μm in diameter, most of which being recognizably arranged in trio groups, were intermingled with from 1 to 8 large ones, whose mean diameter ranged from 59 to 67 μm.

These results were interpreted as follows (Dhouailly, 1976). Upper lip dermis induces the formation of typical whisker follicles in dorsal epidermis. Dorsal dermis induces typical pelage hair follicles in upper lip epidermis. However, under the heterotopic dermal impulsion, the latter also continues to form whisker follicles, apparently because it was already engaged toward constructing vibrissae by the time it was separated from its own dermis. Indeed whisker rudiments arise earliest in development and become visible as soon as at 12 days of gestation. Because of the small size of the mystacial area before 11 days, no upper lip epidermis could be used that was further away from overt whisker morphogenesis than 24 h. Contrarily dorsal epidermis can be easily obtained at 12.5 days, i.e. 48 h prior to pelage hair rudiment formation. So that finally, by the time it is taken from the embryo, upper lip epidermis has probably already been influenced by the underlying mystacial dermis, whereas dorsal epidermis at 12.5 days is still entirely uncommitted.

Nevertheless, it is clear from these experiments that—in mammals,

Figs 1 and 2. Heterotopic recombinants of dorsal 12.5-day mouse dermis and upper lip 11-day mouse epidermis. Formation of pelage hair follicles: central primary follicles after 3 days of culture (Fig. 1), central and (smaller) lateral primary follicles after 5 days of culture on the chick chorioallantoic membrane (Fig. 2).

Fig. 3. Heterotopic recombinant of 12.5-day upper lip mouse dermis and dorsal mouse epidermis. Formation of large whisker follicles in graded sizes.

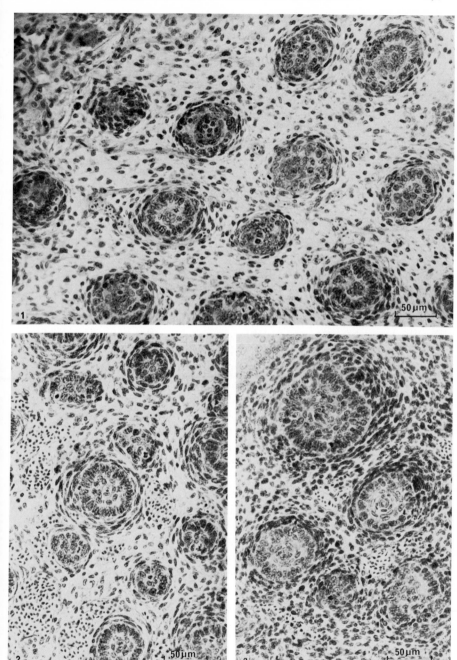

as in birds or reptiles—dermis is responsible for the regional specification of cutaneous appendages. The earlier finding of Kollar (1966), according to which the epidermis contained the region-specific factors, must be rejected and can be explained easily by the fact that it is probably surgically impossible to obtain uncommitted snout epidermis.

The interclass heterospecific recombinations yielded results quite homologous to those that had been obtained with chick dermis. In combination with either dorsal or ventral lizard epidermis, dorsal mouse dermis induced the formation of arrested scale buds, which were of two or three size classes and arranged in a recognizable triad pattern. The larger scale buds probably corresponded to central primary follicles, while the smaller ones corresponded to lateral primary or to secondary follicles. With upper lip mouse dermis, lizard dorsal or ventral epidermis formed giant arrested scale buds whose size and distribution pattern visibly corresponded to those of vibrissae (Dhouailly, 1975).

Likewise recombinants of dorsal mouse dermis with dorsal or tarsometatarsal chick epidermis gave rise to arrested feather filaments, again of two size classes and arranged in more or less well recognizable trio groups. The arrested feather filaments, as compared to fully developed filaments obtained in homospecific or in intraclass (bird/bird) heterospecific recombinations, were characterized by their stoutness, and by their chaotically arranged barb ridges, whose differentiating barbule cells could however be easily recognized (Dhouailly, 1973). When tarsometatarsal, comb or midventral apterium chick epidermis was recombined with upper lip mouse dermis, a set of very large arrested feather filaments were formed, whose arrangement was reminiscent of that of vibrissae.

In summary, heterotopic recombinations of lizard, chick or mouse skin tissues demonstrate that the regional specificity of skin and appendages is determined by the dermis. The epidermis from any prospective appendage-bearing or glabrous region can respond to the dermal region-specific induction, provided it is isolated and reassociated at an early enough stage when it has not yet been irreversibly determined by the dermis of its site of origin. When the epidermis is taken after it has already received a region-specifying induction from the underlying dermis, it may respond to the heterotopic dermis either by producing the appendages of its own original site or both types of appendages, such as, for instance, tarsometatarsal scales *and* feathers, whiskers *and* pelage hairs.

Regarding heterospecific associations, only intraclass recombinants result in the development of fully differentiated appendages. Interclass recombinants lead to the formation of arrested appendage buds.

In the development of the feather, the dermis plays an organizing role on the overlying epidermal cells which brings about the specific plumar architecture. Although it does not participate by any cellular contribution in feather morphogenesis, it imposes on the epidermal cells the specific manner in which they have to become spatially arranged within the feather-forming follicle cylinder. Only the information for barbule morphology and keratin composition is intrinsically contained within the epidermis and cannot be noticeably modified by heterologous dermis.

Heterospecific interclass recombination experiments confirm that the arrangement of appendages in a specific pattern on the surface of the skin is determined by the dermis. Because they lead to the formation of arrested appendages, it must be inferred that only part of the dermal morphogenetic message is effectively transmitted to and correctly understood by the foreign epidermis. Consequently, the dermal morphogenetic induction contains at least two distinct messages. One which triggers off the formation of appendages in a species- and region-specific pattern; this message is effectively understood and interpreted by foreign epidermis. In addition, another message is apparently required for the continuation of appendage morphogenesis, in particular for the acquisition of the species-specific characters of the individual appendages. The second message, which is class-specific, cannot be provided by class-foreign dermis. The formation of fully developed feathers in intraclass chick/duck or duck/chick recombinants, however, indicates that *within* a given class, this second type of dermal message can be transmitted and correctly understood by the epidermis from another species. Moreover, the class-specific type of appendage (scale, feather or hair) is solely dependent on the class-specific origin of the epidermis.

EXPERIMENTS WITH CHIMERIC BISPECIFIC DERMIS

We have seen so far, then, that recombinants of dermis and epidermis from two different classes of amniotes, while they are able to start off the morphogenesis of epidermis-dependent appendages, do not possess the capacity of developing beyond this threshold and cannot therefore give rise to fully developed appendages.

It was recently shown, however, that it was possible to pass this threshold by adding homospecific dermal cells to the foreign dermis (Dhouailly and Sengel, 1975). Experiments were run in the following way, using chick (or quail) and mouse skin tissues. Pieces of chimeric bispecific dermis were prepared by reaggregating dissociated chick (or quail) and mouse dermal cell suspensions in the ratio of 9 : 1 and 1 : 9.

Thus the reconstituted chimeric dermis contained 90% chick cells and 10% mouse cells, or conversely 90% mouse cells and 10% chick (or quail) cells. The 10% cells were obtained from the skin of either appendage-forming (dorsal) regions or glabrous areas (midventral apterium of the chick embryo, foot sole of the mouse embryo). The latter were used in order to test whether dermal cells from those non-appendage-forming regions were totally deprived of any morphogenetic information for the differentiation of cutaneous appendages.

Four types of associations were thus performed and grown for 8 days as chorioallantoic grafts: (1) Dorsal 12.5-day mouse epidermis/chimeric dermis containing 90% 7-day dorsal dermal chick cells and 10% 14.5- or 15.5-day plantar dermal mouse cells. (2) Dorsal 12.5-day mouse epidermis/chimeric dermis composed of 90% 7-day dorsal dermal chick cells and 10% 12.5-day dorsal dermal mouse cells. (3) Dorsal 6-day chick or 5-day quail epidermis/chimeric dermis constituted of 90% 14.5-day dorsal dermal mouse cells and 10% 11- or 12-day dermal chick cells from the midventral apterium. (4) Dorsal 6-day chick epidermis/chimeric dermis composed of 90% 14.5-day dorsal dermal mouse cells and 10% 6-day dorsal dermal quail cells.

Recombinants of types 1 and 2 involving mouse epidermis gave rise to two types of hair follicles (Figs 4 and 5): a large number (from 6 to 24 per explant) of small stage 2 arrested follicles (Fig. 5), and a small number (from 1 to 4 per explant) of large stage 6 fully developed follicles (Figs 4 and 5). At histological examination, it was seen that in close

Figs 4 and 5. Type 1 heterospecific recombinants: dorsal mouse epidermis/chimeric dermis composed of dorsal dermal chick cells (light nuclei) and plantar dermal mouse cells (dark nuclei).

Fig. 4. Stage 6 hair follicle whose dermal papilla is made of mouse cells and whose dermal sheath contains a mixture of chick and mouse cells.

Fig. 5. Stage 2 hair bud (*top*) surrounded by dermal chick cells only, and stage 6 hair follicle (*bottom*) surrounded by a dermal cell population where mouse cells predominate and are in direct contact with the epidermal basement membrane.

Figs 6 and 7. Type 3 heterospecific recombinants: dorsal quail epidermis/chimeric dermis composed of dorsal dermal mouse cells and chick dermal cells from the midventral apterium.

Fig. 6. Base of a long fully developed feather filament whose dermal papilla is composed of a mixture of mouse (M) and chick (C) cells.

Fig. 7. Transverse section through the collar of a long feather filament showing eight well-organized barb ridges and a dermal core containing a majority of chick cells (but no quail cells).

BC, blood cells: BR, barb ridge; DC, dermal core; DP, dermal papilla; DS, dermal sheath; HS, hair shaft.

contact with the arrested follicles only chick cells were found, whereas the fully developed stage 6 follicles were surrounded by a mixed dermal cell population where mouse plantar or dorsal cells predominated and were in immediate contact with the follicle epidermis.

Recombinants of type 3 involving chick or quail epidermis gave rise again to two types of feather filaments: a large number (an average of 20 per explant) of small arrested feather filaments with ill-formed barb ridges, and a few (1 to 5 per explant) of long (3–5 mm) feather

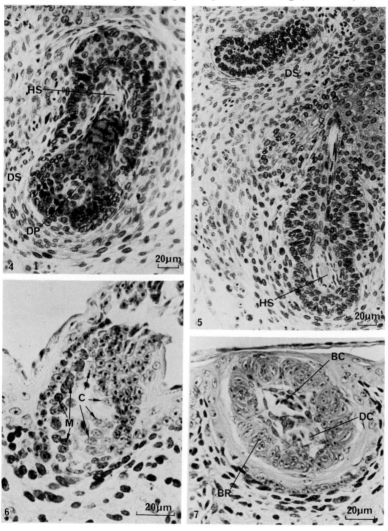

filaments with a typical number of well-organized barb ridges (Figs 6 and 7). At histology, the dermal papilla of the long fully developed filaments comprised a majority of chick cells (from the midventral apterium). In those recombinants involving quail epidermis, no quail dermal cells were found inside the dermal papilla of the filaments, which warranted that these long filaments had not been induced by homospecific dermal cells accidentally contaminating the isolated epidermis. The dermal papilla of the arrested filaments contained mouse cells exclusively or a large majority of them.

Finally recombinants of type 4 involving chick epidermis and a mixed dermis composed of quail (10%) and mouse (90%) cells yielded the same result as type 3, with the additional advantage that even single dermal quail cells, isolated among mouse cells, could be easily detected due to their specific nuclear marker (Le Douarin, 1971). Explants formed a number of arrested feather filaments and a few fully grown ones. The dermal papilla of the latter was composed by a majority of quail cells, while the dermal papilla of the arrested filaments contained either no quail cells or but a very small proportion.

The simultaneous production by the same recombinant of arrested and fully developed appendages demonstrates that the arrest of development in chick–mouse heteroclass recombinants is not due to adverse culture conditions. The differentiation of a feather bud into a feather, and of a hair bud into a hair, requires the presence of avian, resp. mammalian dermal cells. The distribution of these cells in the chimeric dermises in close contact with the appendage-forming epidermis suggests that the necessary morphogenetic information cannot be transmitted at distance but that an immediate cell to cell contact between dermal and epidermal cells is required.

The important finding that dermal cells from non-appendage-forming regions are able to support the continuation of appendage morphogenesis is another argument in favour of the necessity of a two-step induction. Dermal cells from glabrous regions are unable to exert the first initiating induction. Hence the inaptitude of glabrous regions to form cutaneous appendages. However, they possess the necessary class- and species-specific information for the continuation of appendage development (Figs 8 and 9).

CONCLUDING REMARKS

From the reported experiments the following view of appendage morphogenesis in amniotes can be proposed. Dermis and epidermis each play determining roles. Scale, feather and hair morphogenesis appears

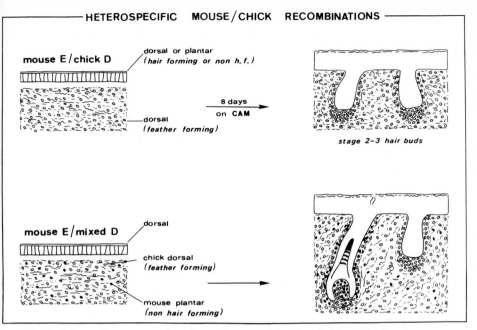

mouse E/chick D

dorsal or plantar
(hair forming or non h. f.)

dorsal
(feather forming)

8 days
on CAM

stage 2-3 hair buds

mouse E/mixed D

dorsal

chick dorsal
(feather forming)

mouse plantar
(non hair forming)

Fig. 8. Schematic representation of experiments with chimeric bispecific dermis, composed of 90% dorsal chick cells and 10% plantar mouse cells.

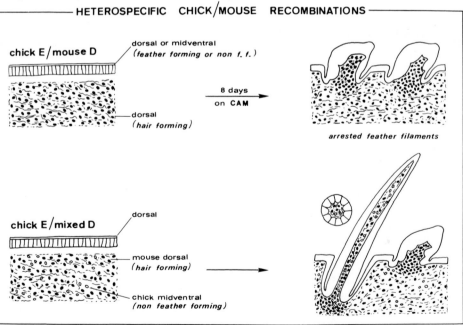

HETEROSPECIFIC CHICK/MOUSE RECOMBINATIONS

chick E/mouse D

dorsal or midventral
(feather forming or non f. f.)

dorsal
(hair forming)

8 days
on CAM

arrested feather filaments

chick E/mixed D

dorsal

mouse dorsal
(hair forming)

chick midventral
(non feather forming)

Fig. 9. Schematic representation of experiments with chimeric bispecific dermis, composed of 90% dorsal mouse cells and 10% chick cells from the midventral apterium.

to occur in at least two steps. During step 1, the dermis exerts a non-class-specific induction, which triggers off the differentiation of epidermal placodes. This induction determines the region-specific size, shape and distribution pattern of the appendages to be formed. Their class-specificity however (scale, feather or hair) resides in the epidermis. The epidermis responds to this step 1 induction by starting to build up a scale, feather or hair primordium according to whether it originates from a reptile, bird or mammalian embryo. The capacity of exerting this step 1 induction is restricted to the dermis of those areas of skin which normally form cutaneous appendages. It is lacking in dermis from glabrous regions.

For the continuation of appendage morphogenesis a step 2 dermal induction is required. This is class-specific and cannot be transmitted across the amniote class barrier. It contains the necessary information for the species-specific shaping of the appendages. It can be transmitted, within the class of birds at least, to an epidermis from another species. Thus the species-specific architecture (of feathers at least) is dermis-dependent, with the exception of morphological details at the cellular (barbules) and molecular (keratins) levels, which remain strictly epidermis-dependent.

REFERENCES

Coulombre, J. L. and Coulombre, A. J. Metaplastic induction of scales and feathers in the corneal anterior epithelium of the chick embryo. *Develop. Biol.* **25**, 464–478 (1971).

Dhouailly, D. Analyse des facteurs de la différenciation spécifique de la plume néoptile chez le canard et le poulet. *J. Embryol. exp. Morph.* **18**, 389–400 (1967).

Dhouailly, D. Déterminisme de la différenciation spécifique des plumes néoptiles et téléoptiles chez le poulet et le canard. *J. Embryol. exp. Morph.* **24**, 73–94 (1970).

Dhouailly, D. Dermo-epidermal interactions between birds and mammals: differentiation of cutaneous appendages. *J. Embryol. exp. Morph.* **30**, 587–603 (1973).

Dhouailly, D. Formation of cutaneous appendages in dermo-epidermal recombinations between reptiles, birds and mammals. *Wilhelm Roux' Arch. Entwickl.-Mech. Org.* **177**, 323–340 (1975).

Dhouailly, D. Regional specification of cutaneous appendages in mammals. *Wilhelm Roux' Arch. Entwikl.-Mech. Org.* **181**, 3–10 (1977).

Dhouailly, D. and Sengel, P. Propriétés phanérogènes des cellules dermiques de peau glabre d'Oiseau ou de Mammifère. *C.R. Acad. Sci. (Paris), Sér. D* **281**, 1007–1010 (1975).

Kato, Y. and Hayashi, Y. The inductive transformation of the chorionic epithelium into skin derivatives. *Exp. Cell. Res.* **31**, 599–602 (1963).

Kieny, M. and Pautou, M. P. Rôle du mésoderme dans l'évolution morphogénétique de la membrane interdigitale du poulet et du canard. *C.R. Acad. Sci. (Paris), Sér. D* **264**, 3030–3033 (1967).

Kollar, E. J. An *in vitro* study of hair and vibrissae development in embryonic mouse skin. *J. invest. Dermatol.* **46**, 254–262 (1966).

Kollar, E. J. The induction of hair follicles by embryonic dermal papillae. *J. invest. Dermatol.* **55**, 374–378 (1970).

Hardy, M. H. The development of pelage hairs and vibrissae from skin in tissue culture. *Ann. N.Y. Acad. Sci* **53**, 546–561 (1951).

Lawrence, I. E. Timed reciprocal dermal-epidermal interactions between comb, mid-dorsal, and tarsometatarsal skin components. *J. exp. Zool.* **178**, 195–210 (1971).

Le Douarin, N. Caractéristiques ultrastructurales du noyau interphasique chez la caille et chez le poulet et utilisation de cellules de caille comme "marqueurs biologiques" en embryologie expérimentale. *Ann. Embryol. Morph.* **4**, 125–135 (1971).

Mizuno, T. Epidermal metaplasia of proamniotic epithelium induced by dorsal skin dermis in the chick embryo. *J. Embryol. exp. Morph.* **27**, 199–213 (1972).

Pautou, M. P. Rôle déterminant du mésoderme dans la différenciation spécifique de la patte de l'Oiseau. *Arch. Anat. microsc. Morph. exp.* **57**, 311–328 (1968).

Rawles, M. E. Tissue interactions in scale and feather development as studied in dermal-epidermal recombinations. *J. Embryol. exp. Morph.* **11**, 765–789 (1963).

Saunders, J. W., Jr and Gasseling, M. T. The origin of pattern and feather tract specificity. *J. exp. Zool,* **135**, 503–527 (1957).

Sengel, P. Recherches expérimentales sur la différenciation des germes plumaires et du pigment de la peau de l'embryon de poulet en culture *in vitro*. *Ann. Sci. nat., Zool.* **11**, 430–514 (1958).

Sengel, P. The organogenesis and arrangement of cutaneous appendages in birds. *Advanc. Morphogenes.* **9**, 181–230 (1971).

Sengel, P. Tissue interactions in skin morphogenesis, M. Balls and M. Monnickendam (ed.), in Organ Culture in Biomedical Research, pp. 111–147, Cambridge Univ. Press, Cambridge, London, New York, Melbourne (1976a).

Sengel, P. Morphogenesis of Skin, Developmental and Cell Biology Series, 277 pp., Cambridge Univ. Press, Cambridge, London, New York, Melbourne (1976b).

Sengel, P., Dhouailly, D., Derminot, J. and Tasdhomme, M. Constitution chimique des kératines des plumes néoptiles chimères issues d'associations ectomésodermiques hétérospécifiques entre le poulet et le canard. *C.R. Acad. Sci.* (*Paris*), *Sér.* D **281**, 885–888 (1975).

Sengel, P., Dhouailly, D. and Kieny, M. Aptitude des constituants cutanés de l'aptérie médio-ventrale du Poulet à former des plumes. *Develop. Biol.* **19**, 436–446 (1969).

Sengel, P. and Pautou, M.P. Experimental conditions in which feather morphogenesis predominates over scale morphogenesis. *Nature* (*Lond.*) **222**, 693–694 (1969).

The Differentiation of the Ganglioblasts of the Autonomic Nervous System Studied in Chimeric Avian Embryos

Nicole Le Douarin

Institut d'Embryologie du Centre National de la Recherche Scientifique et du Collège de France, 49 bis, avenue de la Belle Gabrielle, 94130 Nogent-sur-Marne, France

INTRODUCTION

The autonomic nervous system is composed of two sets of ganglia and nerves synthesizing different neurotransmitters and acting antagonistically upon the same effectors. The orthosympathetic (or sympathetic) ganglia, localized exclusively in the dorsal structures of the cervicotruncal region, act upon target cells through catecholamines (mainly noradrenaline), while the parasympathetic neurons, distributed ventrally in dispersed intravisceral ganglia, produce acetylcholine as a neurotransmitter. The peripheral autonomic ganglia are connected, through cholinergic synapses, with central neurons located in the *medulla oblongata* and lumbosacral spinal cord for the parasympathetic system, and in the dorsal spinal cord for the orthosympathetic (see Fig. 3).

The derivation of the peripheral ganglion cells from the neural crest, although suggested by a variety of experiments consisting in the early ablation of the neural anlage either in amphibians or in higher vertebrates (Muller and Ingvar, 1923; van Campenhout, 1946; Yntema and Hammond, 1947; Nawar, 1956; Strudel, 1953), was not immediately accepted, and various other sources, such as mesenchymal cells (Tello, 1925; Levi-Montalcini, 1947; Keuning, 1944, 1948) or endodermal cells (Masson, 1923; Schack, 1932), were also proposed to account for the formation of the autonomic ganglia. However, since cell markers have been used to trace the fate of neural crest cells, no doubt remains about the neurectodermal origin of these structures. By orthotopic grafting of isotopically labelled neural crests into non-labelled chick embryos, Weston (1963) showed the accumulation of neural cells in the location of the primary sympathetic ganglion chain, and, later on, Johnston demonstrated that grafts of labelled cervical neural crest give rise to the superior cervical ganglia (SCG) (quoted in Weston, 1970).

However, the migration pattern of the autonomic neuroblasts remained unknown. In particular neither the respective sources along the neural axis of the parasympathetic and orthosympathetic neuro-blasts, nor the spatial relationships between the preganglionic neuron they are associated with, had been established. Therefore, using a stable cell marking technique, we have investigated some aspects of the de-velopment of the autonomic nervous system in the avian embryo. In a first step, we have determined the levels of the neural axis from which the adrenomedulla, the sympathetic and enteric ganglia come. Then we investigated the possible role of cell and tissue interactions on the differentiation of autonomic ganglioblasts either into adrenergic or cho-linergic neurons.

EXPERIMENTAL ANALYSIS OF CERVICOTRUNCAL NEURAL CREST CELL MIGRATION AND DEVELOPMENTAL FATE

The cell labelling technique used for this purpose has been described in detail previously (Le Douarin, 1969, 1973a, b). It is based on struc-tural differences in the nucleus of two species of birds: the Japanese quail (*Coturnix coturnix japonica*) and the chick (*Gallus gallus*). In quail cell nuclei, the nucleolar RNP are associated with one or two large masses of heterochromatin, while in the chick very little DNA partici-pates in the nucleolar structure. As a consequence, cells of the two species can readily be identified in chimeric tissues either after applica-tion of the Feulgen–Rossenbeck stain or in electron microscopy (Fig. 1).

Due to the stability of the nuclear characteristics of quail and chick cells, their combination provides a labelling stable, whatever the dura-tion of the association.

Isotopic and isochronic grafts of fragments of quail neural tube into chick embryos were carried out at various developmental stages rang-ing from 6- to 30-somite. The reverse experiment, i.e. graft of chick neural tube into quail was also done as control. No difference was found in timing and pattern of migration in the two grafts. A few hours after

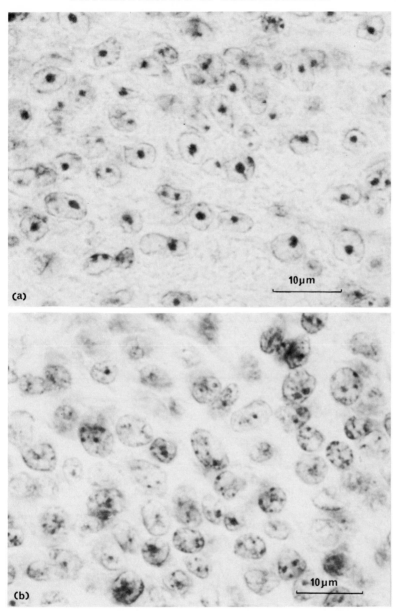

Fig. 1. Sympathetic ganglia stained according to Feulgen–Rossenbeck technique. (a) Ganglion of a $7\frac{1}{2}$-day quail embryo. Large mass of heterochromatin in the nucleus. (b) Ganglion of an $8\frac{1}{2}$-day chick embryo. Small chromocentres; chromatin evenly dispersed in the nucleoplasm.

the graft, the implant is properly incorporated into the axial structures of the host embryo which undergoes normal morphogenesis (Fig. 2). The neural crest cells migrate into the recipient according to the normal pattern and can be recognized on 5 μm serial sections of the trunk after Feulgen–Rossenbeck staining. Differentiated autonomic neurons were identified by a silver staining procedure according to Ungewitter

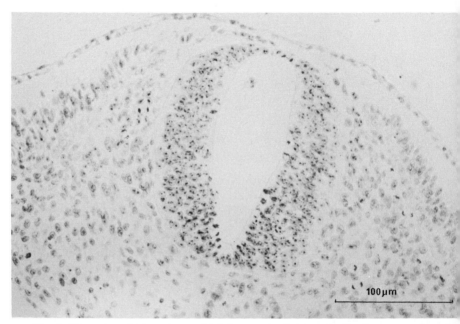

Fig. 2. Three-day chick embryo which has received the graft of a fragment of quail neural tube at the level of the 18th to 24th somites at the 25-somite stage. The grafted neural tube develops normally in the axial structures of the host. The quail neural crest cells migrate in the chick tissues and can be recognized due to the structure of their nucleus (Feulgen–Rossenbeck staining).

(1951) and/or the cytochemical reaction for cholinesterases according to Koelle and Friedenwald (1949) as modified by Karnovsky and Roots (1964). Distinction between cholinergic and adrenergic cells was carried out through the technique of Falck (1962) for catecholamine detection in UV light after formol vapour fixation, which is positive in catecholamine-producing cells and negative in cholinergic neurons.

The heterospecific grafts of the neural primordium were done from the level of the rhombencephalon to the posterior end of the neural axis. The length of the implants corresponded to 4 to 6 somites and the more cranial the graft the younger the embryos were, since neural crest formation begins in the head and progressively extends posteriorly with the closure of the neural tube.

The results of these experiments, which have been reported elsewhere (Le Douarin and Teillet, 1973; Teillet and Le Douarin, 1974; Teillet, 1977), can be summarized as follows:

Sympathetic ganglia and adrenomedullary paraganglia

After the graft of fragments of quail neural tube into chick embryos or inversely, cells of sympathetic ganglia and aortic plexus (Fig. 3) originate from the explant at the corresponding transverse level of the embryo. In fact the origin of the sympathetic ganglion chains extends along the whole spinal neural crest caudad to the level of the 5th somite.

The sympathetic ganglion chains

The paravertebral ganglia have the same segmentary distribution as the corresponding rachidian nerve roots. Due to the morphogenetic movements which affect the dorsal structures during ontogenesis, the sympathetic ganglia of the trunk are drawn caudally. At 7 days of incubation the orthosympathetic ganglia of the trunk are located 2 somites behind the transverse level from which they are actually derived.

The first ganglion of the sympathetic chain, the superior cervical ganglion arises from the neural crest located behind the level of the 5th somite on the length of about 5 segments (Fig. 3). It is of interest to notice that its final localization is anterior to the level of the neural axis from which it originates.

The aortic plexus, formed by ganglion cell and fibres, extends caudally from the level of the suprarenal glands and is located laterally and ventrally around the dorsal aorta. It is derived from the level of the neural crest caudad to the 16th to 18th somites.

The adrenomedullary paraganglia. Cells of implant type were found in the suprarenal gland only when the graft was done at the level of the 18th to 24th segments. If this whole area of a quail neural tube is isotopically grafted into a chick, a chimeric suprarenal gland develops in which cortical cells belong to the host and all adrenomedullary cells are of the graft type (Teillet and Le Douarin, 1974). Dispersed adrenomedullary cells were found in the metanephric blastoma (Le Douarin and Houssaint, 1969; Fontaine and Le Douarin, 1971) and are derived from the posterior levels of the neural crest (Le Douarin, 1969).

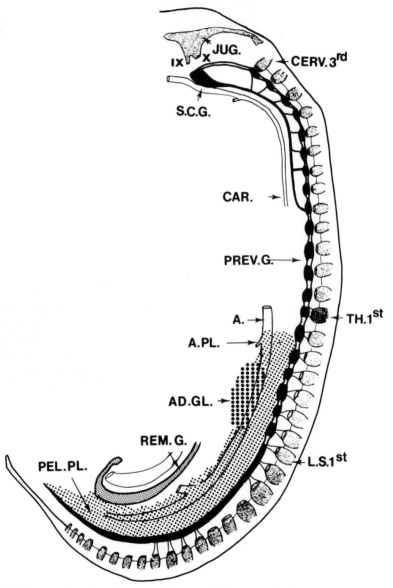

Fig. 3. Diagram showing the general feature of the sympathetic system in the chick. A, aorta; AD.GL., adrenomedullary paraganglion; A.PL., aortic plexus; CAR., internal carotid artery; CERV.3rd = 3rd cervical rachidian nerve; JUG., jugular ganglion; L.S.1st, 1st lumbosacral rachidian nerve; PEL.PL., pelvian plexus; PREV.G., prevertebral ganglia; REM.G., ganglion of Remak; S.C.G., superior cervical ganglion.

Enteric ganglia

At the time the present work was undertaken, although the origin of the enteric ganglioblasts from the primitive neural anlage was admitted, discrepancies arose concerning the level of the neural axis they are derived from. For some authors the whole neural crest gives rise to the enteric ganglia while others consider that they come exclusively from either the vagal or the truncal level of the neural axis (see Le Douarin and Teillet, 1973, for references). Experiments involving isotopic and isochronic grafts of pieces of quail neural primordium into chick (or *vice versa*) were carried out. From the observation of serial sections of the host intestine the following picture emerged. The enteric ganglia appear to have a double origin from the vagal (level of somites 1 to 7) and the lumbosacral (behind the level of the 28th somite) neural primordium. The main source of the enteric neuroblasts is located at the level of somites 1 to 7; it gives rise to ganglion cells which migrate into the whole gut including large intestine, the caeca and the rectum. From the lumbosacral neural primordium, originate some post-umbilical gut ganglion cells and the neurons of the ganglion of Remak. The part of the neural primordium located between the 8th and 28th somite does not participate in the formation of the wall gut ganglia.

The chronology of the enteric neuroblast migration has been followed through the observation of host embryos at successive developmental stages. Most cells of vagal origin leave the neural crest before the 13-somite stage, but a few of them still migrate some time after the 16-somite stage. Those cells which have to reach the hind gut accomplish a long-term migration which can be evaluated as lasting about 6 days. The presumptive neuroblasts of lumbosacral origin are not found in the hind gut before the 7th day of incubation. As a result, the hind gut (from the origin of the caeca to the cloaca) remains aneural until the 5th to 6th day of incubation. When taken from the embryo before this stage and cultivated on the chorioallantoic membrane (CAM) it is completely devoid of enteric ganglion cells and nerve fibres (Fig. 4).

The diagram represented in Fig. 5 summarizes the observations concerning both the ortho- and parasympathetic systems.

The neural primordium appears to be regionalized in several parts with respect to the fate of the presumptive autonomic ganglioblasts they give rise to:

(1) An anterior region located from somites 1 to 7 from which enteric ganglion cells are derived, which can be divided into two parts: one corresponding to the level of somites 1 to 5 gives rise only to enteric

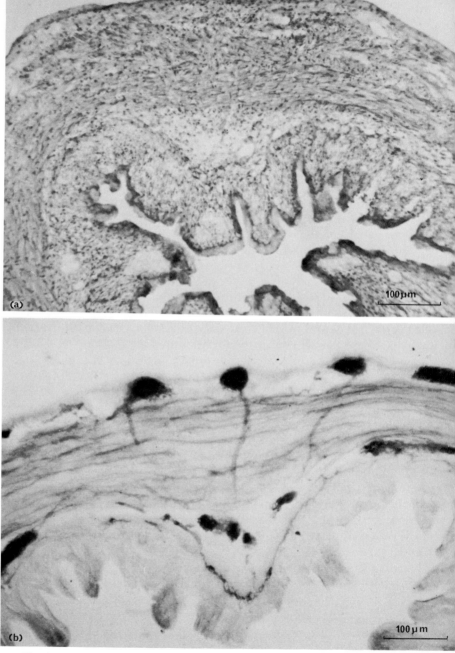

Fig. 4. (a) Transverse section in the intestine of a 5-day chick embryo grafted for 9 days on the CAM. Cytochemical technique for cholinesterase detection. Non-specific cholinesterase activity is present in the epithelium. No enteric ganglia develop; the intestine remains aneural (Hematoxyline staining). (b) Enteric ganglia in the normal gut of a chick embryo. Same treatment as in (a) +action of DFP (di-isopropylfluoro-phosphate): specific cholinesterases in ganglia and nerves. The reaction disappears in the epithelium.

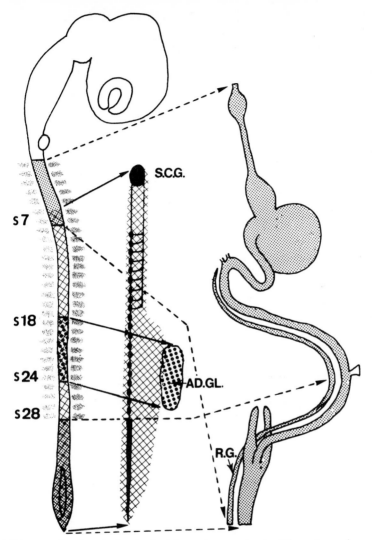

Fig. 5. Diagram showing the levels of origin on the neural axis of the autonomic neurons and adrenomedullary cells. The vagal level of the neural crest (from somites 1 to 7) provides all the enteric ganglia of the pre-umbilical gut and contributes to the innervation of the post-umbilical gut. The lumbosacral level of the neural crest gives rise to the ganglion of Remak (R.G.) and to most of the ganglia of the post-umbilical gut. The orthosympathetic chain derives from the level of the neural crest posterior to the 5th somite and the adrenomedullary cells originate from the level of somites 18 to 24. S.C.G., superior cervical ganglion; AD.GL., adrenomedullary gland; R.G., ganglion of Remak.

cholinergic neurons, the other (level of somites 5 to 7) produces both enteric neurons and some of the adrenergic cells of the SCG.

(2) The part of the neural primordium posterior to the level of the 7th somite, also divided in two: a portion producing only adrenergic cells (levels of somites 8th to 28th) and another (posterior to the level of the 28th somite) which is able to give rise to both adrenergic and cholinergic neurons.

The advantage of the technique used in this work is to disturb as little as possible the normal developmental processes since quail and chick species are closely related in taxonomy and their developmental rate are roughly similar during the first half of the incubation period. The results presented above can therefore be considered as giving a satisfactory view of the normal developmental processess which take place in both species considered.

THE INFLUENCE OF CELL AND TISSUE INTERACTIONS ON AUTONOMIC GANGLIOBLAST DIFFERENTIATION

It was further investigated whether the neuroblasts of the autonomic nervous system are already committed to synthesize either catecholamines (as adrenergic neurons and paraganglion cells) or acetylcholine (as enteric ganglia), when they begin to migrate.

One of the experimental approaches used was to modify artificially the normal pattern of migration and localization of crest cells in the embryo by grafting pieces of the neural primordium into an heterotopic situation along the neural axis. Another was to take the neural crest from the embryo before the onset of migration and associate it in culture (*in vitro* or *in vivo* on CAM) with embryonic rudiments. The latter may or may not be the tissue into which the crest cells migrate and differentiate in normal developmental conditions. The type of differentiation achieved by the ganglioblasts was studied by cytochemical and physiological methods (Le Douarin and Teillet, 1974; Le Douarin *et al.*, 1975).

These experiments can give information on various aspects of the neural crest cell evolution, in particular whether tissue interactions happen promoting or permitting expression of neural crest phenotypes and, in such a case, whether they occur during or after the migration process.

It should be emphasized that the catecholaminergic cells of the autonomic system are restricted to the dorsal trunk region, they never cross the dorso-ventral level of the suprarenal glands nor penetrate into the dorsal mesentery. Laterally, they are found exclusively in the vicinity of the mediodorsal plane and do not overpass at any developmental

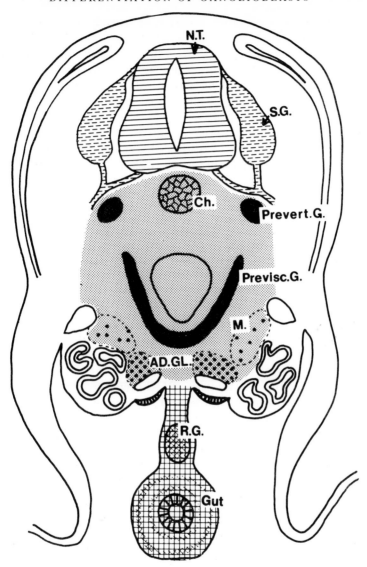

Fig. 6. Scheme showing the distribution in the trunk of a 5-day avian embryo of the adrenergic ganglia and paraganglia ▓ and of the cholinergic parasympathetic ganglia ▦ . The prevertebral (Prevert.G.) and previsceral (Previsc.G.) ganglia and the paraganglia of the suprarenal gland (AD.GL.) and of the metanephritic blastema (M.) are located in the dorsal mesenchyme. In contrast, the enteric ganglia of the gut and the ganglion of Remak (R.G.) are distributed in the splanchnopleural mesenchyme. Ch., notochord; S.G., sensory ganglion.

stage the lateral levels defined by the external sides of the kidney (Fig. 6). Previous work by Cohen (1972) has thrown light on the important role played by the somitic mesenchyme in the differentiation of sympathoblasts and suggested that fundamental changes of developmental significance occur in crest cells during their ventral migration along both the lateral sides of the neural tube and the somite.

According to Norr (1973), however, the somitic mesenchyme is not competent by itself for promoting sympathoblast differentiation, but acquires such a capability from the influence of the complex ventral neural tube–notochord. In order to analyse the possible role of the ventral neural tube, the notochord and the mesenchymal substrate on the autonomic· neuron differentiation, several series of experiments were carried out.

Adrenergic cell differentiation

Mesenchymes non-permissive for adrenergic cell differentiation. Experiment 1 : *Association of the neural primordium (level of somites 18 to 24) with lateral plate derivatives and limb bud mesenchymes)* (Fig. 7)

In the normal development the dorsal region of the neural crest gives rise only to adrenergic cells. This is not the case when it is cultivated in contact with either the gut or the limb bud mesenchymes. In the gut, the neural crest cells differentiate into neurons as shown by the silver procedure and exhibit an acetylcholinesterasic activity (Fig. 8) but are devoid of catecholamines.

Role of the notochord and of the neural tube on adrenergic cell differentiation. Experiment 2: *Association of the neural primordium (level of somites 18 to 24) +notochord with hind gut mesenchyme* (Fig. 7)

In another series, the *neural primordium* taken from the adrenomedullary level of 25-somite quail embryos, was inserted together with the *notochord* into the *hind gut* taken from a 5-day chick embryo and cultivated on the CAM for 8 to 9 days. In 3 out of 7 explants, catecholamine-producing cells developed (Fig. 9).

Experiment 3: *Association of the neural crest with dorsal mesenchyme after removal of the neural tube and the notochord* (Fig. 7)

The role of the neural tube on sympathoblast differentiation was then tested as follows. The neural tube and associated neural crest was removed from 18- to 20-somite chick embryos either at the level of the somites or in the area of the unsegmented somitic mesenchyme. Pieces of quail neural crest from the trunk level were placed into the groove

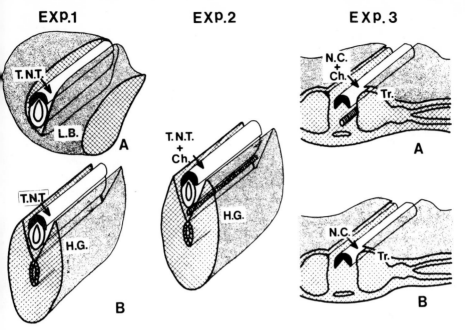

Fig. 7. Diagram showing the various kinds of experiments carried out to investigate environmental requirements for adrenergic cell differentiation.

Exp. 1: (A) Association of the neural primordium taken from a 25-somite quail embryo at the level of somites 18 to 24, with the limb bud of a 3-day chick embryo. Graft of the explant on the CAM of a chick for 9 days. Treatment of the tissues by both the techniques of Falck and of Feulgen-Rossenbeck to study the differentiation of the neural crest derivatives. (B) Association of the same level of the neural tube of a quail embryo as in the experiment 1(A) with the hind gut of a 5-day chick embryo.

Exp. 2: Same experiment as in Exp. 1 (B), but in this case the quail neural tube remains associated with the notochord.

Exp. 3: Transverse trip of a chick embryo at 18- to 20-somite stage, from which the neural tube + neural crest (A) or the neural tube + neural crest + notochord (B) were removed. In a second step, the neural crest taken at the dorsal level from quail embryos was inserted into the groove resulting from the excision.

resulting from the operation and the explant was grafted on the CAM of a chick for 7 days. In these conditions, a large amount of catecholamine-containing cells differentiated.

The same experiment was carried out, but the notochord was also removed. In most cases (7/9) fluorescent quail cells were present in the graft.

These results confirm and extend those of Cohen (1972) showing that various non-somitic mesenchymes are not able to promote sympathoblast differentiation. Our experiments demonstrate, in addition, that

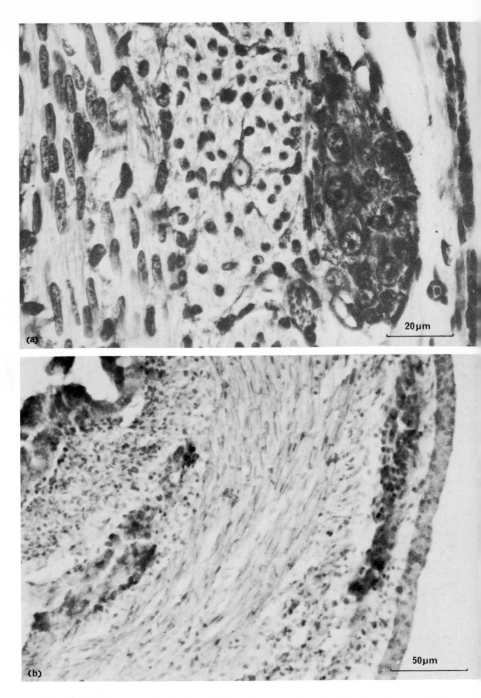

Fig. 8. (a) Hind gut of a 5-day chick embryo associated with the adrenomedullary level of the neural primordium of a 25-somite quail embryo and grafted 9 days on the CAM. Silver stain according to Ungewitter shows an enteric ganglion of the Auerbach's plexus. (b) Same experiment: Auerbach's and Meissner's plexuses develop and ganglion cells show a cholinesterase activity.

Fig. 9. Result of Exp. 2 (see Fig. 7) in which the hind gut of a 5-day chick embryo has been associated with the neural primordium + the notochord of a quail embryo. (a) Differentiation in the gut mesenchyme of an adrenergic ganglion, showing fluorogenic monoamines after the application of the Falck's technique. (b) The adrenergic ganglia (AG) is made up of quail cells. In addition, non-adrenergic enteric ganglia also differentiate (arrow). cm, circular muscle layer of the intestine.

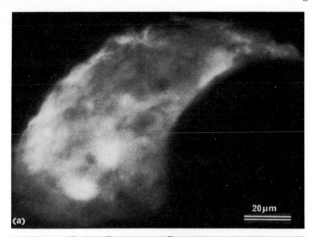

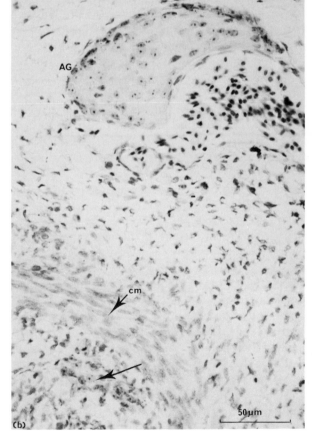

dorsotruncal mesenchyme can promote adrenergic cell differentiation in the absence of both the neural tube and the notochord. That does not exclude, however, a previous influence of the notochord and possibly of the neural tube on the dorsal mesenchyme. It is likely that such an influence has occurred prior to the removal of these structures.

The influence of the preganglionic neuronal input on postganglionic neurons differentiation in the autonomic nervous system has been already demonstrated. Interruption of the spinal cord input at birth prevents maturation of adrenergic neurons of the mouse SCG (Black et al., 1971).

The possible role of the neural tube on sympathoblast differentiation might result from the establishment of a synaptic contact between the preganglionic neuron and the neural crest cell. In such a case, the removal of the neural tube at an early developmental stage would hinder adrenergic cell differentiation. The result of experiment 3 shows that such a synapse is not necessary for the onset of catecholamine synthesis in the presumptive sympathetic neuron, since cells containing large amounts of catecholamine were present in the explants.

On the contrary when comparing the results of experiment 1 vs. experiment 2 the role of the notochord appears of major significance in promoting crest cell differentiation: the presence of this structure results in the differentiation of adrenergic cells in a non-somitic mesenchymal substrate. The primary influence of the neural tube–notochord system on the somitic mesenchyme is thus confirmed, the latter being primed to induce sympathoblast differentiation (Norr, 1973). In addition, this experiment shows that the specificity of the mesenchymal substrate is not fundamental since in this particular case a tissue of splanchnic origin can replace dorsal mesenchyme.

The inductive role of the ventral neural tube and the notochord on cartilage differentiation in the somite is a well-established phenomenon and the similarity between the sympathoblast and the vertebral cartilage systems was already underlined (Norr, 1973). The interacting tissues, the timing of the interactions and the stage at which they occur appear very similar (cf. Strudel, 1953, 1955; Avery et al., 1956; Lash et al., 1957; Lash, 1963, 1967). However, one must admit, from the results reported above, that these two phenomena are in fact independent even though they are concomitant. No cartilage indeed is induced in the gut mesenchyme in experiment 2.

A puzzling developmental process deserves to be mentioned here. Notochord cells appear to be the first cells in the organism with the capability to store catecholamines. Kirby and Gilmore (1972) reported that the notochord shows fluorigenic amines as revealed by the formaldehyde-induced fluorescence technique during a short period of development (2 to 4 days). Recently Strudel *et al.* (1977) confirmed these findings and showed that notochord cells actually have the enzymatic equipment for catecholamine synthesis. Even if no data are available suggesting relationships between the ability of the notochord to synthesize fluorigenic monoamines and its capacity to elicit the same function in autonomic ganglioblasts, it is interesting to confront both these developmental events for further consideration.

Cholinergic cell differentiation

One part of the neural crest never gives rise to enteric ganglia in normal development; it could be wondered whether in this area (level of somites 8 to 28) the presumptive autonomic neurons are early committed to differentiate into adrenergic cells. To study this problem, a piece of the dorsal neural primordium of a quail embryo, taken at the "adrenomedullary" level (somites 18 to 24), was transplanted into a chick embryo at the "vagal" level (Le Douarin *et al.*, 1975). In this case, presumptive adrenergic neuroblasts of the dorsal region migrate into the intestine and give rise to normally located and morphologically developed Auerbach's and Meissner's plexuses enteric ganglia. These cells that in normal development are not attracted by splanchnic structures, when grafted at the hind brain level, are led by a preferential pathway to the developing gut.

Cytochemically the quail neuronal cells which constitute the Auerbach's and Meissner's plexuses of the host embryo intestine are shown to be devoid of formol vapour-induced fluorescence. Furthermore a physiological assay described in detail in a previous paper (Le Douarin *et al.*, 1975) demonstrated that they actually acted as cholinergic neurons.

CONCLUSION

According to the isotopic and isochronic transplantation studies reported above, the neural crest appears regionalized into several areas regarding their ability to produce the various elements of the autonomic nervous system. Some regions of the neural axis are devoted to give rise exclusively to either cholinergic or catecholaminergic neurons while others contain presumptive cells for both the ortho- and parasympathetic systems. However, the neural crest cells do not appear committed at an early stage to differentiate into one or the other cell type, and the microenvironmental conditions they find in the site where they settle is one of the key factors for the expression of their differentiated phenotype.

Therefore, the external conditions which control the neural crest cell migration in the developing embryo appear of great significance for their fate in the sense that they determine the final localization of the cells. Besides their own ability to migrate, little is known either about how the initial orientation and direction of neural crest migration is established and maintained. However, the experiments reported above, involving heterotopic transplantations of pieces of the neural primordium (Le Douarin and Teillet, 1974), have shown that preferential migration pathways for crest cells are present at certain levels of the embryo. It is clear indeed that some of the neural crest cells issued from the vagal level of the neural axis are led to the gut. They migrate massively in a ventrocaudal direction (Le Lièvre and Le Douarin, 1975), and it is likely that the early dispersion of the anterior somite cells facilitates this process. Later on some crest cells are incorporated into the wall of the foregut and their migration then proceeds caudally along the digestive tract.

Another example of pre-established migration routes is given by the way the suprarenal gland is colonized. Although the ability to differentiate into adrenomedullary cells is widespread along the whole length of the neural crest (Sengel and Chevallier, 1971; Le Douarin and Teillet, 1974), only cells coming from the 18th to 24th somitic level migrate into the suprarenal gland in normal developmental conditions. On the other hand, pieces of neural crest originating from other parts of the neural primordium also give rise to the adrenomedulla provided that they are transplanted at the level of the 18th to 24th somites, at a convenient developmental stage.

By submitting selectively the neural crest to various environments we have been able to precise some aspects of the tissue interactions which mediate the differentiation to sympathetic neurons from neural crest cells.

The notochord appears to be the primary source of the developmental stimulus which elicits in the crest cells the ability to synthesize the catecholamines. Notochord can be replaced by somitic mesenchyme which very likely has taken its inductive properties from a previous interaction with the notochord itself. Anyhow, the dorsal mesenchyme is not the only tissular substrate in which adrenergic cells can differentiate, since they appear in the gut when the notochord is present together with the neural primordium.

In the absence of the notochordal stimulus it seems that the presumptive autonomic neurons develop into cholinergic cells, whatever their normal developmental fate may be.

REFERENCES

Avery, G., Chow, M. and Holtzer, H. An experimental analysis of the development of the spinal column. *J. exp. Zool.* **132**, 409–425 (1956).

Black, I. B., Hendry, I. A. and Iversen, L. L. Trans-synaptic regulation of growth and development of adrenergic neurones in a mouse sympathetic ganglion. *Brain Res.* **34**, 229–240 (1971).

Cohen, A. M. Factors directing the expression of sympathetic nerve traits in cells of neural crest origin. *J. exp. Zool.* **179**, 167–182 (1972).

Falck, B. Observations on the possibilities of the cellular localization of monoamines by a fluorescence method. *Acta physiol. scand.* **56**, Suppl. 197, 1–25 (1962)

Fontaine, J. and Le Douarin, N. Mise en évidence par fluorescence de cellules à catécholamines dans le mésenchyme métanéphritique de l'embryon de poulet. *C. R. Acad. Sci. (Paris)* **273**, 1299–1301 (1971).

Karnovsky, M. J. and Roots, L. A "direct-colouring" thiocholine method for cholinesterases. *J. Histochem. Cytochem.* **12**, 219–221 (1964).

Keuning, F. J. The development of the intramural nerve elements of the digestive tract in tissue culture. *Acta neerl. Morph.* **5**, 237–247 (1944).

Keuning, F. J. Histogenesis and origin of the autonomic nerve plexus in the upper digestive tube of the chick. *Acta neerl. Morph.* **6**, 8–48 (1948).

Kirby, M. L. and Gilmore, S. A. A fluorescence study on the ability of the notochord to synthesize and store catecholamines in early chick embryos. *Anat. Rec.* 173, 469–478 (1972).

Koelle, O. B. and Friedenwald, J. S. A histochemical method for localizing cholinesterase activity. *Proc. Soc. exp. Biol. Med.* **70**, 617–622 (1949).

Lash, J. W. Studies on the ability of embryonic mesonephros explants to form cartilage. *Develop. Biol.* **6**, 219–232 (1963).

Lash, J. W. Differential behavior of anterior and posterior embryonic chick somites *in vitro. J. exp. Zool.* **165**, 47–56 (1967).

Lash, J. W., Holtzer, S. and Holtzer, H. An experimental analysis of the development of the spinal cord. VI. Aspects of cartilage induction. *Exp. Cell Res.* **13**, 292–303 (1957).

Le Douarin, N. Particularités du noyau interphasique chez la Caille japonaise *(Coturnix coturnix* japonica). Utilisation de ces particularités comme "marquage biologique" dans les recherches sur les interactions tissulaires et les migrations cellulaires au cours de l'ontogenese. *Bull. biol. Fr. Belg.* **103**, 435–452 (1969).

Le Douarin, N. A biological cell labelling technique and its use in experimental embryology. *Develop. Biol.* **30**, 217–222 (1973a).

Le Douarin, N. A. Feulgen-positive nucleolus. *Exp. Cell Res.* **77**, 459–468 (1973b).

Le Douarin, N. and Houssaint, E. Mise en évidence des cellules phéochromes dans le mésenchyme métanéphritique de poulet évoluant en l'absence de l'uretère. *C.R. Soc. Biol. (Paris)* **163**, 505–508 (1969).

Le Douarin, N., Renaud, D., Teillet, M.-A. and Le Douarin, G. Cholinergic differentiation of presumptive adrenergic neuroblasts in interspecific chimaeras after heterotopic transplantations. *Proc. nat. Acad. Sci. (Wash.)* **72**, 728–732 (1975).

Le Douarin, N. and Teillet, M.-A. The migration of neural crest cells to the wall of the digestive tract in avian embryo. *J. Embryol. exp. Morph.* **30**, 31–48 (1973).

Le Douarin, N. and Teillet, M.-A. Experimental analysis of the migration and differentiation of neuroblasts of the autonomic nervous system and of neurectodermal mesenchymal derivatives, using a biological cell marking technique. *Develop. Biol.* **41**, 162–184 (1974).

Le Lièvre, C. and Le Douarin, N. Mesenchymal derivatives of the neural crest: analysis of chimaeric quail and chick embryos. *J. Embryol. exp. Morph.* **34**, 124–154 (1975).

Levi-Montalcini, R. Ricerche sperimentali sull'origine del simpatico toraco-lombare nell'embrione di pollo. *R. Acad. Naz. dei Lincei* **3** (1947).

Masson, P. Appendicite neurogene et carcinoides. *Ann. Anat. norm. méd.-chir.* **1**, 3–59 (1923).

Müller, E. and Ingvar, S. Über den Ursprung des Sympathikus beim Hühnchen. *Arch. mikr. Anat.* **99**, 650–671 (1923).

Nawar, G. Experimental analysis of the origin of the autonomic ganglia in the chick embryo. *Amer. J. Anat.* **99**, 473–506 (1956).

Norr, S. C. *In vitro* analysis of sympathetic chick neural crest cells. *Develop. Biol.* 34, 16–38 (1973).

Schack, L. Über die gelben Zellen im menschlichen Wurmfortsatz. *Beitr. path. Anat. (Zeiglers Beitrage)* **90**, 441–478 (1932).

Sengel, P. and Chevallier, A. Sur les potentialités de différenciation médullo-surrénalienne des crêtes neurales chez l'embryon de poulet. *C.R. Acad. Sci. (Paris)* **272**, 1301–1304 (1971).

Strudel, G. Conséquences de l'excision de tronçons du tube nerveux sur la morphogenèse de l'embryon de poulet et sur la différenciation de ses organes: contribution à la genèse de l'orthosympathique. *Ann. Sci. nat. Zool.* **15**, 253–329 (1953).

Strudel, G. L'action morphogène du tube nerveux et de la corde sur la différenciation des vertèbres et des muscles vertébraux chez l'embryon de poulet. *Arch. Anat. micr. Morph. exp.* **44**, 209–235 (1955).

Strudel, G., Recasens, M. and Mandel, P. Identification de catécholamines et de Sérotonine dans les chordes d'embryons de poulet. *C. R. Acad. Sci. (Paris)* **284**, 967–970 (1977).

Teillet, M.-A. Le développement du système nerveux sympathique étudié chez les oiseaux par la méthode des greffes interspécifiques de tube neural. In preparation (1977).

Teillet, M.-A. and Le Douarin, N. Détermination par la méthode des greffes hétérospécifiques d'ébauches neurales de Caille sur l'embryon de Poulet, du niveau du névraxe dont dérivent les cellules médullo-surrénaliennes. *Arch. Anat. micr. Morph. exp.* **63**, 51–62 (1974).

Tello, J. F. Sur la formation des chaînes primaire et secondaire du grand sympathique dans l'embryon de poulet. *Trav. Lab. Invest. biol. Univ. Madr.* **23**, 1–28 (1925).

Ungewitter, L. H. A urea silver nitrate method for nerve fibers and nerve endings. *Stain Technol.* **26**, 73–76 (1951).

Van Campenhout, E. Contribution to the problem of the development of the sympathetic nervous system. *J. exp. Zool.* **61**, 295–320 (1930).

Weston, J. A. The migration and differentiation of neural crest cells. *Advanc. Morphogenes* **8**, 41–114 (1970).

Yntema, C. L. and Hammond, W. S. The development of the autonomic nervous system. *Biol. Rev.* **22**, 334–359 (1947).

Tissue Interactions in Tooth Development *in vitro*

Irma Thesleff

*III Department of Pathology, University of Helsinki,
SF-00290 Helsinki 29, Finland*

INTRODUCTION

The tooth is formed from an epithelial and a mesenchymal component. Enamel is formed from the epithelial part, arising as an ingrowth of oral ectoderm, and all other dental tissues derive from the mesenchymal part, seen as a condensation of mesenchymal cells under the epithelial bud. The final differentiation of the tooth-forming cells takes place during the so-called bell stage of development which is reached when the under surface of the epithelial bud invaginates and forms the enamel organ, which then surrounds the mesenchymal dental papilla (Fig. 1a). A row of mesenchymal cells, facing the epithelium, then differentiate into odontoblasts and secrete the collagenous organic matrix of dentine and pre-dentine. The epithelial cells of the inner enamel epithelium, which face the pre-dentine, differentiate into tall columnar ameloblasts and secrete the organic matrix of enamel.

IN VITRO DEVELOPMENT OF THE TOOTH RUDIMENT

In vitro techniques were applied for the study of tooth development already in the 1930s (Glasstone, 1936). Since then a great number of different techniques have been presented, which allow variable degrees of differentiation (Glasstone, 1965; Koch, 1972). The progress of differentiation and the stage of development that can be reached *in vitro* largely depends on the culture conditions. Full functional maturation of ameloblasts with secretion of enamel matrix has been noted only

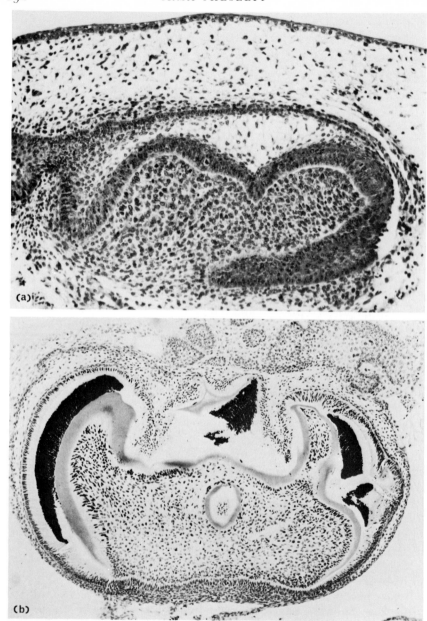

Fig. 1. Photomicrographs of lower first molars of 17-day-old mouse embryos. (a) At the start of cultivation. The epithelial and mesenchymal cells are undifferentiated. (b) After 14 days of cultivation. Large amounts of enamel matrix (black) have been secreted by the ameloblasts. Mallory's phosphotungstic acid–haematoxylin stain.

occasionally in *in vitro* studies, and it has been concluded that a protein-rich medium stimulates this development (Koch, 1972). Thus it is not surprising that the best results on tooth cultivation have been obtained when tooth germs have been grown as transplants at various sites, e.g. in the abdominal wall (Huggins *et al.*, 1934), anterior chamber of the eye (Kollar and Baird, 1970) or on the chick chorioallantoic membrane (Slavkin and Bavetta, 1968).

For our studies on tooth differentiation we have developed culture conditions which regularly allow secretion of considerable amounts of enamel matrix (Thesleff, 1976). This method is largely based on two earlier studies reporting enamel formation *in vitro* (Koch, 1967; Wigglesworth, 1968). The culture medium consists of BGJb (Biggers *et al.*, 1961) supplemented with 20% horse serum and 10% chick embryo extract. When cultured in this medium, undifferentiated bell-staged molar teeth from mouse embryos will start enamel secretion after 1 week of cultivation, and after 2 weeks vast amounts of enamel matrix are seen (Fig. 1).

EPITHELIO-MESENCHYMAL INTERACTIONS IN TOOTH DEVELOPMENT

Sequentiality

Tooth development includes a series of interactive events between the epithelium and mesenchyme, starting probably already during migration of the neural crest cells, which, as demonstrated in amphibian studies, are the predecessors of odontoblasts (Mangold, 1936; Balinsky, 1947). Such studies suggested that migrating neural crest cells interact with the stomatodeal endoderm prior to reaching their final position under the oral epithelium, and that this interaction is a prerequisite for their differentiation into odontoblasts (Sellman, 1946). Subsequent interaction between the mesenchymal cells and oral ectoderm results in the formation of the dental bud, and finally the differentiation of odontoblasts and ameloblasts and their secretion of the respective organic matrices also involves epithelio-mesenchymal interactions (Gaunt and Miles, 1967; Slavkin, 1974). These final interactions which take place during the bell stage of tooth development have been more extensively studied than the preceding ones.

Differentiation in the bell-staged tooth occurs in definite sequence. The mesenchymal cells will first differentiate into odontoblasts, and this requires an interaction with the inner enamel epithelium. It is only after the odontoblasts have started their secretion of pre-dentine that the cells of the inner enamel epithelium differentiate into ameloblasts and start secretion of enamel matrix. Again this differentiation results from an epithelio-mesenchymal interaction, and it has frequently been shown that differentiation of ameloblasts never takes place in the absence of differentiated, secreting odontoblasts (Huggins *et al.*, 1934; Gaunt and Miles, 1967; Kollar, 1972; Koch, 1972; Slavkin, 1974).

Specificity

The mesenchymal component of the tooth rudiment is specific for tooth formation, whereas the epithelial part is not. Already the results of the early studies on amphibian tooth development suggested that the ability to form teeth resides in the mesenchymal component derived from the neural crest (Henzen, 1957; Gaunt and Miles, 1967). Later studies, in which mammalian tissues have been cultivated in heterotypic recombinations, have conclusively demonstrated that the dental papilla mesenchyme is the only mesenchymal tissue in which odontoblast differentiation can be induced, whereas differentiation of ameloblasts can be induced also in other than dental epithelium (Kollar and Baird, 1970; Ruch *et al.*, 1973). When the dental papilla from a mouse molar is combined with epithelium from the diastema region (the region between incisors and molars where mice do not normally develop teeth), and cultivated *in vitro*, odontoblasts differentiate and secrete pre-dentine and induce ameloblast differentiation in the heterotopic epithelium (Fig. 2a). Similar development has been obtained also when dental papilla has been transplanted in combination with epithelium from dorsal skin or foot pads (Kollar and Baird, 1970; Ruch *et al.*, 1973).

When enamel organ epithelium has been cultivated in combination with mesenchymal tissues other than dental papilla, neither odontoblast nor ameloblast differentiation has been seen. Dental epithelium keratinizes when cultured *in vitro* with mesenchyme from the diastema region (Fig. 2b), and forms vibrissae when transplanted with snout mesenchyme (Kollar and Baird, 1970).

The above studies and the information derived from experiments on amphibian tooth formation suggest that in the bell-staged tooth rudiment the mesenchymal cells are pre-determined and that the epithelio-mesenchymal interaction leading to differentiation of odontoblasts is of a "permissive" type. On the contrary, the differentiation of ameloblasts seems to involve a "directive" type of interaction between the inner enamel epithelium and the pre-dentine secreting odontoblasts.

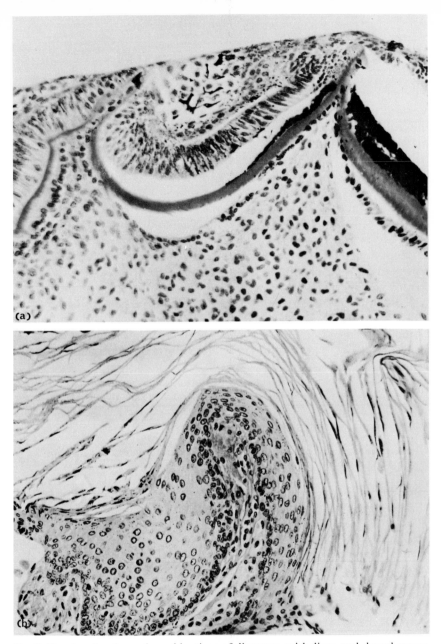

Fig. 2. Results of reciprocal combinations of diastema epithelium and dental mesenchyme (a), and dental epithelium and mesenchyme from the diastema region (b). In the former ameloblasts have differentiated and secreted enamel matrix. In the latter the dental epithelium shows squamous metaplasia with keratinization, but no differentiation towards ameloblasts. Mallory's phosphotungstic acid–haematoxylin stain.

TRANSMISSION OF THE INDUCTIVE SIGNALS

Alternative hypotheses

Various different mechanisms have been suggested to be involved in the transmission of signals in morphogenetic tissue interactions. *Signal substances* transmitted over considerable extracellular distances are believed to be mediators in primary embryonic induction (Saxén, 1961; Tiedemann, 1971; Toivonen *et al.*, 1976). Transmission of signals via *interaction between cells and extracellular matrix* (ECM) has been reported to be involved in, for example, salivary gland morphogenesis (Bernfield *et al.*, 1972; Kallman and Grobstein, 1966), and in corneal differentiation (Meier and Hay, 1975). The definition of ECM varies from cell-surface-associated macromolecules to organic matrices, e.g. in cartilage and dentine. Consequently the mechanism of transmission of inductive signals via ECM may include long-range transmission of molecules as well as close contacts between the interacting cells and matrix. Actual *cell-to-cell contacts* have been suggested to be mediators of the inductive signal in the development of kidney tubules (Wartiovaara *et al.*, 1974; Lehtonen, 1976; Saxén *et al.*, 1976; Nordling *et al.*, this volume). Close cell-to-cell contacts can be thought to mediate the inductive signal by formation of specialized junctions allowing passage of small molecules, or by interaction of complementary molecules of the cell surface (Saxén, this volume).

Ultrastructure of the epithelio-mesenchymal interface in the developing tooth

Ultrastructure of the interface between the inner enamel epithelium and the dental papilla mesenchyme during differentiation of odontoblasts has been described in several electron-microscopical studies (Pannese, 1962; Reith, 1967; Frank and Nalbandian, 1967; Slavkin *et al.*, 1969; Kallenbach, 1971; Silva and Kailis, 1972; Slavkin and Bringas, 1976). From this information something of the nature of the epithelio-mesenchymal interaction in odontogenesis can be conjectured.

In the undifferentiated bell-staged tooth germ a basal lamina separates the epithelium from the mesenchyme, and it seems to be intact during differentiation of odontoblasts. Tufts of fibrils have been reported to be attached to the mesenchymal side of the basal lamina, and their appearance has been correlated with the alignment of odontoblasts beneath the lamina and their subsequent differentiation. Thus it seems that because of the intact basal lamina, the inductive signal leading to odontoblast differentiation is not transmitted via actual cell-to-cell contacts between the heterotypic cells.

Prior to ameloblast differentiation pre-dentine is secreted. This is seen as accumulation of collagen fibres in the epithelio-mesenchymal interface. At this time numerous matrix vesicles have also been observed in the interface (Slavkin et al., 1969). The basal lamina is disrupted by epithelial cell processes prior to ameloblast differentiation, thus making close cell contacts possible with the odontoblast processes which have remained close to the basal lamina in spite of the migration of the odontoblast cell bodies away from the basal lamina during secretion of pre-dentine. In fact, desmosomes have been reported between odontoblasts and the cells of the inner enamel epithelium (Pannese, 1962). Close cell contacts between odontoblast processes and the epithelial cells, between which the processes had infiltrated, were also reported recently, and they were suggested to play a role in the differentiation of ameloblasts (Slavkin and Bringas, 1976; Slavkin et al., this volume).

Role of the extracellular matrix in tooth differentiation

Prior to odontoblast differentiation the extracellular matrix at the epithelio-mesenchymal interface is mainly composed of the basal lamina. Experimental disruption of this lamina by agents interfering with collagen deposition has usually been reported to prevent the differentiation of odontoblasts. Based on such studies and the electron-microscopical evidence (above) it has been suggested that the basal lamina guides the differentiation of odontoblasts (Slavkin, 1974; Hetem et al., 1975).

After the differentiation of odontoblasts the ECM between epithelium and mesenchyme consists mostly of pre-dentine secreted by the odontoblasts. If agents that act on collagen deposition interfere with tooth development at this stage, secretion of pre-dentine is affected, and ameloblast differentiation is prevented. Thus pre-dentine has been suggested to play a significant role in the differentiation of ameloblasts (Ruch et al., 1974; Slavkin, 1974). The matrix vesicles, seen in the ECM prior to ameloblast differentiation (above), have been shown to contain RNA, and these vesicles have also been suggested to play a role in transferring information from the mesenchymal to the epithelial cells (Slavkin, 1974).

Transfilter experiments on tooth differentiation

The transfilter technique has been widely used in studies on morphogenetic tissue interactions (Grobstein, 1953). The cultivation of the
interacting tissue components on opposite sides of a filter allows the examination of the mechanism of interaction between the inductor and
target tissue.

In tooth development the transfilter technique was first applied by
Koch (1967). He cultivated the epithelial and mesenchymal components of mouse incisors with an interposed Millipore filter with pore
sizes of 0.45 and 0.35 μm, and showed that the entire series of interactions, from odontoblast differentiation to secretion of enamel matrix
by the differentiated ameloblasts, occurred in this experimental set-up.
In light-microscopical examination he noticed cytoplasmic material in
the filters, but since a part of the thickness of the filter seemed to remain
unpenetrated, he did not come to any clear conclusion whether close
cell contacts or some transmissible extracellular material was responsible for the passage of the inductive signal.

We have used Millipore filters with a nominal pore size of 0.8 μm
for similar experiments. When the epithelial and mesenchymal components of molars from 17-day-old mouse embryos (Fig. 1a) are grown on
opposite sides of such a Millipore filter, the first signs of differentiation
are seen after 3 days of cultivation, when the mesenchymal cells align
on the upper filter surface (Fig. 4). After 4 days of cultivation odontoblasts have differentiated and secreted pre-dentine (Fig. 3a), and after
1 week ameloblasts on the opposite filter surface have polarized and
started enamel secretion. After 2 weeks of cultivation larger amounts
of enamel matrix are seen on the filter surface and between the ameloblasts as well as deep in the pores of the filter (Fig. 3b).

Ultrastructural examination of the filter after 3 days of cultivation,
i.e. at the time of odontoblast differentiation, reveals penetration of
cytoplasmic material mainly from the mesenchymal side. Some of this
material can be seen at all levels of the filter (Fig. 4).

Because of the spongy structure of the Millipore filters their ultrastructural examination is difficult. Nuclepore filters, which have
straight channels, thus allowing the examination of the growth of one
single cell process in the pore, have therefore largely replaced Millipore
filters in studies of this kind (Wartiovaara et al., 1972; Meier and Hay,
1975; Saxén et al., 1976; Toivonen et al., 1976). Nuclepore filters were
thus also used for more detailed studies on cytoplasmic penetration into
the filters during transfilter tooth differentiation (Thesleff et al., 1977).

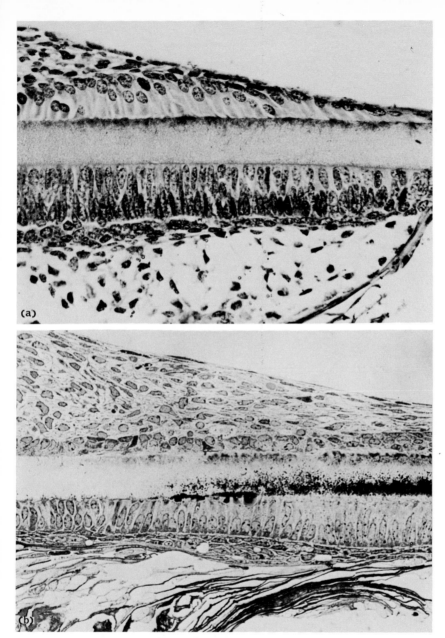

Fig. 3. Photomicrographs of dental epithelium and mesenchyme cultivated on opposite sides of a Millipore filter with a nominal pore size of 0.8 μm. (a) Explant fixed after 4 days of cultivation showing polarized odontoblasts (*top*) and pre-dentine secreted by them (dark). The epithelial cells below show beginning of ameloblast differentiation but no secretion of enamel matrix. (b) A similar explant fixed after 14 days of cultivation showing secreted enamel matrix (black) at the epithelium-filter interface and in the filter. Mallory's phosphotungstic acid–haematoxylin stain.

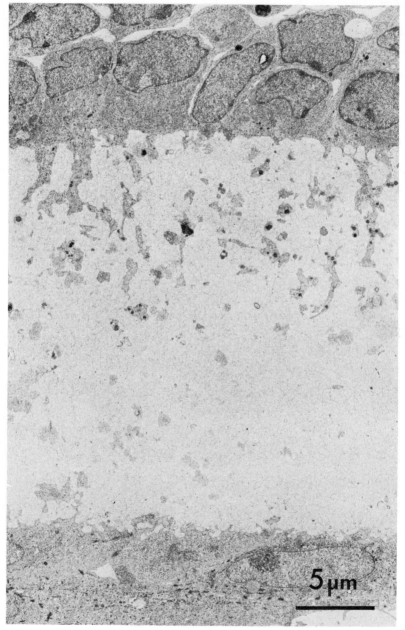

Fig. 4. An electron micrograph of dental mesenchyme (*above*) and epithelium separated by a Millipore filter (Fig. 3) and cultivated for 3 days. Abundant cytoplasmic penetration is seen from the mesenchymal side and similar material is seen throughout the filter. Glutaraldehyde and osmium fixation, post-staining with uranyl and lead.

Fig. 5. Photomicrographs of dental mesenchyme and epithelium separated by Nucle-pore filters with 0.6 μm pore size. (a) After 3 days of cultivation the mesenchymal cells are seen to align on the upper filter surface and the pores of the filter are filled with stainable material. (b) After 14 days of cultivation odontoblasts (*above*) have secreted homogeneous pre-dentine. Dark stained enamel matrix secreted by the ameloblasts is seen on the lower filter surface. Epon-embedded specimens stained with Mallory's phosphotungistic acid haematoxylin (from Thesleff *et al.*, 1977).

When Nuclepore filters with pore size of 0.6 or 0.2 μm were interposed between the dental epithelium and mesenchyme, the development occurred as in the case of Millipore filter: the exact sequence of
events was here also seen to start with differentiation of odontoblasts,
and ended with secretion of enamel matrix by ameloblasts (Fig. 5).
Ameloblasts never differentiated unless the odontoblasts on the opposite surface had started their secretion of pre-dentine.

The progress of differentiation correlated with the pore size of the
filter, being more rapid in the larger μm filter. Ultrastructural examination of the filters showed that penetration of cell processes into the filter
pores also occurred more rapidly in the larger pore size filter. Thus
the progress of differentiation was related to the ingrowth of cell processes into the filter channels. With 0.6 μm pore size filters differentiation of odontoblasts was seen after 3 days of cultivation, and here also
cytoplasmic material was seen at all levels of the filter (Fig. 5). With
0.2 μm filters no differentiation was seen after 3 days of cultivation and
electron-microscopical examination showed that penetration of cytoplasmic process into the filter pores was only minimal, whereas after
6 days of cultivation, when differentiation of odontoblasts had started,
cytoplasmic material was seen throughout the filter.

Nuclepore filter with pore size of 0.1 μm always prevented differentiation. Even after 2 weeks of cultivation neither the mesenchymal

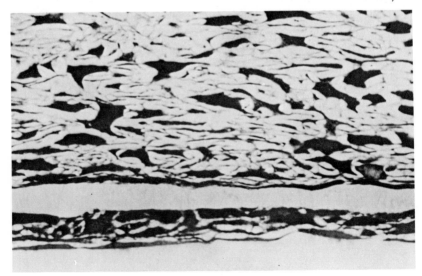

Fig. 6. Photomicrograph of dental mesenchyme (*above*) and epithelium separated by
Nuclepore filter with 0.1 μm pore size and cultivated for 14 days. Neither morphological
differentiation nor specific secretory products are seen on either side of the filter.

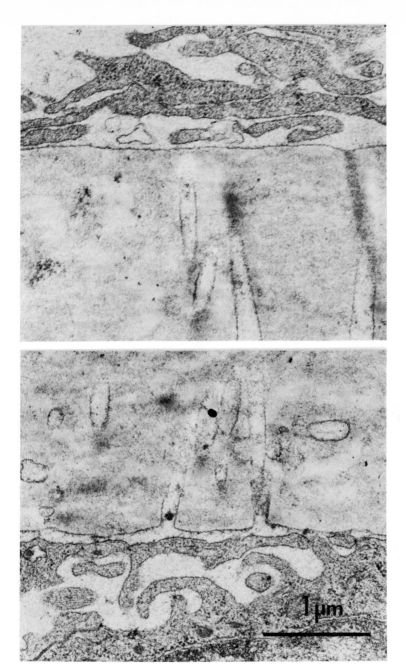

Fig. 7. Electron micrographs of the tissue–cell interface of the explant in Fig. 6. No penetration of cell processes into the 0.1 μm filter pores is observed either from the mesenchymal (*above*) or the epithelial side. Glutaraldehyde and osmium fixation, post-staining with uranyl and lead.

nor the epithelial cells had aligned on the filter surface (Fig. 6). Electron-microscopical examination on the filters showed that cytoplasmic material did not penetrate into the filter pores from either side of the filter (Fig. 7). Bearing in mind the sequentiality of differentiative events in odontogenesis, the differentiation of odontoblasts always preceding that of the ameloblasts, it was concluded that by using the 0.1 μm pore size filter the differentiation of odontoblasts was prevented. The results also indicated that penetration of cell processes into the filter pores was a prerequisite for differentiation of these cells to occur.

CONCLUSIONS

Based on our experiments, not much can be concluded about the mechanism of ameloblast differentiation, because with the 0.1 μm pore size filter we blocked the chain of interactive events already at the stage of mesenchymal cell differentiation to odontoblasts. The observation that ameloblasts did not differentiate if the odontoblasts on the opposte filter surface had not already differentiated and secreted pre-dentine is in accordance with the prevailing hypothesis that secretion of pre-dentine by the odontoblasts is a prerequisite for the differentiation of ameloblasts. Because agents interfering with deposition of pre-dentine prevent their differentiation, it is probable that pre-dentine plays an important role in the differentiation of ameloblasts. It is, however, difficult to conclude whether the function of pre-dentine is to serve as a structural basis for the differentiating ameloblasts, or whether it also mediates the inductive signal leading to ameloblast differentiation. The observations on close heterotypic cell contacts between odontoblasts and undifferentiated epithelial cells, however, retain the possibility that here transmission of the inductive signal could be contact-mediated.

Concerning the mechanism of the differentiation of odontoblasts, our results indicated that a close apposition of the interacting tissues is a prerequisite for the differentiation to occur, and accordingly, they cannot be regarded to support the idea of long-range transmission of signal substances. However, earlier ultrastructural observations of the intact basal lamina separating the inner enamel epithelium and the dental papilla mesenchyme seem to exclude the possibility of the formation of actual heterotypic cell contacts. Previous experimental studies have suggested that the extracellular matrix at the epithelio-mesenchymal interface mediates the inductive signal. Thus we suggest that by using the 0.1 μm pore size Nuclepore filter we prevented the mesenchymal cells making contact with the ECM under the epithelial cells on the opposite filter surface. The experimental situation would then resemble that in the Nuclepore filter studies by Meier and Hay (1975), where the differentiation of corneal cells was shown to depend on contact with a collagenous matrix on the opposite filter surface.

The mechanism involved in the epithelio-mesenchymal interaction leading to odontoblast differentiation would thus involve a cell–ECM interaction, not mediated by long-range transmission of ECM molecules, but by a close contact between the pre-odontoblasts and the basal lamina-associated extracellular material.

REFERENCES

Balinsky, B. J. Korrelation in der Entwicklung der Mund- und Kiemenregion und des darmkanals bei Amphibien. *Wilhelm Roux' Arch. Entwickl.-Mech. Org.* **143**, 365–395 (1947).

Bernfield, M. R., Banerjee, S. D. and Cohn, R. H. Dependence of salivary epithelial morphology and branching morphogenesis upon acid mucopolysaccharide-protein (proteoglycan) at the epithelial surface. *J. Cell Biol.* **52**, 674–689 (1972).

Biggers, J. D., Gwatkin, R. B. L. and Heyner, S. The growth of avian and mammalian tibiae on a relatively simple chemically defined medium. *Exp. Cell Res.* **25**, 41–58 (1961).

Frank, R. M. and Nalbandian, J. Ultrastructure of amelogenesis, in A. E. W. Miles (ed.), Structural and Chemical Organization of Teeth, pp. 399–466, Academic Press, New York (1967).

Gaunt, W. A. and Miles, A. E. W. Fundamental aspects of tooth morphogenesis, in A. E. W. Miles (ed.), Structural and Chemical Organization of Teeth, Vol. 1, p. 151, Academic Press, New York (1967).

Glasstone, S. The development of tooth germs *in vitro. J. Anat. (Lond.)* **70**, 260–266 (1936).

Glasstone, S. The development of tooth germs in tissue culture, in E. N. Willmen (ed.), Cells and Tissues in Culture, Vol. 2, pp. 273–283, Academic Press, New York (1965).

Grobstein, C. Morphogenetic interaction between embryonic mouse tissues separated by a membrane filter. *Nature (Lond.)* **172**, 869–871 (1953).

Henzen, W. Transplantation zur Entwicklungs-physiologischen Analyse der Larvalen Mundorgane bei Bombinator und Triton. *Wilhelm Roux' Arch. Entwickl.-Mech. Org.* **149**, 387–442 (1957).

Hetem, S., Kollar, E. J., Cutler, L. S. and Yaeger, J. A. The effect of α,α'-dipyridyl on the basement membrane of tooth germs in vitro. *J. dent. Res.* **54**, 783–787 (1975).

Huggins, C. B., McCarroll, H. R. and Dahlberg, A. A. Transplantation of tooth germ elements and the experimental heterotopic formation of dentine and enamel. *J. exp. Med.* **60**, 199–210 (1934).

Kallenbach, E. Electron microscopy of the differentiating rat incisor ameloblast. *J. Ultrastruct. Res.* **35**, 508–531 (1971).

Kallman, F. and Grobstein, C. Localization of glucosamine-incorporating materials at epithelial surfaces during salivary epithelio-mesenchymal interaction *in vitro. Develop. Biol.* **14**, 52–67 (1966).

Koch, W. E. *In vitro* differentiation of tooth rudiments of embryonic mice. I. Transfilter interaction of embryonic incisor tissues. *J. exp. Zool.* **165**, 155–170 (1967).

Koch, W. E. Tissue interaction during *in vitro* odontogenesis, in H. C. Slavkin and

L. A. Bavetta (ed.), Developmental Aspects of Oral Biology, pp. 151–164, Academic Press, New York (1972).

Kollar, E. J. and Baird, G. Tissue interactions in embryonic mouse tooth germs. II. The inductive role of the dental papilla. *J. Embryol. exp. Morph.* **24**, 173–186 (1970).

Kollar, E. J. The development of the integument: Spatial, temporal and phylogenetic factors. *Amer. Zool.* **12**, 125–135 (1972).

Lehtonen, E. Transmission of signals in embryonic induction. *Med. Biol.* **66**, 275–291 (1976).

Mangold, O. Experimente zur Analyse der Zusammenarbeit der Keimblätter. *Naturwissenschaften* **24**, 753–760 (1936).

Meier, S. and Hay, E. D. Stimulation of corneal differentiation by interaction between cell surface and extracellular matrix. *J. Cell Biol.* **66**, 275–291 (1975).

Pannese, E. Observations on the ultrastructure of the enamel organ. III. Internal and external enamel epithelia. *J. Ultrastruct. Res.* **6**, 186–204 (1962).

Reith, E. J. The early stage of amelogenesis as observed in molar teeth of young rats. *J. Ultrastruct. Res.* **17**, 503–526 (1967).

Ruch, J. V., Fabre, M., Karcher-Djuricic, V. and Stäubli, A. The effects of L-Azetidine-2-carboxylic acid (analogue of proline) on dental cytodifferentiations *in vitro*. *Differentiation* **2**, 211–220 (1974).

Ruch, J. V., Karcher-Djuricic, V and Gerber, R. Les déterminismes de la morphogenèse et des cytodifférenciations des ébauches dentaires de souris. *J. Biol. Buccale* **1**, 45–56 (1973).

Saxén, L. Transfilter neural induction of amphibian ectoderm. *Develop. Biol.* **31**, 140–152 (1961).

Saxén, L., Lehtonen, E., Karkinen-Jääskeläinen, M., Nordling, S. and Wartiovaara, J. Morphogenetic tissue interactions: Mediation by transmissible signal substances or through cell contacts?, *Nature (Lond.)* **259**, 662–663 (1976).

Sellman, S. Some experiments on the determination of the larval teeth in *Ambyostoma mexicanum. Odont. T.* **54**, 1–128 (1946).

Silva, D. G. and Kailis, D. G. Ultrastructural studies on the cervical loop and the development of the amelo-dentinal junction in the cat. *Arch. oral Biol.* **17**, 279–289 (1972).

Slavkin, H. C. Embryonic tooth formation. A tool for developmental biology. *Oral Sci. Rev.* **4**, 1–136 (1974).

Slavkin, H. C. and Bavetta, L. A. Organogenesis: Prolonged differentiation and growth of tooth primordia on the chick chorio-allantois. *Experientia* **24**, 192–194 (1968).

Slavkin, H. C. and Bringas P. Epithelial-mesenchyme interactions during odontogenesis. IV. Morphological evidence for direct heterotypic cell–cell contacts. *Develop. Biol.* **50**, 428–442 (1976).

Slavkin, H. C., Bringas, P., LeBaron, R. D., Cameron, J. C. and Bavetta, L. A. The fine structure of the extracellular matrix during epithelio-mesenchymal interactions in the rabbit embryonic incisor. *Anat. Rec.* **165**, 237–243 (1969).

Thesleff, I. Differentiation of odontogenic tissues in organ culture. *Scand. J. dent. Res.* **84**, 353–356 (1976).

Thesleff, I., Lehtonen, E., Wartiovaara, J. and Saxén, L. Interference of tooth differentiation with interposed filters. *Develop. Biol.* in press (1977).

Tiedemann, H. Extrinsic and intrinsic information transfer in early differentiation of amphibian embryos, in D. D. Davies and M. Balls (ed.), Control Mechanisms of Growth and Differentiation, pp. 223–234, Cambridge Univ. Press, Cambridge (1971).

Toivonen, S., Tarin, D. and Saxén, L. Transmission of morphogenetic signals from amphibian mesoderm to ectoderm in primary induction. *Differentiation* **5**, 49–55 (1976).

Toivonen, S., Tarin, D., Saxén, L., Tarin, P. J. and Wartiovaara, J. Transfilter studies on neural induction in the newt. *Differentiation* **4**, 1–7 (1975).

Wartiovaara, J., Lehtonen, E., Nordling, S. and Saxén, L. Do membrane filters prevent cell contacts? *Nature (Lond.)* **238**, 407–408 (1972).

Wartiovaara, J., Nordling, S., Lehtonen, E. and Saxén, L. Transfilter induction of kidney tubules: Correlation with cytoplasmic penetration into Nucleopore filters. *J. Embryol. exp. Morph.* **31**, 667–682 (1974).

Wigglesworth, D. J. Formation and mineralisation of enamel and dentine by rat tooth germs *in vitro*. *Exp. Cell Res.* **49**, 211–215 (1968).

Epigenetic Regulation of Enamel Protein Synthesis during Epithelial-Mesenchymal Interactions

Harold C. Slavkin, Gary N. Trump, Steven Schonfeld, Anna Brownell, Nino Sorgente and Victor Lee-Own

Laboratory for Developmental Biology, Ethel Percy Andrus Gerontology Center and Departments of Biochemistry and Microbiology-Immunology, School of Dentistry, University of Southern California, Los Angeles, California 90007, USA

INTRODUCTION

The regulatory processes operating during epithelial-mesenchymal interactions associated with tooth development are as yet obscure. The value of the tooth organ lies in its high degree of developmental programming for sequential synthesis of several unique extracellular matrix proteins derived from either the inner enamel epithelia or the adjacent mesenchymal cells. During differentiation, histogenesis, and organogenesis a program of changing protein synthetic and secretion patterns characterize tooth morphogenesis. Understanding the biochemical basis underlying such developmental processes would be facilitated by the isolation and comparison of the messenger RNAs from those cells producing a number of different extracellular matrix proteins (e.g. the protein constituents of dentine and enamel matrix formation). The kinetics of specific protein synthesis and secretion (e.g. dentine collagen, dentine phosphoprotein, dentine proteoglycans, enamel proteins), if correlated with the kinetics of synthesis and translation of the corresponding messenger RNAs, would facilitate the identification of regulatory controls of information flow in epithelial-mesenchymal interactions.

The basic objective of this discussion is to describe recent experiments to identify and monitor the quality of specific gene products which characterize the phenotype of the inner enamel epithelia, preamelo-blasts and ameloblasts during embryonic epithelial-mesenchymal interactions. These experiments employ immunological methods and a variety of morphological and biochemical techniques in an attempt to study the nature of epigenetic influences upon ameloblast cell differentiation, and the mechanisms by which epithelial cells respond to inductive signals derived from mesenchyme.

THE DEVELOPMENTAL PATTERN OF ENAMEL PROTEIN SYNTHESIS AND SECRETION

General embryological features

During mammalian embryogenesis (9.5–10th day of gestation in mice), the dental lamina is observed as a band of thickening of ectodermally derived oral epithelium which indicates the future position of each dental arch. The dental lamina forms an ingrowth of oral epithelial cells into the adjacent ectomesenchyme (i.e. cranial neural crest-derived cells) of the jaw (see reviews by Gaunt and Miles, 1967; Slavkin, 1974). Both tissues are separated by a continuous basal lamina associated with the˙ undersurface of the epithelia (a metachromatic basement membrane is evident in the light microscope). Presumably, through cell proliferation, the lamina first forms a bud which subsequently becomes invaginated and forms a bell-shaped epithelial structure called the "enamel organ epithelia" in association with a dental papilla mesenchyme. Through continued epithelial-mesenchymal interactions, the enamel organ increases in size and morphological complexity until the basic configuration of the tooth organ is determined with the onset of odontoblast cell differentiation and dentine matrix deposition. Following the initiation of dentine mineralization, enamel protein synthesis and secretion are observed in secretory amelogenesis.

General cytological features

The histological characteristics of amelogenesis in a large number of mammalian animals have been described (see extensive reviews by Gaunt and Miles, 1967; Slavkin, 1974). Light-microscopic autoradiographic studies have employed tritiated thymidine, various amino acids, and hexosamines to describe unique features of cell divison and protein synthesis characteristic of the differentiation of secretory ameloblast cells from proliferating inner enamel epithelium (e.g. see Blumen

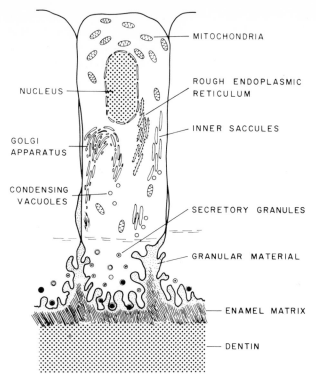

MITOCHONDRIA

ROUGH ENDOPLASMIC
RETICULUM

NUCLEUS

INNER SACCULES

GOLGI
APPARATUS

CONDENSING
VACUOLES

SECRETORY GRANULES

GRANULAR MATERIAL

ENAMEL MATRIX

DENTIN

Fig. 1. Diagrammatic presentation of the embryonic mouse secretory ameloblast. The available evidence indicates that the terminally differentiated, non-dividing, secretory ameloblast is a merocrine-type cell. These cells are elongated and tall columnar in appearance and possess a prominent Golgi apparatus located in the supranuclear region. The majority of mitochondria are located in the infranuclear region of the cell. Enamel protein synthesis is on membrane-associated polysomes (rough endoplasmic reticulum). The translated gene product is glycosylated in the Golgi apparatus, condensed in the inner saccules of the smooth endoplasmic reticulum, packaged into condensing vacuoles, and then transported through the cytoplasm as secretory granules. The granules release enamel protein materials from the lateral and apical surfaces of Tome's process; each ameloblast has one Tome's process. During secretion, phosphokinases mediate phosphorylation of the proteins. Also during secretion inter- and intramolecular disulphide linkages are formed. Newly secreted enamel matrix appears as a granular material which is then transformed into the enamel matrix *per se*. (Ultrastructural descriptions are based upon electron-microscopic autoradiographic studies published by Weinstock and LeBlond, 1971; Weinstock, 1972; Slavkin *et al.*, 1976; and the morphological descriptions of Reith, 1967, and Kallenbach, 1971, 1976.)

and Merzel, 1972; Greulich and Slavkin, 1965; Slavkin *et al.*, 1976; Weinstock, 1972). Ultrastructural studies have described the sites and pathways for the synthesis and secretion of enamel protein (e.g. see Frank, 1970; Kallenbach, 1971, 1976; Reith, 1967; Warshawsky, 1966; Weinstock and LeBlond, 1971; Weinstock, 1972).

In situ high resolution electron-microscopic autoradiographic descriptions of enamel protein synthesis and secretion indicated that this process required 30 min (Slavkin *et al.*, 1976; Weinstock, 1972). The enamel proteins appear to be synthesized on the rough endoplasmic reticulum, glycosylated in the Golgi apparatus (i.e. studies using glucosamine or fucose isotopic precursors), packaged in condensing vacuoles derived from the inner saccules, transported in secretory vesicles to the secretory region of the ameloblast termed Tomes' process, and secreted as a granular material which is subsequently transformed into the enamel matrix *per se* (see review by Weinstock, 1972). The earliest extracellular matrix form of enamel protein is the granular material described initially by Reith (1967), and then Slavkin *et al.* (1976) and Kallenbach (1976) (Fig. 1).

HISTOCHEMICAL FEATURES OF EMBRYONIC EXTRACELLULAR ENAMEL MATRIX

Histochemically, available reports have concluded that newly secreted enamel protein matrix is quite analagous to keratin (Fraser *et al.*, 1972), i.e. positive reactions for sulphydryls, tyrosine and basic amino acids, intense anionic dye binding, and a highly critical electrolyte concentration (Everett and Miller, 1974).

SUMMARY OF EMBRYONIC ENAMEL PROTEIN BIOCHEMICAL CHARACTERISTICS

Most of the modern methods for studying the chemical properties of protein molecules were developed for soluble proteins. Enamel proteins are naturally insoluble under physiological conditions in that they are constituents of a thixotropic gel in association with forming calcium hydroxyapatite crystals (Eastoe, 1960). Very little if any protein can be extracted unless the enamel is first demineralized and a high proportion of the disulphide linkages have been broken. Like keratins, unless the enamel proteins are further modified they will reform when the extraction medium is removed or diluted. These requirements are crucial since the original molecular species will be converted into a heterogeneous mixture of polypeptides and many of these polypeptides will aggregate, thereby presenting erroneous information of the actual number of enamel protein gene products which characterize the secretory ameloblast phenotype. To circumvent some of these technical difficulties it is imperative to consider how to demineralize the forming enamel matrix, which denaturing agents to use (e.g. urea), and how to solubilize the proteins in the presence of an appreciable number of inter- and intramolecular disulphide linkages.

The task of preparing embryonic enamel matrix and purifying soluble enamel proteins has proved to be extremely difficult and replete with technical difficulties (see review edited by Fearnhead and Stack, 1971). The following discussion will briefly summarize recent methods and results which offer the possibility of preparing extracellular matrix gene products characteristic of embryonic amelogenesis.

TABLE I

Summary of embryonic enamel protein characteristics

Characteristic	Reference
Methionine and leucine are the major N-terminal amino acids	Fukae and Shimizu (1974)
No detectable hydroxyproline or hydroxylysine in enamel protein	Elwood and Apostolopoulos (1975)
Content of enamel protein decreases from approximately 30% to 2% of wet weight during mineralization in early stages of development	Fukae and Shimizu (1974)
Organic phosphate is attached to the enamel protein as phosphoserine	Seyer (1972)
Hexosamines (glucosamine and galactosamine) are present in the enamel matrix ($>75\%$ is galactosamine)	Seyer and Glimcher (1969)
L-fucose, D-xylose, D-mannose, D-arabinose, N-acetylglucosamine, N-acetylgalactosamine, N-acetylmannosamine and glucuronic acid are present in the enamel organic matrix	Elwood and Apostolopoulos (1975)
Embryonic enamel proteins aggregate by disulphide linkages	Mechanic (1971); Everett and Miller (1974)
Enamel protein is antigenic	Nikiforuk and Gruca (1971); Elwood and Apostolopoulos (1975); Schonfeld (1975)
Enamel proteins are phosphorylated	Stavropoulos and Glimcher (1972); Seyer (1972); Guenther et al. (1976)
Proline, glutamic acid, leucine, histidine and serine (order of decreasing concentration) represent 64% of the total amino acid residues	Levine and Glimcher (1965); Elwood and Apostolopoulos (1975)

Varying degrees of heterogeneity amongst the enamel matrix proteins have been reported in a number of conflicting reports. Much of the available literature is replete with reports which have indicated that the enamel matrix consists of 9–22 different polypeptides of different molecular weights, charge distribution, degrees of phosphorylation and glycosylation and amino acid compositions (e.g. see Eggert *et al.*, 1973; Elwood and Apostolopoulos, 1975a, b, c; Fincham, 1971; Mechanic, 1971; Seyer and Glimcher, 1971). A summary of enamel protein biochemical characteristics is presented in Table 1.

IDENTIFICATION AND PARTIAL CHARACTERIZATION OF NEWLY SECRETED EMBRYONIC RABBIT ENAMEL PROTEIN

Several laboratories, including our own, have recently "revisited" the question of determining the number of gene products which characterize the extracellular matrix phenotype of enamel matrix during embryonic tooth morphogenesis (Chrispens *et al.*, 1977; Guenther *et al.*, in press).

Our experimental strategy was to first identify the enamel proteins within newly secreted embryonic enamel extracellular matrix in 26-day New Zealand White incisor and molar tooth organs. Subsequently, we initiated studies to identify and partially characterize the major detectable poly-A containing mRNAs in comparable embryonic molar tooth organs (Lee-Own *et al.*, 1977).

Our initial investigations identified four proteins within the enamel matrix of embryonic New Zealand White rabbit molar and incisor tooth organs. We noted that newly secreted enamel matrix was approximately 78% mineral in the form of calcium hydroxyapatite. To ensure a maximum yield of enamel proteins from our extraction procedure, we used 5% cold TCA (trichloroacetic acid) for demineralization, a methanol wash to remove the TCA, and then extracted the enamel proteins with 6M urea in 100mM Trisborate buffer at pH 8.6 at 60°C (Guenther *et al.*, 1975). It is essential to reduce the urea extracts with either 400 mM mercaptoethanol or dithiothreitol before electrophoresis on either 7.5% urea polyacrylamide gel electrophoresis (PAGE) or sodium dodecylsulphate (SDS)-PAGE. The reduction of sulphydryl groups, presumed to be intra- and intermolecular disulphide linkages formed between cystine residues, was essential in these extraction procedures.

SDS-PAGE methods were used to provide further information about the degree of protein heterogeneity within the enamel matrix, and to also determine relative molecular weight values for the enamel polypeptides. Newly secreted enamel matrix proteins were found to consist of four polypeptides: (1) 65 000, (2) 58 000, (3) 22 000, and (4) 20 000 daltons, respectively (Guenther *et al.*, in press). The first and third enamel proteins were both phosphorylated (Fig. 2). Following a 6 h incubation *in vitro*, all four proteins were labelled with isotopic glucosamine, proline, leucine and tryptophan. Puromycin, present in the culture medium at a concentration of 2mm, caused a 95% inhibition of labelled amino acid and hexosamine incorporation into enamel proteins. The inhibitory effects of puromycin on a labelled precursor incorporation suggested that continuing protein synthesis was a prerequisite for enamel protein secretion. Preliminary pulse-chase labelling experiments indicated a precursor-product relationship in which the 65 000 molecular weight enamel protein was synthesized and secreted *prior to* the appearance of the smaller enamel polypeptides. Shimokawa and Suzaki (1975) have reported that bovine secretory ameloblasts synthesized and secreted one major enamel protein (greater than 50 000 daltons) *in vitro* which was enzymatically degraded by an enamel matrix

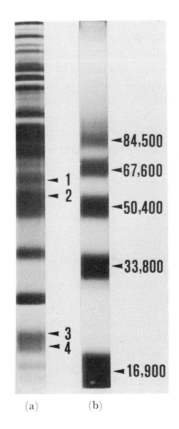

Fig. 2. SDS-polyacrylamide gel electrophoresis protein patterns of (a) a 6m urea buffered extraction of 26-day-old embryonic New Zealand White rabbit incisor tooth organ extracellular matrix containing both dentine and newly secreted enamel protein matrix, and (b) myoglobin chains (monomers, dimers, trimers, tetramers and pentamers) used as molecular weight markers. Four enamel proteins have been identified having molecular weights of (1) 65 000, (2) 58 000, (3) 22 000 and (4) 20 000 daltons, respectively. Of these polypeptides, all four were glycosylated and enamel proteins (1) and (3) were phosphorylated (from Guenther *et al.*, in press).

(a)   (b)

protease into a 25 000 dalton polypeptide and then into increasingly smaller polypeptides (Fukae, personal communication).

The immunological characteristics of enamel proteins have recently been found to be extremely interesting. Upon secretion, enamel proteins become incorporated into a thixotropic gel in which calcium hydroxyapatite crystals are rapidly forming (Eastoe, 1960). Therefore, newly secreted embryonic enamel proteins are essentially sequestered molecules within a mineralizing extracellular matrix and are not "seen" by the host's immune system. Recently, our laboratory began a series of studies to investigate the possibility that enamel matrix might be immunogenic and also capable of producing either an alloantisera or autoantisera against embryonic rabbit molar tooth enamel matrix. Schonfeld (1975) produced an alloantibody directed against embryonic enamel matrix. Embryonic New Zealand White rabbit 26-day molar tooth organ extracellular matrices were used as antigens in an immunization schedule in which the immunogen in Freund's adjuvant was injected into young female New Zealand White rabbits over a 6-week period. This procedure resulted in a high titre of antibody which was found by indirect fluorescence microscopy to be specific for the newly secreted enamel matrix. Subsequent studies using immunoelectrophoresis demonstrated that of the three major electrophoretic constituents visualized on 7.5% urea-PAGE characteristic for embryonic enamel matrix, two of these electrophoretic constituents were antigenic and reacted with the anti-enamel matrix antisera (Schonfeld *et al.*, in press).

On the basis of these as yet preliminary studies, we now consider that one major gene product characterizes the extracellular matrix

TABLE II

Enamel protein biosynthesis and metabolism

Molecular event	Mediating enzymes
Selection of structural gene(s)	
Transcription	
Translation of mRNA(s)	
Glycosylation	Galactosyl transferases
	Glucosyl transferases
Phosphorylation	Protein phosphokinase
Disulphide linkages	Copper-dependent oxidation
Pro-enamel/enamel conversion	Enamel peptidases
Enamel matrix formation/calcification	Phosphatases
Enamel maturation (degradation)	Enamel proteases

phenotype associated with embryonic ameloblast cell differentiation. The available data further indicate that this protein has appreciable post-translational modifications during the course of biosynthesis, secretion, extracellular matrix formation, mineralization and degradation (Table II). It is also evident that the heterogeneity reported

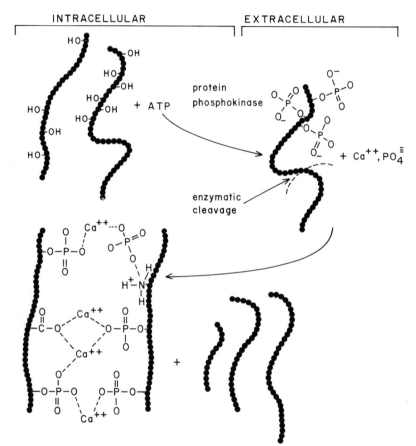

Fig. 3. A postulated scheme for the conversion of newly secreted enamel protein into increasingly smaller polypeptides associated with the initiation of calcium hydroxyapatite crystal formation in the matrix is presented. We assume that the phosphorylation of the precursor enamel protein is mediated by phosphokinases *prior to* secretion from the ameloblast cells. Thereafter, an extracellular enamel protease (as described by Fukae, personal communication) cleaves the protein into numerous smaller polypeptides resulting in molecular weights of 25 000 to 3000 daltons. This process is coincident with calcium hydroxyapatite crystal formation which is initiated by these "calcium-binding" polypeptides in a thixotropic gel (Eastoe, 1960). The collective process is termed enamel maturation.

amongst the polypeptides in mature and older enamel reflected protease-mediated degradation associated with enamel maturation. One interesting aspect of the initial development of the enamel extracellular matrix is to consider the possible role of the enamel protein in initiating calcium hydroxyapatite crystal formation. In this context, we have considered enamel protein to be a "calcium-binding" protein and, on the basis of amino acid composition, ultrastructural characteristics, solubility properties, histochemistry, and physical properties in general, we have postulated a mechanism by which newly secreted enamel protein participates in the initiation of mineralization and also becomes degraded concomittant with enamel maturation *in vivo* (Fig. 3).

IDENTIFICATION OF THE mRNAs FROM DEVELOPING EMBRYONIC RABBIT MOLAR TOOTH ORGANS

As discussed above, we identified four enamel proteins of 65 000, 58 000, 22 000 and 20 000 daltons, respectively. We assume that only one enamel gene product represents the extracellular enamel matrix phenotype prior to the complex post-translational modifications described in the preceding discussion. Therefore, we anticipated that the molecular weight determinations from SDS-PAGE for enamel protein(s) were somewhat higher than the actual molecular weight of the polypeptide representing the nascent translation gene product.

For this discussion we will only briefly outline the experimental protocol and the preliminary results obtained since details of these studies are currently in preparation for publication elsewhere. Four litters of 26-day embryonic New Zealand White rabbit molar tooth organs (approximately 360 tooth organs) were found to give a yield of 540 μg $\pm$ 40 μg total cellular RNA. We purified this material using oligo-dT column chromatography to isolate an enriched poly-A-containing RNA preparation. Aliquots of the poly-A-containing RNA (0.25 μg/ml) was assayed for translation activity using a wheat germ cell-free protein synthesis system. Approximately 200 000–250 000 counts per minute of ^{3}H-proline were incorporated into 5% TCA precipitable counts. One third of these counts was bacterial collagenase-labile. In other studies we fractionated the poly-A preparations using a sucrose density gradient and then assayed each fraction for translational activity and poly-A content using a tritiated poly-U assay system. We assumed that the total poly-A-containing preparations contained mRNAs for procollagen polypeptide chains (145 000 m.w.), alkaline phosphatase (80 000 m.w.), enamel protein(s), and dentine phosphoproteins ($\sim$35 000 m.w.), in addition to "housekeeping" proteins found in all somatic cells. Table III summarizes the data describing the degree of heterogeneity (seven major fractions) amongst the poly-A-containing mRNAs isolated from the 26-day embryonic rabbit molar tooth organ. On the basis of these preliminary studies we predict that the major gene product for enamel protein, prior to post-translational modifications (e.g. glycosylation and phosphorylation), will be 50 000–53 000 daltons and will have a corresponding mRNA of approximately 460 000 daltons (Lee-Own et al., 1977).

TABLE III

Summary of messenger ribonucleic acids isolated from embryonic rabbit molar tooth organs

Major fractions (sucrose density gradient)	Poly-A-containing RNAs	mRNAs[1]	Calculated protein sizes[2]	In vitro translation products using 7.5% SDS-gels
1	Yes	160 000	17 700	20 000
2	Yes	300 000	33 000	32 000
3	Yes	460 000	51 000	53 000 (enamel protein)
4	Yes	960 000	106 000	85 000
5	Yes	1 200 000	130 000	105 000
6	Yes	1 390 000	154 300	142 000 (procollagen α chain)
7	Yes	1 740 000	193 000	—

[1] Estimated RNA molecular weights based upon sucrose density gradient fractionation and formamide gel electrophoresis of fractionations using 28S, 18S and 4S markers.

[2] Our calculations assumed that there were three bases for each amino acid and that each amino acid had an approximate molecular weight of 300.

A MODEL FOR EPIGENETIC REGULATION OF ENAMEL PROTEIN SYNTHESIS AND SECRETION

In almost all published investigations of epithelial-mesenchymal interactions, histogenesis and morphogenesis have been used as the criteria to assess mesenchymal specificity (see reviews by Grobstein, 1967, 1975; Saxen et al., 1968; Slavkin, 1974). In a defined system, no evidence as yet has been reported which unequivocally demonstrated mesenchymal induction of new gene transcription, translation and post-translational modifications in a responding epithelium. Therefore, one of the central issues is to design experiments which might obtain evidence capable of discriminating between (1) instructive developmental signals derived from the mesenchymal cells, and (2) permissive and/or growth rate-limiting factors which might amplify or retard the synthesis and secretion of already determined epithelial gene products (Trump *et al.*, 1977).

The following discussion presents a series of indirect observations which, if taken collectively, support the thesis that developmental signals required to initiate ameloblast cell differentiation and the synthesis and secretion of enamel protein reside (in part) within mesenchymal cell surface molecules. The transmission, the receiving of a signal by epithelia, and the translation of the signal into a specific phenotype appear to be mediated by close-range interactions between outer surfaces of mesenchymal cell processes, matrix vesicles and a responsive inner enamel epithelia.

During tooth morphogenesis, mesenchymal cells produced matrix vesicles which accumulated along the undersurface of the inner enamel epithelia adjacent to a continuous basal lamina (Slavkin *et al.*, 1972; Silva and Kailis, 1972; Croissant, 1975; Croissant *et al.*, 1975; Slavkin and Bringas, 1976). Matrix vesicles were deposited *prior to* degradation of the basal lamina (Croissant, 1975). Croissant (1975) postulated that matrix vesicles and/or mesenchymal cell processes served to degrade the basal lamina and, thereby, "unmasked" the outer cell surfaces of the epithelia (Croissant *et al.*, 1975). Following lamina degradation, direct heterotypic cell–cell contact has been described in several different mammalian species during tooth development (Pannesse, 1962; Kallenbach, 1971, 1976; Slavkin and Bringas, 1976).

We designed several experiments to test Croissant's postulate (1975) that matrix vesicles provided a cell-regulated mechanism for the transport of proteases for functions within the extracellular matrix microenvironment. Since we assumed that matrix vesicles and/or mesenchymal cell processes mediated the degradation of the basal lamina, and since the lamina was allegedly formed of Type IV collagen and proteoglycans, we assayed for mammalian collagenase activity associated with physically isolated matrix vesicles. Fractions (matrix vesicles and mesenchymal cell membranes) were isolated from embryonic New Zealand White rabbit incisor tooth organs in a void volume effluent from Bio-Gel A-50 M columns by procedures previously reported by our laboratory (Slavkin *et al.*, 1972). For the collagenase activity assay, ^{14}C-labelled Type I rabbit skin collagen was reconstituted in fibrillar form and then incubated, in the presence of various protease inhibitors, with aliquots of matrix vesicles for 18 h at 25°C. At the termination of the incubation, experimental and control groups were processed for scanning electron microscopy, 200 μl aliquots were assayed for radioactive counts released into the reaction mixture supernatant, and the remaining supernatant was dialysed, lyophilized and then analysed using 5% sodium dodecylsulphate-polyacrylamide gel electrophoresis (SDS-PAGE) to determine the degree of heterogeneity and relative molecular weights of the reaction products. The 5% SDS-PAGE patterns of the reaction products resulting from the collagenase activity of the matrix vesicles demonstrated two major products (P_1 and P_2) having molecular weights of 67 000 and 32 000 daltons, respectively. The matrix vesicles demonstrated a mammalian collagenase-like activity characterized by the enzymatic cleavage of native collagen into 3/4 and 1/4 fragments (Brownell et al., 1976; Sorgente et al., 1977).

Following basal lamina degradation, possibly mediated by matrix vesicle-associated collagenase activity, direct mesenchymal-epithelial cell contacts were observed prior to the initiation of enamel protein secretion from preameloblasts (Pannesse, 1962; Croissant, 1975; Croissant et al., 1975; Slavkin and Bringas, 1976). These direct heterotypic cell–cell contacts persisted during the initial stages of precursor enamel protein secretion (Kallenbach, 1971, 1976; Slavkin et al., 1976) (Fig. 4). Further, indirect data which also implicated close-range, direct heterotypic cell–cell interactions in tooth morphogenesis have been obtained from transfilter embryonic induction studies. Koch (1967) demonstrated that embryonic mouse incisor tooth organ tissues could be successfully dissociated and cultured in juxtaposition to a 0.45 μm pore size Millipore filter; both types of tissues expressed histogenesis and deposited extracellular matrices in vitro. If cellophane was interposed between tissue interactants, advanced differentiation was not observed. More recently, Thesleff (this volume) reported that 0.1 μm pore size Nucleopore filters restricted differentiation using embryonic mouse molar tissues cultured transfilter. The transfilter studies and the various electron-microscopic descriptions of epithelial-mesenchymal interactions in situ support the thesis that mesenchymal cell processes and/or matrix vesicles (approximately 0.1 μm in diameter) signal epithelia to initiate enamel protein synthesis and secretion.

The following sequence of events possibly describe the as yet unknown mechanism of epithelial-mesenchymal interactions during embryonic tooth morphogenesis: (1) the fibrous material associated with the basal lamina in the proliferative zone is synthesized and secreted by the inner enamel epithial cells prior to preodontoblast differentiation; (2) the fibrous material is collagen and serves as a template to signal adjacent mesenchymal cells to secrete Type I dentine collagen

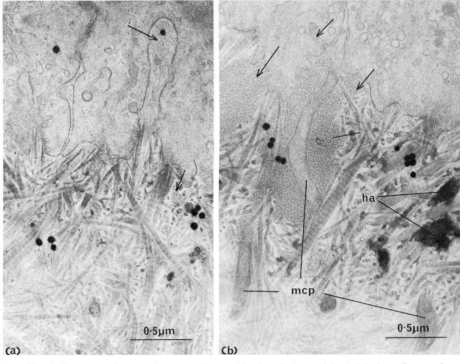

Fig. 4. Electron-microscopic autoradiographic descriptions of the initiation of enamel matrix secretion using ³H-tryptophan as an isotopic precursor to study the kinetics of enamel protein biosynthesis and secretion in a newborn mouse incisor tooth organs (after studies reported by Slavkin *et al.*, 1976). (a) Enamel precursor protein matrix is first secreted as a granular material (arrows). *In vivo* synthesis and secretion of the granular material requires 30 min. (b) Increased secretion is observed in adjacent cells (arrows) in proximity to mesenchymal cell processes (mcp). Coincident with direct cell–cell heterotypic contacts and enamel protein secretion (black dots) in the formation of calcium hydroxyapatite crystals (ha) in the adjacent dentine matrix.

(analogous to the synthesis and secretion of Type I collagen by the embryonic chick corneal epithelial cells *prior to* the appearance of mesenchymal cell synthesis and secretion of the primary corneal stroma) (Hay and Revel, 1969); (3) the inner surface cells of the enamel organ epithelia (i.e. inner enamel epithelium) synthesize and secrete Type IV basal lamina collagen *and* Type I collagen *prior to* the differentiation of preodontoblast cells (Trelstad and Slavkin, 1974); (4) matrix vesicles are formed by "budding" from the plasma membrane of the preodontoblast cells and are sequestered along the undersurface of the basal lamina (Slavkin *et al.*, 1976); (5) the basal lamina degradation is the result of matrix vesicle and/or mesenchymal cell process protease activity; and (6) the degradation of the lamina "unmasks" the epithelial cell surface receptors which form direct contacts with the outer surfaces of the mesenchymal cell processess and, thereby, function to mediate the initiation of enamel protein synthesis and secretion.

ACKNOWLEDGEMENTS

The authors wish to acknowledge the excellent technical assistance of Pablo Bringas, Jr, P. Matosian, P. Wilson and W. Mino. The research investigations reviewed in this manuscript were supported by US Public Health Service Research Grants DE-02848, DE-03569, DE-03513 and Training Grant DE-00094 from the National Institute of Dental Research. The authors wish to express their appreciation to Mrs Joanne Leynnwood for typing the manuscript.

REFERENCES

Blumen, G. and Merzcl, J. The decrease in the concentration of organic material in the course of formation of the enamel matrix. *Experientia* **28**, 545–548 (1972).

Brownell, A., Sorgente, N. and Slavkin, H. C. Collagenase activity associated with extracellular matrix vesicle-enriched fractions. *J. Cell Biol.* **70**, 271 (1976).

Chrispens, J. B., Weliky, B. G. and Slavkin, H. C. Phylogenetic variations amongst extracellular matrix macromolecules characteristic of murine and lagomorph odontogenesis. *J. dent. Res.* **56**, A136 (1977).

Croissant, R. D. Induction, extracellular matrix vesicles, and RNA: A study of epithelial-mesenchymal interaction during late embryonic rabbit incisor development. Ph.D. Thesis, University of Southern California (1975).

Croissant, R. D., Guenther, H. and Slavkin, H. C. How are embryonic preameloblasts instructed by odontoblasts to synthesize enamel? In H. C. Slavkin and R. C. Greulich (ed.), Extracellular Matrix Influences on Gene Expression, pp. 515–521, Academic Press, New York (1975).

Eastoe, J. E. Organic matrix of tooth enamel. *Nature* (*Lond.*) **187**, 411–412 (1960).

Eggert, R. M., Allen, G. A. and Burgess, R. C. Amelogenins: purification and partial characterization of proteins from developing bovine dental enamel. *Biochem. J.* **131**, 471–484 (1973).

Elwood, W. K. and Apostolopoulos, A. X. Analysis of developing enamel of the rat. I. Fractionation: Protein and calcium content. *Calcif. Tiss. Res.* **17**, 317–326 (1975a).

Elwood, W. K. and Apostolopoulos, A. X. Analysis of developing enamel of the rat. II. Electrophoretic and amino acid studies. *Calcif. Tiss. Res.* **17**, 327–336 (1975b).

Elwood, W. K. and Apostolopoulos, A. X. Analysis of developing enamel of the rat. III. Carbohydrate, DEAE-Sephadex, and immunological studies. *Calcif. Tiss. Res.* **17**, 337–347 (1975c).

Everett, M. M. and Miller, W. A. Histochemical studies on calcified tissues. I. Amino acid histochemistry of foetal calf and human enamel matrix. *Calcif. Tiss. Res.* **14**, 229–244 (1974).

Fearnhead, R. W. and Stack, M. V. (ed.), Tooth Enamel: Its Composition, Properties and Fundamental Structure, Vol. II, John Wright & Sons Ltd., Bristol (1971).

Fincham, A. G. Experiments on the DEAE-cellulose chromatography of bovine foetal enamel matrix and the isolation of a low molecular weight fraction, in R. W. Fearnhead and M. V. Stack (ed.), Tooth Enamel: Its Composition, Properties and Fundamental Structure, pp. 79–87, John Wright & Sons Ltd., Bristol (1971).

Frank, R. M. Autoradiographique quantitative de l'amelogenèse en microscopique électronique a l'aide de la proline tritiée chez le chat. *Arch. oral Biol.* **15**, 569–581 (1970).

Fraser, R. D. B., MacRae, T. P. and Rogers, G. E. Keratins: Their Composition Structure and Biosynthesis, Charles C. Thomas, Springfield, Ill. (1972).

Fukae, M. and Shimizu, M. Studies on the proteins of developing bovine enamel. *Arch. oral Biol.* **19**, 381–386 (1974).

Gaunt, W. A. and Miles, A. E. W. Fundamental aspects of tooth morphogenesis, in A. E. W. Miles (ed.), Structural and Chemical Organization of Teeth, Vol. I, pp. 151, Academic Press, New York (1967).

Grobstein, C. Mechanisms of organogenetic tissue interaction. *Nat. Cancer Inst. Monogr.* **26**, 279–299 (1967).

Grobstein, C. Developmental role of intercellular matrix, in H. C. Slavkin and R. C. Greulich (ed.), Extracellular Matrix Influences on Gene Expression, pp. 9–16 Academic Press, New York (1975).

Greulich, R. C. and Slavkin, H. C. Amino acid utilization in the synthesis of enamel and dentine matrices as visualized by autoradiography, in C. P. LeBlond and K. B. Warren (ed.), The Use of Radioautography in Investigating Protein Synthesis, pp. 199–214, Academic Press, New York (1965).

Guenther, H., Croissant, R. D., Schonfeld, S. and Slavkin, H. C. Enamel proteins: Identification of epithelial-specific differentiation products, in H. C. Slavkin and R. C. Greulich (ed), Extracellular Matrix Influences on Gene Expression, pp. 387–398, Academic Press, New York (1975).

Guenther, H., Croissant, R. D., Schonfeld, S. and Slavkin, H. C. Identification of four extracellular matrix enamel proteins during embryonic rabbit tooth organ development. *Biochem. J.* (1977).

Hay, E. D. and Revel, J. P. Fine structure of the developing avian cornea, in A. Wolsky and P. S. Chen (ed.), Monographs in Development Biology, p. 30, Karger, Basel (1969).

Kallenbach, E. Electron microscopy of the differentiating rat incisor ameloblast. *J. Ultrastruct. Res.* **35**, 508–531 (1971).

Kallenbach, E. Fine structure of differentiating ameloblasts in the kitten. *Amer. J. Anat.* **145**, 283–317 (1976).

Koch, W. E. *In vitro* differentiation of tooth rudiments of embryonic mice. I. Transfilter interaction of embryonic tissues. *J. exp. Zool.* **165**, 155–170 (1967).

Lee-Own, V., Zeichner, M., Benveniste, K., Denny, P., Paglia, L. and Slavkin, H. C. Cell-free translation of messenger RNAs of embryonic tooth organs: Synthesis of the major extracellular matrix proteins. *Biochem. biophys. Res. Commun.* **74**, 849–856 (1977).

Levine, P. T. and Glimcher, M. J. The isolation and amino acid composition of the organic matrix and neutral soluble proteins of developing rodent enamel. *Arch. oral Biol.* **10**, 753–756 (1965)

Mechanic, G. L. The multicomponent re-equilibrated, protein system of bovine embryonic enamelin. (Dental Enamel Protein.) Chromatography in deaggregating solvents, in R. W. Fearnhead and M. V. Stack (ed.), Tooth Enamel II: Its Composition, Properties and Fundamental Structure, pp. 88–92, The Williams & Wilkins Co., Baltimore (1971).

Nikiforuk, G. and Gruca, M. Immunological and gel-filtration characteristics of bovine enamel protein, in R. W. Fearnhead and M. V. Stack (ed.), Tooth Enamel II: Its Composition, Properties and Fundamental Structure, pp. 95–98, The Williams & Wilkins Co., Baltimore (1971).

Pannesse, E. Observations on the ultrastructure of the enamel organ. III. Internal and external enamel epithelia. *J. Ultrastruct. Res.* **6**, 186–204 (1962).

Reith, E. J. The early stage of amelogenesis as observed in molar teeth in young rats. *J. Ultrastruct. Res.* **17**, 503–526 (1967).

Saxén, L., Koskimies, O., Lahti, A., Miettinen, H., Rapola, J. and Wartiovaara, J. Differentiation of kidney mesenchyme in an experimental model system, in M. Abercrombie, J. Brachet and T. J. King (ed.), Advances in Morphogenesis, pp. 251–294, Academic Press, New York (1968).

Schonfeld, S. Demonstration of an alloimmune response to embryonic enamel matrix proteins. *J. dent. Res.* **54**, 72–77 (1975).

Schonfeld, S., Trump, G. N. and Slavkin, H. C. Immunogenicity of two naturally occurring solid-phase enamel proteins. *Proc. Soc. Exp. Biol.* (N.Y.) (1977) in press.

Seyer, J. Evolution of mineralizing tissues, in H. C. Slavkin (ed.), The Comparative Molecular Biology of Extracellular Matrices, pp. 276–289, Academic Press, New York (1972).

Seyer, J. and Glimcher, M. J. The content and nature of the carbohydrate compounds of the organic matrix of embryonic bovine enamel. *Biochem. biophys. Acta (Amst.)* **184**, 509–522 (1969).

Seyer, J. and Glimcher, M. J. The isolation of phosphorylated polypeptide components of the organic matrix of embryonic bovine enamel. *Biochem. biophys. Acta (Amst.)* **236**, 279–291 (1971).

Shimukawa, H. and Sasaki, S. Biosynthesis of enamel protein *in vitro. J. dent. Res.* **54**, A107 (1972).

Silva, D. and Kailis, D. (1972) Ultrastructural studies on the cervical loop and the development of the amelo-dentinal junction in the cat. *Arch. oral Biol.* **17**, 279–289 (1972).

Slavkin, H. C. Embryonic tooth formation: a tool for developmental biology, in A. H. Melcher and G. A. Zarb (ed.), Oral Sciences Reviews, Vol. 4, Munksgaard, Copenhagen (1974).

Slavkin, H. C. and Bringas, P. Epithelial-mesenchymal interactions during odontogenesis. IV. Morphological evidence for direct heterotypic cell–cell contacts. *Develop. Biol.* **50**, 428–442 (1976).

Slavkin, H. C., Bringas, P. and Croissant, R. D. Epithelial-mesenchymal interactions during odontogenesis. II. Intercellular matric vesicles. *Mech. Age Develop.* **1**, 139–161 (1972).

Slavkin, H. C., Croissant, R. D. and Bringas, P. Epithelial-mesenchymal interactions during odontogenesis. III. A simple method for the isolation of matrix vesicles. *J. Cell Biol.* **53**, 841–849 (1972).

Slavkin, H. C., Croissant, R. D., Bringas, P., Matosian, P., Wilson, P., Mino, W. and Guenther, H. Matrix vesicle heterogeneity: possible morphogenetic functions for matrix vesicles. *Fed. Proc.* **35**, 127–134 (1976).

Slavkin, H. C., Mino, W. and Bringas, P. The biosynthesis and secretion of precursor enamel protein by ameloblasts as visualized by autoradiography after tryptophan administration. *Anat. Rec.* **185**, 289–312 (1976).

Slavkin, H. C., Trump, G. N., Brownell, A. and Sorgente, N. Epithelial-mesenchymal interactions: Mesenchymal specificity, in J. Lash and M. Burger (ed.), Cell and Tissue Interactions, Raven Press, New York, in press.

Sorgente, N., Brownell, A. G. and Slavkin, H. C. Basal lamina degradation: The identification of mammalian-like collagenase activity in mesenchymal-derived matrix vesicles. *Biochem. biophys. Res. Commun.* **74**, 448–454 (1977).

Trelstad, R. L. and Slavkin, H. C. Collagen synthesis by the enamel organ epithelia of the embryonic rabbit tooth. *Biochem. biophys. Res. Commun.* **59**, 443–449 (1974).

Trump, G., Bringas, P. and Slavkin, H. C. Embryonic mammalian amelogenesis: *In vitro* organ culture methods. *J. dent. Res.* **56**, A58 (1977).

Warshawsky, H. Steps in secretion of enamel matrix protein, as shown by electron microscope radioautography of the ameloblasts of rat incisors following tyrosine-H[3] injection. *Anat. Rec.* **154**, 438–439 (1966).

Weinstock, A. Matrix development in mineralizing tissues as shown by radioautography: Formation of enamel and dentine, in H. C. Slavkin and L. A. Bavetta (ed), Developmental Aspects of Oral Biology, pp. 202–242, Academic Press, New York (1972).

Weinstock, A. and LeBlond, C. P. Elaboration of the matrix glycoprotein of enamel by the secretory ameloblasts of the rat incisor as revealed by radioautography after galactose-^{3}H-injection. *J. Cell Biol.* **51**, 26–37 (1971).

Studies on Primary Embryonic Induction

D. Tarin

*Department of Histopathology, Royal Postgraduate Medical School,
Du Cane Road, London W12, England*

INTRODUCTION

In vertebrate embryos the formation of the nervous system by the surface ectoderm is elicited by a stimulus emanating from the dorsal mesoderm. This effect of the dorsal mesoderm in evoking nervous system formation in ectoderm which would otherwise remain undifferentiated is referred to as neural induction. Not only is this process responsible for neural development but also for the laying down of the embryo's cranio-caudal axis and it is therefore frequently referred to more generally as primary embryonic induction. The use of the term primary also distinguishes this event from subsequent inductive phenomena in other organs which are referred to as secondary.

The significance and power of this process in development were first shown by the experiments of Spemann and Mangold (1924) who transplanted a second dorsal blastopore lip (which contains invaginating dorsal mesoderm) into amphibian gastrulae. The spectacular result was the formation of embryos with two nervous systems and two cranio-caudal axes.

The position, orientation and size of the secondary nervous system was directly influenced by the position and orientation of the grafted dorsal mesoderm. It was further shown by experiments involving grafting of dorsal lip tissue from a non-pigmented embryo into a pigmented one that the tissues of the induced (second) neural tube and the secondary embryonic axis were only partially derived from the graft. The remainder of the cells required for building the organs of the neuraxis were derived from the host. This established that the cells in the vicinity of the dorsal lip of the amphibian blastopore do not consist of a mosaic of predetermined organs but are able to influence and coordinate other cells to cooperate with them in building harmoniously proportioned structures.

Subsequently it was demonstrated that the phenomenon of primary induction occurs in other species e.g. birds (Waddington and Schmidt, 1933) and fish (Oppenheimer, 1936), and there is good reason to believe it applies in all vertebrate embryos. Understanding of the mechanisms involved would clearly provide insight into

(1) how different groups of cells communicate to produce orderly patterns of differentiation, and
(2) the pathogenesis of developmental neural abnormalities.

Despite much further research the mechanisms by which the dorsal chordamesoderm alters the fate and developmental behaviour of the overlying ectoderm remains obscure and there is confusion about the interpretation of the apparently contradictory experimental findings. For instance, Spemann (1938, pp. 262–268) showed that implantation of the dorsal lip of the blastopore from early gastrulae results in induction of secondary brain and head structures in the hosts whereas the dorsal lip from late gastrulae induce tail formations. This indicates an "instructive" quality in the inductive stimulus. On the other hand, inductive effects were shown to be exerted by tissues from a variety of sources and even killed tissues. This suggests that the inductive stimulus has a "releasing" or permissive quality, but as Spemann (1938, p. 369) remarked: "A dead 'organiser' is a contradiction in itself".

Against this background it was decided to reinvestigate the mechanism of primary induction by a programme of studies using modern methods of analysis, so far as possible, in the intact embryo. The studies were all performed on amphibian embryos—predominantly *Xenopus laevis* because the eggs can be obtained at will in any season. The findings are outlined below and discussed in the context of pertinent studies using other techniques.

Fig. 1 (*inset*). Survey view of sagittal section, stage 12 embryo, showing the yolk plug (Y) protruding between the dorsal and ventral lips of the blastopore (L_1 and L_2). A, archenteric cavity; C, blastocoel; H, endoderm.

Fig. 1. Dorsal part of the same, stage 12 embryo. Note the thickening of the ectoderm and the presence of columnar cells in this region. The rostral extent of mesodermal invagination is vague, but it probably corresponds with the position marked by the asterisk. The ectomesodermal junction is clearly defined and the gap between the two layers varies in size.

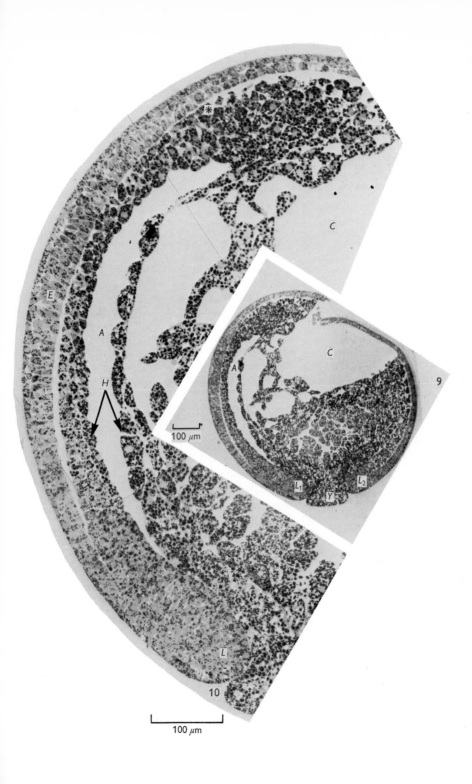

E

*

C

A

H

100 μm

A

C

9

L₁

Y

L₂

100 μm

L

10

100 μm

DEVELOPMENT OF THE NEURAXIS

The histological observations (Tarin, 1971) revealed a fairly clear-cut sequence of events. Initial thickening occurs in the mid-dorsal ectoderm at about stage $11\frac{1}{2}$ of Nieuwkoop and Faber (1967) and spreads over the rest of the dorsal surface by stage $12-12\frac{1}{2}$ (Fig. 1). This is rapidly followed by segregation of brain and spinal regions clearly recognizable by stage 13 (Fig. 2).

There is next a period of relative quiescence during which the neural

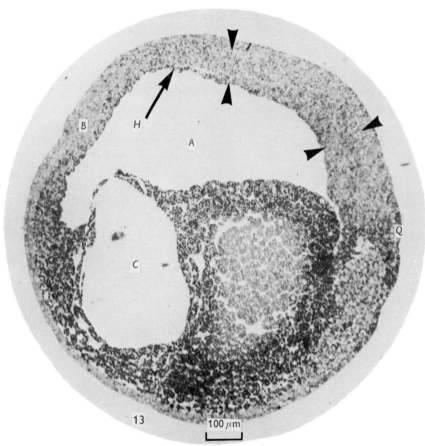

Fig. 2. Sagittal section, stage 13 embryo, survey picture. The brain (B) has segregated from the more caudal neural plate which will form spinal cord. The arrow heads indicate the gradual decrease in dorsal mesodermal thickness between the blastopore (Q) and the cephalic end.

plate continues to thicken and mechanical changes bring about folding around a longitudinal axis. Then, between stages 17 and 19 regionalization of the brain begins with segregation of its substance into three main masses separated from each other by deep constrictions (Fig. 3).

These observations show that the morphological events comprising neural induction are complex and gradual. Thus the nervous system does not appear *ab initio* as a pre-formed unit with inherent structural differences in various regions but instead as a diffuse thickening of the dorsal ectoderm (interpreted as "activation") which subsequently becomes regionally modified to form different parts of the CNS (interpreted as "transformation"). One therefore expects that the inductive processes responsible for these changes will prove to be similarly complex and sequential rather than dependent on a single rapid event. This conclusion is similar to those reached by other investigators using grafting and other experimental manipulations (Nieuwkoop *et al.*, 1952; Takaya, 1955; Eyal-Giladi, 1954) and supports in principle the activation-transformation hypothesis advanced by Nieuwkoop *et al.* (1952) and the similar scheme proposed by Sala (1955).

It seems established by the work of several investigators that the segregation of the neural plate into different regions (regionalization) is effected by the underlying mesoderm (Spemann, 1931; Holtfreter, 1933; Alderman, 1935; Deuchar, 1953; Takaya, 1955; see also review by Saxén and Toivonen, 1962). What is not agreed, however, is whether the mesoderm itself is a mosaic of separate areas each responsible for the evocation of a different part of the central nervous system or whether regionalization is the result of variation in the length of exposure to cumulative influences exerted by the mesoderm as it slides under the ectoderm. As might be expected in such a situation there are various items of experimental evidence which are considered to support or oppose each of these interpretations. For instance, Waddington and Deuchar (1952) substituted the neural plate of late gastrulae with vitally stained ectoderm from early ones and found that in the embryos which survived, a whole normal neural axis was formed from the marked transplant. This suggests that regionalization is not produced by differential effects exerted during mesodermal invagination and tends to favour the alternative possibility of regional differences in the mesoderm. On the other hand, Eyal-Giladi (1954) removed portions of ectoderm from the dorsal surface at different stages during gastrulation and transplanted them to the ventral aspect of the embryos. She noted the development of progressively more caudal neural structures in the grafts taken at successively later stages of mesodermal invagination, and attributed this sequence to differences in the length of

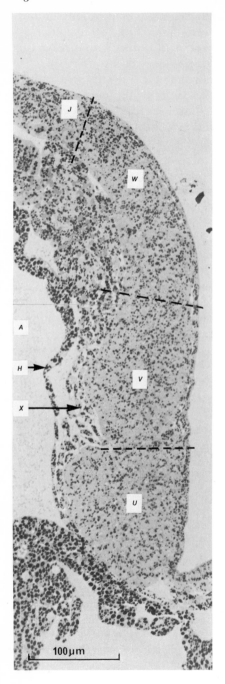

Fig. 3. Detail of brain of stage 19 embryo, sagittal section. The brain is divided into 3 regions (U, V, W) by ventral grooves. The rhombencephalon (W) merges imperceptibly with the spinal cord (J) and the prosencephalon (U) has a sharp boundary with the ectoderm of the stomadeal membrane. The prosencephalon (U) and mesocephalon (V) are underlain by a very thin sheet of mesoderm—the prechordal plate (X). (A, archenteron; H, endoderm.)

exposure to cumulative effects exerted by the mesoderm sliding underneath the dorsal surface.

There are still other experiments whose results are difficult to reconcile with either of these hypotheses: Waddington and Yao (1950) found that antero-posterior reversal of the organization centre in young *Triturus alpestris* embryos did not affect development and normal embryos were obtained. Even at the late gastrula stage (Harrison stage 13) regulation of the antero-posterior alignment of the embryonic axis was found to be possible. If at this stage an extra archenteron roof (i.e. dorsal chorda-mesoderm) is placed in reversed polarity between the ectoderm and archenteron roof of the gastrula normal morphogenesis is observed.

In order to exclude possible influences from host tissues Deuchar (1953) used pieces of archenteron roof and tested their inductive activity in explantation experiments. She found that, providing the mass of inducing tissue was adequate, regional differences could be demonstrated, but also, if the archenteron roof was disaggregated as a whole, its inductive effect after reaggregation did not differ significantly from normal non-disaggregated tissue. It would seem that regional differences in the inductive capacity of the archenteron roof are reversible and can be re-established after gross experimental disturbance.

The histological observations presented above do not permit a clear interpretation of these data. However, it is suggested that the further grafting experiments which will be needed to understand how regionalization occurs should be performed at suitable stages selected on the basis of the histological changes.

The mechanism of folding of the neural plate to form a neural tube is a problem related to the study of neural induction. In contrast to the newt, where the neural plate is only one cell thick (Burnside and Jacobson, 1968), in *Xenopus* it is composed of several cell layers which all participate in the folding process. Thus, it seems unlikely that the deformation in the shape of the surface ectodermal cells described by some investigators (Waddington and Perry, 1966; Baker and Schroeder, 1967; see also Wren and Wessels, 1969) and suggested as a possible mechanism of neurulation is alone an adequate causal factor in this species because it is difficult to see how this could achieve more than the production of a surface dimple or wrinkle. Although such a mechanism might initiate the formation of a neural groove it could be expected to have little effect on the deeper layers of the ectoderm. It is therefore imagined that to accomplish folding of the full thickness of the neural plate other factors must be in operation. In fact, it seems highly likely that the formation of a longitudinal gutter, by the sinking of the notochord and the relatively fast dorsal expansion of the somite masses (Figs 4, 5 and 6) between stages 13 and 19 (Tarin, 1971), makes a significant contribution to this process. These histological observations confirm those of Schroeder (1970), who drew similar conclusions and argued that this mechanism augmented the effects of cellular re-arrangement and alteration of cell shape.

Although this histological work established when the thickening of the dorsal ectoderm and its regionalization take place in *Xenopus* neural induction, the timing of the stimuli which trigger these responses remains unknown. However, the work of Spemann (1938) and his associates shows that this is not of particular importance because inductive stimuli are present for long after the neural tube is formed and it is the ability of the ectoderm to respond which decides when induction begins. Thus, we should consider (1) whether there are any features which persist for long periods during induction and might represent the agents responsible, and (2) whether there are any changes in the ectoderm prior to induction which indicate when the cells become competent to respond to the inductive stimulus.

CHANGES AT THE ECTO-MESODERMAL INTERFACE

In a histological study it is of course only possible to identify visible features, and the only one of these which persisted for long enough to represent a possible inductive agent was amorphous material with metachromatic staining properties (Fig. 7) lying in the intercellular spaces (Tarin, 1971). It first appeared between the dorsal mesodermal cells at the beginning of gastrulation (stage 12) and persisted there and at the ectomesodermal boundary until the thickening of the dorsal ecto-derm resulted in the formation of the neural plate. After stage 14 the metachromatic material rapidly declined in quantity, beginning in the trunk region, although some persisted in the vicinity of the head until segregation of the various parts of the brain occurred. Similar material was also observed in the ventral lip of the blastopore when mesodermal invagination began in this site and remained there until beyond stage 15. Electron-microscopical studies (Tarin, 1972) showed that this material was composed of masses of tightly packed small dense granules (Fig. 8) of uniform size (mean diameter 28.9 nm, SD 3.5 nm). The two

Figs 4–6. Coronal sections of stage 15, stage 18 and stage 19 embryos respectively, showing that the progressive dorsoventral expansion of the somite masses (K) relative to the diameter of the notochord creates a dorsal longitudinal gutter, into which the folding neural plate sinks as it folds to form the neural tube. Asterisks indicate neural groove and neural plate.

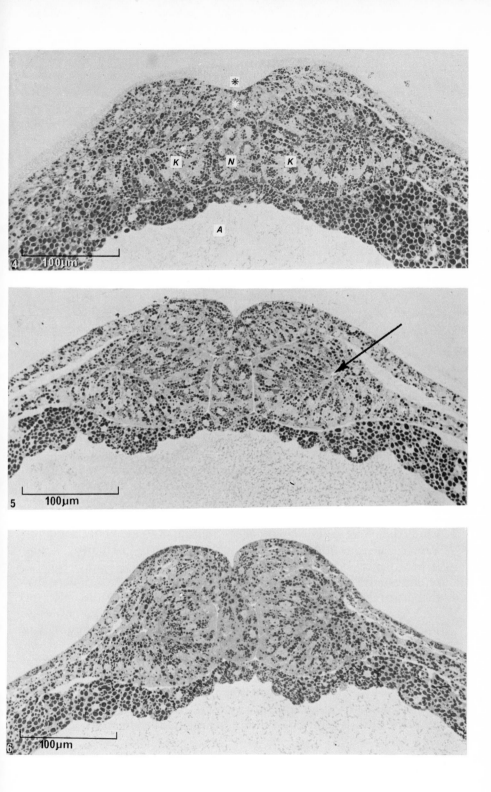

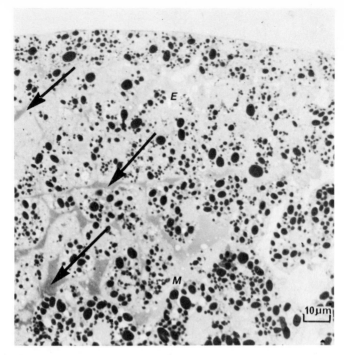

Fig. 7. Mid-dorsal region, stage 11½ embryo. Metachromatic material (arrows) lies between the cells in the mesoderm and, to a lesser extent, in the ectoderm and at the ectomesodermal boundary.

substances with metachromatic staining properties and granular composition most likely to be presented in these embryos are glycogen and/or RNA.

The arrangement of the granules was mostly irregular, but areas in which they were grouped in paracrystalline array were not uncommon. Ribosomes from other animals have been reported to show this tendency (Byers, 1967; Birks and Wheldon, 1971).

Light-microscopical histochemical studies (Tarin, 1973) showed that the distribution of pyroninophilic, ribonuclease digestible, material (RNA) corresponded exactly to that of the granular aggregates and the metachromatic material. Tests for glycogen with PAS staining and amylase digestion gave strong positive results in all the embryonic tissues, but because of the marked intensity of the cellular staining it was not possible, at this level of resolution, to convincingly exclude the presence of glycogen in the extracellular space. There is, however, no doubt that its location is predominantly intracellular.

At the ultrastructural level, staining methods claimed to be specific

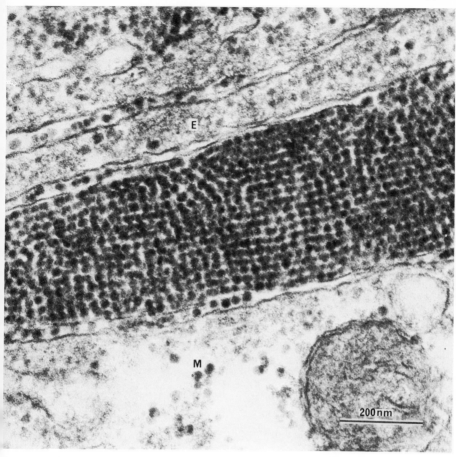

Fig. 8. Dorsal ectomesodermal junction, stage 14. Collections of granules lie in the spaces between cells in a distribution corresponding exactly to that of the metachromatic material in Fig. 7. The granules are sometimes very regularly arranged in rows. A portion of an ectodermal cell (E) lies in the upper part of the picture and a portion of a mesodermal cell (M) is seen below the mass of granules.

for these substances gave disappointing and equivocal results (Tarin, 1973). However, it was found that embryos fixed in potassium permanganate, which is reported to preserve glycogen granules but not ribosomes, had no granules in their extracellular spaces. The persistence of a little amorphous material in this site is believed to indicate that other substances, perhaps glycogen, are normally also present.

To summarize, these findings constitute strong circumstantial evidence for the accumulation of ribosomes in the extracellular spaces of the *Xenopus* embryo. The tests applied cannot exclude the presence of glycogen in the same site nor can they convincingly establish it. For the time being therefore it is proposed to reserve judgement on this issue and accept that glycogen might co-exist with RNA in the extra-cellular space.

Experiments to assess the functional significance of this material are described below (see p. 240).

In the sequential ultrastructural studies on neural induction (Tarin, 1972) it was also found that, as the extracellular granules decreased in number in the trunk region, fine fibrils about 6 nm in diameter were

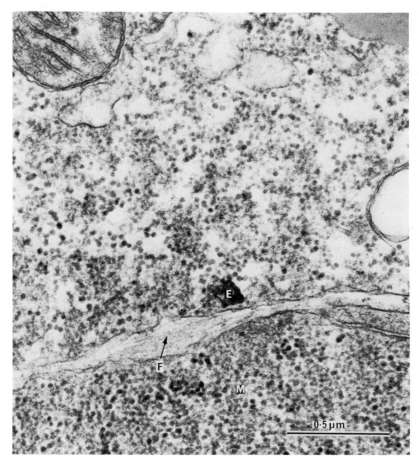

Fig. 9. Dorsal ectomesodermal junction, stage 16. Fine fibrils (F) lie in the small inter-cellular spaces between the notochordal cell (M) and a neural plate cell (E). They are straight, unstriated and irregularly arranged to form a loose meshwork.

present at the interface between the germ layers. These were long, straight and unstriated, and were irregularly orientated to form a loose meshwork (Fig. 9). They first appeared at stage 14 in the trunk region and their distribution as seen in coronal sections at this time is shown in Fig. 10. The fibrils were present not only between the ectoderm and the mesoderm but also between the mesoderm and the endoderm. Similar fibrils were observed between the notochordal and paraxial segments of the dorsal mesoderm. Fibrils were only found in the spaces between different groups of cells and never in the small intercellular spaces within these cell masses: in both the ectomesodermal and the endomesodermal spaces they were not present further laterally than the edge of the neural plate.

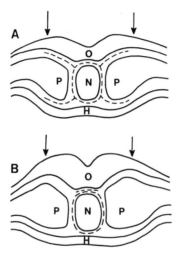

Fig. 10. Diagram illustrating the distribution of the fine fibrils in embryos (A) at stage 14 and (B) at stage 17. The broken lines represent the fine fibrils and the arrows indicate the edges of the neural plate. N, notochord; O, neural plate; P, paraxial mesoderm; H, endoderm.

In older embryos the form of the fibrils did not change, but their distribution was different. At stages 16 and 17 they were observed both in the head region and the trunk. In coronal sections, however, they were present only around the notochord and no longer between the paraxial mesoderm and adjacent tissues; there appeared to be more of them between the notochord and the neural plate than between the notochord and the paraxial mesoderm or the endoderm.

It was found by experiments involving the injection of minute quantities of enzymes such as collagenase and hyaluronidase into the dorsal ectomesodermal space with a calibrated micropipette that the fine fibrils are composed mainly of glycosaminoglycans (Tarin, 1973).

The distribution of these fibrillae and their exact correspondence with the lateral spread of the neural plate suggested they may play a role in neural induction. Similar fibrils have been observed at the interface between interacting tissues in other developing organs, e.g. the pancreas (Kallman and Grobstein, 1965), the teeth (Slavkin *et al.*, 1969) and the lung (Wessells, 1971), and although they do not seem to be responsible for *initiating* induction in these sites, they certainly exercise a morphogenetic function demonstrable by the disturbances which arise after their removal with enzymes (Grobstein and Cohen, 1965; Wessells and Cohen, 1968).

Experiments analysing the significance of these interfacial materials in neural induction are considered below.

With regard to the matter mentioned earlier (item 2, p. 234 of when the ectodermal cells acquire the competence to respond to the neural inductive stimulus, all means of investigation—histological, electron-microscopical and histochemical—failed to identify any specific features in the ectodermal cells which would serve as an indicator of this stage of readiness. Grafting and transplantation techniques therefore remain the only means of recognizing when reactivity to activating and regionalizing stimuli is acquired and lost.

DEVELOPMENTAL SIGNIFICANCE OF THE INTERFACIAL MATERIALS

These experiments were done *pari passu* with the studies on the chemical composition of the extracellular materials described above. Each batch of embryos injected with a given enzyme or other agent was split into two groups, one of which was used to assess the effect of the injected agent on the extracellular materials and one to assess its effects on the development of the living embryo. This permitted direct comparison of the results since both groups were injected simultaneously with the same agent under the same conditions. Assessment of the effects of removal of the extracellular materials on the development of the central nervous system and cephalocaudal axis was performed by the examination of embryos incubated in the enzyme solution for 24 h. At the end of this period, the embryos were carefully inspected with a dissecting microscope and the percentage viability and general stage of development were compared with untreated controls from the same batch of eggs. Particular attention was paid to the stage and perfection of development and, if there was any delay or deformity, the embryos were killed, and histologically examined. Functional activity was assessed by stimulation of the embryos with a fine glass needle to show whether they responded, at least in terms of reflex activity, at all points along their bodies including the treated areas.

It was found that the extracellular granules present in the early stages of induction disappeared after injection of ribonuclease. Surprisingly they also disappeared after injection of amylase or plain saline. This could indicate either high solubility or lysis by enzymes released from damaged cells, but injection of ribonuclease antagonists, such as polyvinyl sulphonate, did not result in the preservation of granules. The removal of the granules had no effect either on general development or on neural induction (Tarin, 1973).

Injection of collagenase into slightly older embryos had only a slight effect on fibrillar size and quantity, but injection of hyaluronidase caused rapid and complete disappearance of fibrils in 100% of the specimens. Once again removal of the interfacial materials had no significant effect on neural induction or on embryonic development in general (Tarin, 1973).

It was therefore important to establish whether the granules and fibrils are reconstituted during the incubation period after injection. It was found that both reappear and that reconstruction takes about 3 h for granules and 6 h for fibrils (Tarin, 1973). In the intervening time neural development proceeded normally and there was no delay compared with untreated controls.

These findings suggest that the electron-optically visible materials at the ectomesodermal junction are not essential for neural induction. This does not necessarily mean that they have no role in this process in the undisturbed embryo, for it is known that in some developmental systems two or more mechanisms operate synergistically to achieve a given effect, but that each can achieve it alone. This is termed the principle of double assurance (see Spemann, 1938, p. 92). For practical purposes, however, one can assume that the extracellular materials do not exercise an important role in the induction of the central nervous system because in their absence there is not even a delay in development of the organ, nor a reduction in its size.

It is also possible that although the enzymic activity breaks down the extracellular structures, it may produce smaller compounds which retain their inductive activity. Even so, this argument does not affect the validity of the conclusion that the morphological features observed at the ectomesodermal junction are not essential mediators of neural induction.

EXPERIMENTS WITH INTERPOSED POROUS MEMBRANES

Currently it is believed that there remain two other plausible means (Saxén, 1972) of transmitting short-range activating and coordinating signals between different tissues in development:

(1) transmission of soluble materials from one to the other,
(2) direct contact between the plasma membranes of the interacting tissues.

Whether there is any requirement for direct cellular contact between the ectoderm and the mesoderm in neural induction was tested by interposing porous barrier membranes between the interacting tissues (Toivonen *et al.*, 1975). In early experiments the barriers were inserted between the dorsal mesoderm and the ectoderm of living embryos, but these proved unsuccessful because of displacement of the membranes by the tissue movements which occurred during healing after the operation. The method developed by Saxén (1961) (see Fig. 11) for previous transfilter culture studies on neural induction was therefore used. After exposure of the dorsal ectoderm to inductive action by dorsal chordamesoderm for 24 h the culture assemblies were either fixed intact for electron-microscopical examination or dismantled and the ectoderm isolated for further culture (8 days) to establish whether neural development ensued. It was found that even when the pores of the interposed membranes were only 0.1 μm in diameter (the range of pore sizes used was 0.1 μm to 8 μm) inductive action could be transmitted across the interposed membrane. Control cultures in which ectoderm was not exposed to mesoderm or in which the membrane between the reacting tissues was composed of cellophane showed no neural development. Electron-microscopical examination of the pores in the membranes did

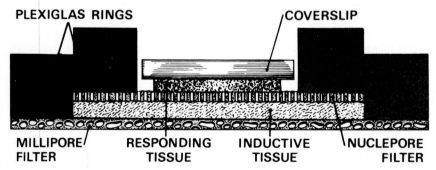

PLEXIGLAS RINGS   COVERSLIP

MILLIPORE   RESPONDING   INDUCTIVE   NUCLEPORE
FILTER   TISSUE   TISSUE   FILTER

Fig. 11. The transfilter method used in the series of experiments with Nuclepore filters (after Saxén, 1961).

not reveal any cytoplasmic processes traversing them nor any evidence of contact between the interacting tissues in any of the specimens examined whatever the pore size.

It is therefore concluded (Toivonen et al., 1975; see also Saxén, 1961, and Vainio et al., 1962) that neural development can be initiated in competent *early* gastrular ectoderm by an inductive stimulus transmitted from the underlying mesoderm without demonstrable cytoplasmic contact between the two tissues.

It is important to emphasize that these results apply strictly to the *initiation* of neural development in uncommitted gastrular ectoderm. Very recent results (Toivonen and Wartiovaara, 1976) suggest that there may be substantial changes in the mode of cellular signalling in the later stages of primary induction, when the transformation of the neural plate to successively more caudal-parts of the central nervous system takes place. In these experiments using neural and archenteron roof from older embryos, where segregation of the different parts of the CNS is occurring, cytoplasmic processes were observed to enter the pores of the filters. Unfortunately, transformation of the neural tissues to successively more caudal parts of the CNS was not seen in these explants. One therefore cannot be sure whether the cytoplasmic processes observed in transfilter cultures of older tissues have a role in the late inductive interactions influencing the regionalization of the CNS and this possibility requires more detailed study.

CONCLUSIONS

This information from transfilter culture experiments indicates that direct membrane-to-membrane contact between the inducing and reacting tissues is not required at least for the initiation of neural induction but may be for later stages in which regionalization takes place. This is supported by the findings of Gallera et al. (1968), that in birds neural induction can occur across a Millipore filter, providing it is less than 25 μm thick. As the extracellular materials at the interface between the mesoderm and ectoderm are also not essential for induction it seems very likely that the transmission of the initial stimulus in neural induction is by the diffusion of soluble factors. This interpretation is supported by other experimental results with autoradiographic and fluorescent antibody tracing methods and with the use of cell-free extracts (see Toivonen et al., 1976, for review) which suggest that there is a transfer of soluble and diffusible molecules, which contain an inductive principle, from one tissue to the other. This of course raises the intriguing question of how the action of this agent *in vivo* is so localized and precise. Although the whole of the gastrular ectoderm is capable of responding to it the boundaries of the developing nervous system are always sharp and distinct.

Comparison of this state of knowledge about primary neural induction with recently obtained information on other inductive systems shows that the means of transmitting the inductive stimulus varies considerably. For instance, when mouse metanephrogenic mesenchyme is induced by the spinal cord to form kidney tubules, contacts are consistently found between the inducing and responding cells (Wartiovaara *et al.*, 1974; Lehtonen *et al.*, 1975). Since, in transfilter culture studies, the time required for tubule formation is also closely related to that for contacts to be established between processes, there is good reason for believing that intercellular contact is the means for transmitting the inductive signal in this system (Wartiovaara *et al.*, 1974; Lehtonen *et al.*, 1975). On the other hand, studies on inductive interactions in rabbit tooth development suggest that transmission of the signal between epithelial and mesenchymal cells is effected by small vesicles in the intercellular matrix (Slavkin, 1972). In yet other developing organs, such as the mouse salivary gland, there is evidence that extracellular materials deposited at the interface between the interacting tissues play an important role in inductive relationships between epithelium and mesenchyme (Bernfield and Wessells, 1972; Grobstein, 1967).

The significance of this variation is that although inductive phenomena in many different organs possess several features in common, they need not necessarily depend on identical mechanisms.

CONCLUDING REMARKS

Considerable amounts of information, some of it apparently contradictory, are now available on the process of neural induction, but there is as yet no clear understanding of how it takes place or of what the crucial factors are. This is probably because of the complexity of the process. The information that has been accumulated indicates that this is not an all or none trigger event but rather a kinetic multiphasic process involving at least the following separate but interlocked steps: initial activation, cytodifferentiation, histodifferentiation, formation and regionalization of the neural tube. Even the initial activation process depends on the prior development of competence and of inductive power which in some systems have been shown to be in turn dependent on specific earlier associations of the tissues involved with other tissues elsewhere in the embryo (see Jacobson, 1966; Sudarwati and Nieuwkoop, 1971). Much of the confusion and contradictory evidence in the field may be due to incorrectly assuming that evidence which is in fact correct and valid applies to the overall process when in reality it applies to only a segment of the process. Thus, data which provide insight into the mechanisms (see above) of initial activation *in vivo* or of initial neural cytodifferentiation in culture should not be considered to explain how the whole orchestrated process of neural induction is triggered and driven.

REFERENCES

Alderman, A. L. The determination of the eye in the anuran Hyla regilla. *J. exp. Zool.* **70**, 205–232 (1935).

Baker, P. C. and Schroeder, T. E. Cytoplasmic filaments and morphogenetic movements in the amphibian neural tube. *Develop. Biol.* **15**, 432–450 (1967).

Bernfield, M. R. and Wessels, N. K. Intra- and extracellular control of epithelial morphogenesis. *Develop. Biol.* Suppl. **4**, 195–249 (1970).

Birks, R. L. and Wheldon, P. R. Formation of crystalline ribosomal arrays in cultured chick embryo dorsal root ganglia. *J. Anat. (Lond.)* **109**, 143–156 (1971).

Burnside, M. B. and Jacobson, A. G. Analysis of morphogenetic movements in the neural plate of the newt, *Taricha torosa. Develop. Biol.* **18**, 537–552 (1968).

Byers, B. Structure and formation of ribosome crystals in hypothermic chick embryo cells. *J. mol. Biol.* **26**, 155–167 (1967).

Deuchar, E. M. The regional properties of amphibian organiser tissue after disaggregation of its cells in alkali. *J. exp. Biol.* **30**, 18–43 (1953).

Eyal-Giladi, H. Dynamic aspects of neural induction in amphibia. *Arch. Biol. Liège* **65**, 179–259 (1954).

Gallera, J., Nicolet, G. and Bannan, M. Neural induction in birds through a millipore filter: Study by optical and electron microscopy. *J. Embryol. exp. Morph.* **19**, 439–450 (1968).

Grobstein, C. Mechanisms of organogenetic tissue interaction. *Nat. Cancer Inst. Monogr.* **26**, 279–298 (1967).

Grobstein, C. and Cohen, J. Collagenase: Effect on the morphogenesis of embryonic salivary epithelium *in vitro. Science* **150**, 626–628 (1965).

Holtfreter, J. Organisierungstuffen nach regionaler Kombination von Entomesoderm mit Ektoderm. *Biol. Zbl.* **53**, 404–431 (1933).

Jacobson, A. G. Inductive processes in embryonic development. *Science* **152**, 25–34 (1966).

Kallman, F. and Grobstein, C. Source of collagen at epithelio-mesenchymal interfaces during inductive interaction. *Develop. Biol.* **11**, 169–183 (1965).

Lehtonen, E., Wartiovaara, J., Nordling, S. and Saxén, L. Demonstration of cytoplasmic processes in Millipore filters permitting kidney tubule induction. *J. Embryol. exp. Morph.* **33**, 187–203 (1975).

Nieuwkoop, P. D. *et al.* Activation and organisation of the central nervous system in amphibians I, II and III. *J. exp. Zool.* **120**, 1–108 (1952).

Nieuwkoop, P. D. and Faber, J. Normal Table of *Xenopus laevis* (Daudin), North Holland Publishing Co., Amsterdam (1967).

Oppenheimer, J. M. Transplantation experiments on developing teleosts (Fundulus and Perca) *J. exp. Zool.* **72**, 409–437 (1936).

Sala, M. Distribution of activating and transforming influences in the archenteron roof during the induction of the nervous system in amphibians. *Proc. kon. ned. Akad. Wet. Ser. C.* **58**, 635–647 (1955).

Saxén, L. Transfilter neural induction of amphibian ectoderm. *Develop. Biol.* **3**, 140–152 (1961).

Saxén, L. Interactive mechanisms in morphogenesis, in D. Tarin (ed.), Tissue Interactions in Carcinogenesis, pp. 49–80, Academic Press, London (1972).

Saxén, L. and Toivonen, S. Primary Embryonic Induction, Logos Press, London (1962).

Schroeder, T. E. Neurulation in *Xenopus laevis*. An analysis and model based upon light and electron microscopy. *J. Embryol. exp. Morph.* **23**, 427–462 (1970).

Slavkin, H. C. Intercellular communication during odontogenesis, in H. C. Slavkin and L. A. Bavetta (ed.), Developmental Aspects of Oral Biology, pp. 165–199, Academic Press, New York, London (1972).

Slavkin, H. C., Bringas, P., LeBaron, R., Cameron, J. and Bavetta, L. A. The fine structure of the extra-cellular matrix during epithelio-mesenchymal interactions in the rabbit embryonic inciser. *Anat. Rec.* **165**, 237–256 (1969).

Spemann, H. Uber don Anteil von Implantat und Wirtskeim an der Orientierung und Beschaffenheit der induzierten Embryonalanlage. *Wilhelm Roux' Arch. Entwickl.-Mech. Org.* **123**, 390–517 (1931).

Spemann, H. Embryonic Development and Induction. Yale Univ. Press (1938).

Spemann, H. and Mangold, H. Uber Induktion von Embryonalanlagen durch Implantation artfremder Organisatoren. *Arch. mik. Anat. Entwickl.-Mech.* **100**, 599–638 (1924).

Sudarwati, S. and Nieuwkoop, P. D. Mesoderm formation in the anuran *Xenopus laevis*. *Wilhelm Roux' Arch. Entwickl.-Mech. Org.* **166**, 189–204 (1971).

Takaya, H. Formation of the brain from the prospective spinal cord of amphibian embryos. *Proc. imp. Acad. Japan* **31**, 360–385 (1955).

Tarin, D. Histological features of neural induction in *Xenopus laevis*. *J. Embryol. exp. Morph.* **26**, 543–570 (1971).

Tarin, D. Ultrastructural features of neural induction in *Xenopus laevis*. *J. Anat. (Lond.)* **111**, 1–28 (1972).

Tarin, D. Histochemical and enzyme digestion studies on neural induction in *Xenopus laevis*. *Differentiation* **1**, 109–126 (1973).

Toivonen, S., Tarin, D. and Saxén, L. The transmission of morphogenetic signals from amphibian mesoderm to ectoderm in primary induction. *Differentiation* **5**, 49–55 (1976).

Toivonen, S., Tarin, D., Saxén, L., Tarin, P. J. and Wartiovaara, J. Transfilter studies on neural induction in the newt. *Differentiation* **4**, 1–7 (1975).

Toivonen, S. and Wartiovaara, J. Mechanisms of cell interaction during primary embryonic induction studied in transfilter experiments. *Differentiation* **5**, 61–66 (1976).

Vainio, T., Saxén, L., Toivonen, S. and Rapola, J. The transmission problem in primary embryonic induction. *Exp. Cell Res.* **27**, 527–538 (1962).

Waddington, C. H. and Deuchar, E. The effect of type of contact with the organiser on the nature of the resulting induction. *J. exp. Biol.* **29**, 496–512 (1952).

Waddington, C. H. and Perry, M. M. A note on the mechanisms of cell deformation in the neural folds of the amphibian. *Exp. Cell Res.* **41**, 691–693 (1966).

Waddington, C. H. and Schmidt, G. A. Induction by heteroplastic grafts of the primitive streak in birds. *Wilhem Roux' Arch. Entwickl.-Mech. Org.* **128**, 522–563 (1933).

Waddington, C. H. and Yao, T. Studies on regional specificity within the organisation centre of the Urodeles. *J. exp. Biol.* **27**, 126–144 (1950).

Wartiovaara, J., Nordling, S., Lehtonen, E. and Saxén, L. Transfilter induction of kidney tubules: correlation with cytoplasmic penetration into Nuclepore filters. *J. Embryol. exp. Morph.* **31**, 667–682 (1974).

Wessells, N. D. Mammalian lung development: Interactions in formation and morphogenesis of tracheal buds. *J. exp. Zool.* **175**, 455–466 (1971).

Wessells, N. K. and Cohen, J. H. Effect of collagenase on developing epithelia *in vitro*: lung ureteric bud and pancreas. *Develop. Biol.* **18**, 294–309 (1968).

Wren, J. T. and Wessells, N. K. An ultrastructural study of lens invagination in the mouse. *J. exp. Zool.* **171**, 359–369 (1969).

Kidney Tubule Induction:
Physical and Chemical Interference

STIG NORDLING, PETER EKBLOM, EERO LEHTONEN, LAURI SAXÉN
and
JORMA WARTIOVAARA

Third Department of Pathology, University of Helsinki,
SF-00290 Helsinki 29, Finland

INTRODUCTION

In principle, a tissue could influence another tissue either from a distance or when in contact with, or very close to, the other tissue. At the present stage of our knowledge, the ability to exert morphogenetic influence on another tissue from distance implies that diffusible substances are at work. If close proximity or cell contacts are required, a cell could act on another cell by means of complementary molecules in a manner similar to antigen-antibody (Grossberg, this volume) or enzyme-substrate interactions, or by intercytoplasmic bridges permitting transfer of substances from one cytoplasm to another without the necessity of crossing the plasma membrane. Close proximity could also be required because of a limited survival time of the morphogenetic signal in the extracellular milieu. In a narrow space between cells the extracellular milieu is almost certainly different from that in the environmental bulk-phase, and might tend towards that of the interior milieu of the cell.

Concepts about the relative importance of long- or short-range morphogenetic interactions have changed during the years. Spemann (1905) observed that in some experiments with apposition of presumptive lens ectoderm and optic cup, the lens failed to develop and a thin layer of mesenchymal cells was seen between the two tissues. Spemann interpreted this as an evidence for the requirement of cell contacts between the interacting tissues. This observation was later confirmed by McKeehan (1951), who interposed a cellophane membrane between optic cup and ectoderm, resulting in inhibition of lens formation. However, interposition of Nuclepore filters with a pore size of 0.1 µm and with a thickness of about 10 µm does not prevent the passage of an inductive signal from the optic cup to responsive ectoderm (Karkinen-Jääskeläinen, unpublished).

Spemann's view that morphogenetic tissue interactions were contact-mediated remained unchallenged until the thirties, when Bautzmann (1932) noticed that in the gastrula, killed blastoporal lip will induce competent ectoderm to form a neural plate, and Mangold (1932) found that non-cellular material in agar had the same effect.

The diffusibility of the morphogenetic factors in primary induction in amphibia has since been repeatedly elaborated (review, Holtfreter, 1955). Saxén (1961) reported that the neuralizing factor would pass through a Millipore filter approximately 25 µm thick, with a nominal pore size of 0.8 µm. Subsequent electron-microscopic examination of these filters failed to demonstrate cytoplasmic material in the pores, and this was taken as an indication for the diffusibility of the inductive agent (Nyholm *et al.*, 1962). Later this experiment was repeated using filters with different pore sizes and with a more uniform structure. Nuclepore filters with pore sizes ranging from 0.1 to 1.0 µm would allow the morphogenetic factor to pass. In electron micrographs, no cytoplasmic processes were seen to penetrate these filters (Toivonen *et al.*, 1975; Tarin, this volume). Thus at present, it appears that the inductive signal initiating primary induction in Amphibia is mediated without cytoplasmic contacts. It should be emphasized, however, that in some recent studies scanning electron microscopy has shown its superiority to transmission EM in the search of cytoplasmic processes in filter cultures (Saxén *et al.*, 1976; Lash *et al.*, this volume). Until such examination has confirmed Tarin's negative findings, caution is warranted in their interpretation.

PHYSICAL INTERFERENCE WITH KIDNEY TUBULE INDUCTION

Hypothesis of matrix interaction

Can the results obtained in primary induction be extrapolated to secondary induction? Grobstein (1953, 1956, 1957) found that the presence of ureteric bud was required for the morphogenesis of mouse metanephric mesenchyme, but that it could be replaced with spinal cord. The morphogenetic signal from spinal cord could pass a Millipore filter with a thickness of approximately 25 μm and with a nominal pore size of 0.45 μm. It could also pass through a stack of two filters, occasionally three, but never four filters, so that the limiting distance between the two interacting tissues was about 75 μm. This distance is too short to fit general concepts about the behaviour of stable diffusible substances. The signal would also occasionally pass thin Millipore filters with a pore size of only 0.1 μm (Grobstein and Dalton, 1957). Cytoplasmic material was visualized in filters with a pore size of 0.45 μm but not in those with a pore size of 0.1 μm, when the standard fixative for electron microscopy at that time, chrome-osmium, was used. This observation therefore apparently ruled out contact-dependent induction. Grobstein was thus left with the observation that the morphogenetic signal was neither freely diffusible nor contact-dependent; leading him to postulate his matrix interaction hypothesis of "something intermediate between full cellular contact and free diffusion" (Grobstein, 1956b), a possibility discussed in biophysical terms by Weiss (this volume, p. 280). The main emphasis of investigation was thereafter focussed not on cellular contacts but on extracellular inductive factors.

Hypothesis of contact-mediated interactions

The question of free diffusion versus contact as the mechanism in embryonic induction was reopened in 1970 when Crick published a theoretical paper on the feasibility of diffusion as the underlying mechanism in establishing morphogenetic gradients in embryonic development. Crick further suggested that a small molecule such as cyclic AMP might be a good candidate. Although diffusion mechanism can operate in primary induction, the extremely slow rate of transmission of the morphogenetic signal in transfilter induction of kidney tubules seemed to contradict long-range diffusion in this system (Nordling et al., 1971). Metanephric mesenchyme is induced to form secretory tubules by spinal cord when the electron-dense membranes are separated by a narrow interspace of 10 nm or less, and as no thick, basement-membrane-like material was seen to separate the cells, we believed that close contacts between the cells might be needed for the transmission of inductive signals, and, hence, a systemic study employing various types of membrane filters was undertaken.

Experiments with Nuclepore filters

In an experimental series where polycarbonate (Nuclepore) filters with straight pores of different sizes were used, a good correlation was found between ingrowth of cytoplasmic processes and transmission of an inductive signal (Wartiovaara *et al.*, 1974). Filters with a pore size of 0.2 μm and larger regularly permitted the passage of inductive signals. Electron microscopy of these filters invariably showed cytoplasmic processes from both the spinal cord cells and the mesenchymal cells penetrating deep into the filter. At places these processes met and were separated by distances not exceeding 10 nm between the electron-dense, linear membranes (elsewhere in this volume Weiss, p. 282, discusses the problems associated with interpreting separation distances of this magnitude). Filters with a nominal pore size of 0.1 μm (0.15 measured) only occasionally permitted induction, and in transmission electron-microscopic examination only occasional processes were seen. However, in scanning electron microscopy of these filters on which spinal cord had been grown on one side, processes were often seen extending through to the opposite side of the filter.

Experiments with Millipore filters

The possibility existed that the presence of cytoplasmic processes inside our filters was only coincidental and not positively correlated with the passage of the inductive signals, and was somehow dependent on the special type of filter used. Therefore, similar experiments to those using Nuclepore filters were made, but using Millipore filters of the same type as those employed by Grobstein and Dalton (1957).

The results on the passage of the inductive signals were essentially the same as those obtained by Grobstein and Dalton (*op. cit.*). Millipore filters with a nominal pore size of 0.1 μm only occasionally permitted the passage of inductive signals, whereas filters with larger pores always did so (0.45 μm pores in Grobstein's studies, 0.22 and 0.8 μm in ours) (Lehtonen *et al.*, 1975). When these filters were examined by electron microscopy after the same chrome-osmium fixation as used by Grobstein and Dalton (*op cit.*) only filters with 0.8 μm pores were observed to contain cytoplasmic material, whereas when a double fixation with glutaraldehyde and osmium tetroxide was used, cytoplasmic material was visualized throughout filters with 0.22 μm pores. The evidence for the interdependence between ingrowth of cytoplasmic processes and induction of kidney tubules was further strengthened by the observation

Fig. 1. Whole mount of metanephric mesenchyme. After 72 h of culture the spinal cord is scraped off and the mesenchyme fixed in 4% buffered formaldehyde and then stained with Weigert's haematoxylin-eosin stain. (a) Culture on low porosity filter, 0.5% pores. (b) Culture on high porosity filter, 4.2% pores. Both types of filter have the same pore size (0.6 μm).

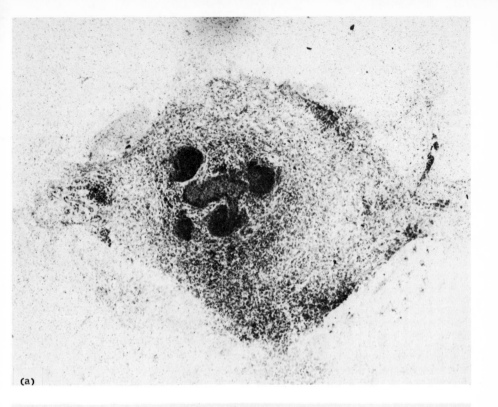

(a)

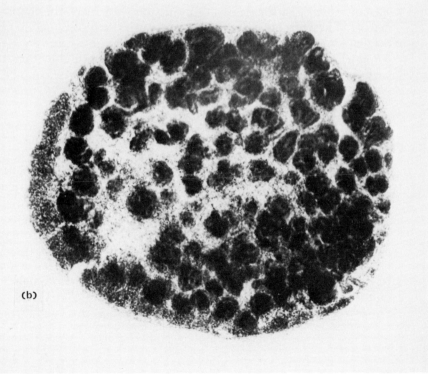

(b)

that induction of metanephric mesenchyme by salivary mesenchyme (another potent inductor of kidney tubule formation) required much larger pores (Saxén *et al.*, 1976). Nuclepore filters with 0.1 μm pores occasionally permitted passage of the inductive signal from spinal cord, but 3 μm pores were required for the signal to pass from salivary mesenchyme. Ingrowth of cytoplasmic material was regularly seen in 3 μm filters, but only in four out of seven filters with 0.6 μm pores, and then only at places. Thus it seems that although ingrowth of cytoplasmic processes is a prerequisite for induction, ingrowth is no guarantee of a successful induction.

Role of contact area

A possible explanation for the finding that ingrowth is not invariably followed by induction is provided by recent observations in another inductive system. Meier and Hay (1975) demonstrated a direct correlation between the pore size (and possibly the pore density) of filters interposed between corneal epithelium and lens capsule and the synthesis of epithelial glycosaminoglycans. They suggested that the contact area of the interactants is of importance.

When Nuclepore filters with different numbers of pores per unit were interposed between spinal cord and metanephric mesenchyme, there was an increase in the degree of induction with increasing filter pore density. The degree of induction was measured as the number of tubules per explant (Fig. 1). The density of pores was in all instances high enough for each mesenchymal cell to be above a hole in the filter. These very recent experiments support the hypothesis that the contact area is of importance for induction.

CHEMICAL INTERFERENCE WITH KIDNEY TUBULE INDUCTION

It seems with reasonable certainty that at least in the kidney tubule system induction requires close apposition of parts of cell peripheries, although at present little is known about the mechanisms involved. Theoretically contacts could be required because of a rapid destruction of the inducing agent, which thus only can cross a very narrow extracellular space of the order of several nanometres. Another possibility is that the inducing substance acts inside the target cell, but is unable to cross the plasma membrane, and induction is therefore dependent on the formation of specialized junctions between the interacting cells. However, the type of junction discussed by Gilula (this volume, p. 326) only permits the passage of small (ca. 10^3 daltons) molecules. Finally induction could be the result of an interaction between complementary molecules on the two tissues.

So far no inductive signal substance operative in tubule induction has been found. In the present system no specialized junctions have yet been identified, which obviously does not rule out their existence, since their identification inside the filter is difficult. Attempts have failed to prevent induction by adding antisera against the major histo-compatibility antigens on the interacting cells or by adding limulus polyphaemus haemolymph, which should cover sialic acid moieties on the tissues (unpublished). However, nothing is known about the fate of these covering agents on the cell surfaces at the time of induction, which seems to take place during the first 12 to 24 h after the transfilter culture has been started.

In an experimental series metabolic inhibitors were used to shed some light on the molecular requirement of the inductive event.

Inhibition of DNA-synthesis

Treatment of the inductor, spinal cord with the irreversible inhibitor Mitomycin C at a concentration of 50 μg/ml did not abolish its capacity to induce. No tubules formed in a mesenchyme which had been treated with the same concentration.

Inhibition of RNA-synthesis

Both ethidium bromide (EB) (Table I) and proflavin (Table II) which are reversible inhibitors of RNA synthesis, prevented the induction of

TABLE I

Effect of proflavine on the induction of kidney tubules by spinal cord

Proflavine (μM)	Number of positive explants/total	
	Spinal cord removed after 24 h	Spinal cord left
0	19/19	25/25
2	9/12	10/10
3	2/23	2/16
4	0/12	0/7

After 24 h cultures were transferred to normal medium. The total cultivation time was always 72 h.

TABLE II

Effect of ethidium bromide (EB) on the induction of kidney tubules by spinal cord

EB (μM)	Number of positive explants/total	
	Spinal cord removed after 24 h	Spinal cord left
0	19/19	25/25
5	7/7	2/2
10	2/7	4/6
15	0/7	0/11

After 24 h cultures were transferred to normal medium. The total cultivation time was always 72 h.

kidney tubules. However, it seems that these agents irreversibly damaged the tissues, as no tubules would form if after 24 h the explant was transferred to a normal medium without removal of spinal cord. Furthermore, pretreatment of either spinal cord or mesenchyme with the inhibitor for 24 h, followed by combination with fresh tissue, did not result in the formation of tubules (Table III). A general toxicity of EB was ruled out by the observation that if EB was added to a trans-filter culture after 24 h of cultivation, tubules would form normally. At 24 h induction has already taken place but no tubules have yet formed. The results thus suggest that the interactive process is sensitive to RNA-inhibitors during the "critical" induction period.

TABLE III

Effect of precultivation with RNA synthesis inhibitor on the induction of kidney tubules by spinal cord

Proflavin (μM)	EB (μM)	Number of positive explants/total	
		Precultivated spinal cord fresh metanephric mesenchyme	Precultivated metanephric mesenchyme fresh spinal cord
0	0	12/12	15/15
3	0	0/4	0/4
4	0	0/4	0/4
0	10	0/4	0/4
0	15	0/4	0/4

Inhibition of protein synthesis

Cycloheximide (CH) was an efficient inhibitor of kidney tubule induction (Table IV). After removal of the inhibitor, tubules formed, provided the spinal cord remained, even if the explant had been exposed to a concentration of CH which was three times higher than that which usually inhibits induction. Hence, protein synthesis during the induction period seems to be a prerequisite for this type of intercellular communication.

TABLE IV

Effect of cycloheximide (CH) on the induction of kidney tubules by spinal cord

CH (μM)	Number of positive explants/total	
	Spinal cord removed after 24 h	Spinal cord left
0	19/19	25/25
0.2	6/7	—
0.4	3/3	—
0.5	2/17	3/3
1.0	0/19	—
1.5	0/30	17/17

After 24 h cultures were transferred to normal medium. The total cultivation time was always 72 h. —, not tested.

Inhibition of the synthesis of hexosamine-containing compound

Some encouraging results have been obtained by using the glutamine analogue (6-diazo-5-oxo-L-norleucine) (DON). At a concentration of 20 μM present for the first 24 h, DON completely inhibits the subsequent formation of kidney tubules if the spinal cord is removed at the same time as DON (Table V). In control cultures without DON tubules develop if the spinal cord is scraped off after 24 h. Addition of a 500-fold excess of glutamine nearly completely prevents the inhibitory effect of DON. This indicates that the inhibitory effect of DON is due to its nature of a glutamine analogue and not to some other mechanism (possibly toxicity). A general toxic effect is also ruled out by the finding that kidney tubules will form if the spinal cord is left after removal of DON after 24 h of culture. Pretreatment of spinal cord with as high a concentration as 100 μM for 24 h did not prevent its inducing capacity.

DON affects cellular metabolism in many ways (Green and Pratt, 1976). It prevents glycosaminoglycan synthesis because it inhibits the

TABLE V

Effect of DON on the induction of kidney tubules by spinal cord

Addition[1]				Number of positive explants/total	
DON (μM)	GlNH$_2$ (mM)	GluNH$_2$ (mM)	AIC (mM)	Spinal cord removed after 24 h	Spinal cord left
0	0	0	0	19/19	25/25
10	0	0	0	6/8	2/2
20	0	0	0	0/15	6/6
30	0	0	0	0/9	3/3
20	10	0	0	8/10	—
20	0	10	0	4/10	—
20	0	0	5	0/10	5/5
20	0	10	5	2/9	5/5
0	0	0	5	3/10	7/7

[1] in GlNH$_2$ free medium.

After 24 h cultures were transferred to normal medium. The total cultivation time was always 72 h.

DON, diazooxonorleucin; GlNH$_2$, glutamine; GluNH$_2$, glucosamine; AIC, aminoimidazol carboxamide; —, not tested.

formation of glucosamine and it inhibits nucleotide synthesis. The inhibition of glycosaminoglycan synthesis can be overcome by adding large amounts of glucosamine. Nucleotide synthesis can be partially inhibited by adding aminoimidazol carboxamide (AIC). When glucosamine was added together with DON the inhibitory effect was only partly abolished. AIC could not prevent the inhibitory effect of DON, but AIC alone to some extent diminished induction (Table V). These results suggest that glycosaminoglycans or glycoproteins may be involved in the establishment of the heterotypic cell contact, or in the exchange of messages between the interacting cells.

None of these inhibitors prevented ingrowth of cytoplasmic processes into the filter as examined both in scanning and in transmission electron microscopy (Fig. 2). Therefore the inhibitory effect is likely to be at the molecular level. However, since these inhibitors have several different effects it seems premature to make final conclusions on the basis of these results.

CONCLUSIONS

The induction of kidney tubules in metanephric mesenchyme seems to require close contacts between the interacting tissues. A correlation

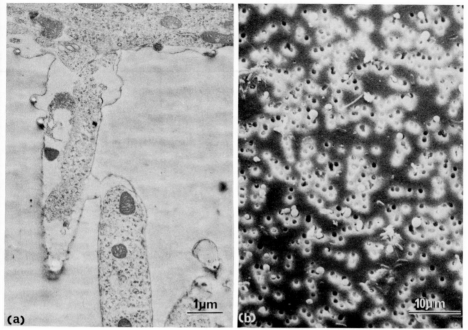

Fig. 2. Electron micrographs of 1.0 μm Nuclepore filters from cultures with 1.5 μM cyclo-heximide in the medium. (a) Transmission electron micrograph of explant with metanephric mesenchyme above the filter and spinal cord under the filter. (b) Scanning electron micrograph of filter on which spinal cord has been cultivated on opposite side.

exists between the degree of induction and the contact area between the tissues. Induction does not require DNA-synthesis or cell division in the inducing tissue, but this requirement exists for the expression of induction. Both inhibitors of RNA and protein synthesis prevent the inductive event from taking place. The inhibitors of RNA-synthesis seem to do so irreversibly, whereas the inhibitor of protein synthesis reversibly prevented induction. A reversible inhibition of induction was also obtained by interfering with glycosaminoglycan synthesis using the glutamine analogue DON. These inhibitors do not seem to interfere with the establishment of contacts.

ACKNOWLEDGEMENTS

The work reviewed in this paper has been supported by the Sigrid Jusé-lius Foundation and the Academy of Finland.

REFERENCES

Bautzmann, H., Holtfreter, J., Spemann, H. and Mangold, O. Versuche zur Analyse der Induktionsmittel in der Embryonalentwicklung. *Naturwissenschaften* **20**, 971–974 (1932).

Crick, F. Diffusion in embryogenesis. *Nature (Lond.)* **225**, 420–422 (1970).

Green, R. M. and Pratt, R. M. Inhibition by diazo-oxo-norleucine (DON) of rat palatal glycoprotein synthesis and epithelial cell adhesion *in vitro*. *J. Cell Biol.* in press.

Grobstein, C. Morphogenetic interaction between embryonic mouse tissues separated by a membrane filter. *Nature (Lond.)* **172**, 869–871 (1953).

Grobstein, C. Inductive interaction in the development of the mouse metanephros. *J. exp. Zool.* **130**, 319–340 (1955a).

Grobstein, C. Tissue interaction in the morphogenesis of mouse embryonic rudiments in vitro, in D. Rudnick (ed.), Aspects of Synthesis and Order in Growth, pp. 233–256, Princeton Univ. Press, Princeton (1955b).

Grobstein, C. Transfilter induction of tubules in mouse metanephrogenic mesenchyme. *Exp. Cell Res.* **10**, 424–440 (1956).

Grobstein, C. Some transmission characteristics of the tubule-inducing influence on mouse metanephrogenic mesenchyme. *Exp. Cell Res.* **13**, 575-587 (1957).

Grobstein, C. and Dalton, A. J. Kidney tubule induction in mouse metanephrogenic mesenchyme without cytoplasmatic contact. *J. exp. Zool.* **135**, 57–73 (1957).

Holtfreter, J. Studies on the diffusibility, toxicity and athogenic properties of "inductive" agents derived from dead tissues. *Exp. Cell Res.* Suppl. **3**, 188–209 (1955).

Lehtonen, E., Wartiovaara, J., Nordling, S. and Saxén, L. Demonstration of cytoplasmic processes in Millipore filters permitting kidney tubule induction. *J. Embryol. exp. Morph.* **33**, 187–203 (1975).

Mangold, O. Ist das Induktionsmitteldiffusionsfahig? In H. Bautzmann, J. Holtfreter, H. Spemann and O. Mangold (ed.), Versuche zur Analyse der Induktionsmittel in Embryonalentwicklung. *Naturwissenschaften* **20**, 974 (1932).

McKeehan, M. S. Cytological aspects of embryonic lens induction in the chick. *J. exp. Zool.* **117**, 31–64 (1951).

Meier, S. and Hay, E. D. Stimulation of corneal differentiation by interaction between cell surface and extracellular matrix. I. Morphometric analysis of transfilter "induction". *J. Cell Biol.* **66**, 275–291 (1975).

Nordling, S., Miettinen, H., Wartiovaara, J. and Saxén, L. Transmission and spread of embryonic induction. I. Temporal relationships in transfilter induction of kidney tubules *in vitro*. *J. Embryol. exp. Morph.* **26**, 231–252 (1971).

Nyholm, M., Saxén, L., Toivonen, S. and Vainio, T. Electron microscopy of transfilter neural induction. *Exp. Cell Res.* **28**, 209–212 (1962).

Saxén, L. Transfilter neural induction of Amphibian ectoderm. *Develop. Biol.* **3**, 140–152 (1961).

Saxén, L., Lehtonen, E., Karkinen-Jääskeläinen, M., Nordling, S. and Wartiovaara, J. Are morphogenetic tissue interactions mediated by transmissible signal substances or through cell contacts? *Nature (Lond.)* **259**, 622–663 (1976).

Spemann, H. Uber Linsenbildung nach experimenteller Entfernung der primären Linsenbildungszellen. *Zool. Anzeiger* **28**, 419–432 (1905).

Spemann, H. Zur Entwicklung des Wirbeltierauges. *Zool. Jb. Abt. Allgem. Zool. Physiol. Tiere* **32**, 1–98 (1912).

Toivonen, S., Tarin, D., Saxén, L., Tarin, P. J. and Wartiovaara, J. Transfilter studies on neural induction in the newt. *Differentiation* **4**, 1–7 (1975).

Wartiovaara, J., Lehtonen, E., Nordling, S. and Saxén, L. Do membrane filters prevent cell contacts? *Nature (Lond.)* **238**, 407–408 (1972).

Wartiovaara, J., Nordling, S., Lehtonen, E. and Saxén, L. Transfilter induction of kidney tubules; Correlation with cytoplasmic penetration into Nucleopore filters. *J. Embryol. exp. Morph.* **31**, 667–682 (1974).

Stimulation of Chondrogenic Differentiation with Extracellular Matrix Components: An Analysis using Scanning Electron Microscopy

James W. Lash, Elizabeth Belsky and N. S. Vasan

Department of Anatomy, School of Medicine G3, University of Pennsylvania, Philadelphia, Pennsylvania 19104, USA

INTRODUCTION

The use of porous filters has provided significant information concerning the nature of tissue interactions. Millipore filters have been used to place artificial barriers between interacting tissues, and it has now been accepted that the tissue masses need not be directly apposed to one another for regulative interactions to take place (Grobstein, 1957; Lash *et al.*, 1957; Wartiovaara *et al.*, 1974; Meier and Hay, 1975; Nordling *et al.*, this volume). There is a direct correlation in that the thicker the filter, or the smaller the pore size, the weaker is the stimulation received by the responding tissue (Grobstein, 1956; Wartiovaara *et al.*, 1974; Meier and Hay, 1975). Because of the tortuous channels which provide the porosity of the Millipore filter (Grobstein, 1956, 1957; Lehtonen *et al.*, 1975; Meier and Hay, 1975), it was never clearly established whether penetrating cellular processes make contact with similar processes from the tissues on the other side of the filter, even though the tissue masses are separated.

With the advent of Nuclepore filters, in which the pores are straight channels through 10 μm thick polycarbonate discs (Porter, 1974; Wartiovaara *et al.*, 1974; Meier and Hay, 1975), evidence is mounting that contact between cell processes does occur, and indeed, most tissue interactions may require such contact (Wartiovaara *et al.*, 1974; Slavkin and Bringas, 1976; Nordling *et al.*, this volume). There is some difference of opinion as to a precise definition of cell contact (cf. Grobstein, 1975; Weiss, this volume), and the issue will be skirted here. Of more immediate concern is a closely related issue of whether molecular products of one tissue may be used to effect the differentiation of another itssue. There are occasional reports of "factors" which can evoke a differentiative response from another tissue, and there are increasing instances where identifiable and characterizable tissue products have been reported to be capable of promoting differentiation. The most notable instances (other than hormonal activity) are those involving the components of the extracellular matrix (Kosher *et al.*, 1973; Meier and Hay, 1974, 1975; Lash and Kosher, 1975; Kosher and Church, 1975; Galbraith and Kollar, 1976). The implications of these experiments are that cell contact is not a prerequisite for induction, and that cell and tissue interactions may in some instances be mediated by certain classes of molecules.

Considerable work has been done investigating the tissue interactions between the embryonic notochord and somites during somite chondrogenesis (Lash, 1963, 1968). It has recently been established that the embryonic notochord produces collagen Type II (Linsenmayer *et al.*, 1973; Miller and Mathews, 1974) and proteoglycans (Kosher and Lash, 1975; Lash and Vasan, 1977), and that these molecules may be implicated in the promotion of somite chondrogenesis (Kosher *et al.*, 1973; Kosher and Lash, 1975; Kosher and Church, 1975). Both proteoglycans and collagen have been shown to effectively stimulate somite explants to form cartilage (Kosher *et al.*, 1973; Kosher and Church, 1975; Lash and Vasan, 1977).

FILTER EXPERIMENTS

In attempts to correlate the penetration of cell processes through Nuclepore filters with chondrogenic stimulation by exogenous proteoglycans in the medium beneath the filter, it was noticed in preparations fixed for scanning electron microscopy (SEM) that extracellular matrix products could be observed forming on the underside of the Nuclepore filter. Thus scanning electron microscopy provided a method whereby the normal deposition of cartilage matrix could be observed as well as deposition occurring after stimulation with proteoglycans.

Previous work has shown that proteoglycans produce a marked biochemical response in explanted somites with respect to the synthesis of new proteoglycans, as well as an increase in the amount of cartilage seen histologically (Kosher *et al.*, 1973). By exploiting the unusual deposition of cartilage matrix on the underside of the Nuclepore filter, it was possible to analyse the deposition of proteoglycans and collagen, and their eventual amalgamation into a hyaline cartilage matrix.

Somites were dissected from stage 17 chick embryos ($2\frac{1}{2}$ days of incubation, staging series of Hamburger and Hamilton, 1951), according to methods previously described (Gordon and Lash, 1974). Ten to fifteen somites were clustered upon the centre of a 2 mm² piece of Nuclepore filter (0.8 μm pore size), "rough" side up (cf. Wartiovaara et al., 1974). The filter rested upon a Nitex grid (656 μm mesh) which in turn floated upon 1.0 ml of nutrient medium in a Falcon Organ Culture dish. The nutrient medium used was F12X (Marzullo and Lash, 1970). After various intervals in culture, the filter containing the explant was removed and fixed in Karnovsky's fixative, post-fixed in osmium, stained with uranyl acetate, and coated with gold. Preparations were observed in a model JSM-U-3 Jeol Scanning Electron Microscope. Biochemical analyses of glycosaminoglycan synthesis were determined according to the methods published by Kosher et al., 1973. (Details of the methods are published elsewhere by Belsky et al., 1977.)

Matrix production

Non-stimulated somites

Observations at 24, 48, 60 and 72 h of culture create the following pattern of matrix deposition by non-stimulated somites. After 24 h of culture, the under surface of the Nuclepore filter shows cell processes emerging, and in the vicinity of the processes fibres and sparse granules are seen (Fig. 1). On occasions, the fibres show indications of periodicity. The periodicity, created by granules of proteoglycans adhering to collagen fibres, is approximately 0.1 μm (1000 Å) (Belsky et al., 1977). This figure includes the gold coating of approximately 100–200 Å thickness. As the age of the culture increases, the fibres increase greatly in numbers, and the number of granules also increases (Figs 2 and 3). After $2\frac{1}{2}$ days in culture, the fibres and granules become transformed into an amorphous matrix (Fig. 4). In some instances, discrete transitions can be seen between the fibrous-granular regions and the amorphous matrix (Fig. 5).

Stimulated somites

When proteoglycan aggregate (PGA), extracted from 13-day embryonic chick sterna (Lash and Vasan, 1977), is added to the nutrient medium (200 μg/ml), striking differences are seen in the pattern of matrix deposition. After 24 h in culture, fibres and granules are seen in greater abundance in the vicinity of the cell processes (Fig. 6).

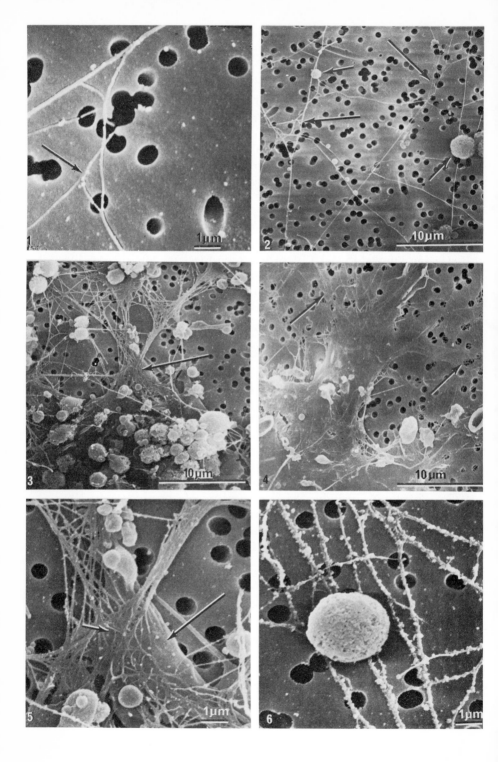

The same transition occurs in the matrix of these explants as in the control explants with one important exception; the fibres and granules are more numerous (Figs 7, 8 and 9). Thus the addition of PGA to the medium enhances greatly the rate at which fibres and granules appear. This enhancement is in agreement with the results reported by Kosher *et al.* (1973) in their analysis of GAG synthesis by "chondro-mucoprotein" (PGA) stimulated explants of somites on nutrient agar. Biochemical analyses of the sulphated GAG production in comparable explants are given in Table I.

TABLE I

Stimulation of GAG synthesis in Nuclepore filter explants of somites. The proteoglycan extract used in these experiments was a different batch from those used in similar cultures for SEM analysis. DPM/DNA ratios represent DPM of radioactive sulphate labelled GAG per ng of DNA (see Kosher *et al.*, 1973). The data are from duplicate cultures, representing comparable results obtained from 10 separate experiments

Explant	Age (days)	DPM/DNA
Control	3	0.30
Control	3	0.31
PGA-stimulated	3	0.71
PGA-stimulated	3	0.84

Fig. 1. Fibres secreted by 24 h control somite explant. Cell processes are not seen in this photograph. Sparse proteoglycan granules (arrow) can be seen on some of the fibres.

Fig. 2. Fibres secreted by 48 h control somite explant. Cell processes (short arrows) and an increased number of fibres (long arrows) are shown. Granules and periodicity barely discernible at this magnification.

Fig. 3. Matrix formation in 60 h control somite explant. An amorphous matrix is seen (arrow) forming in the vicinity of the cell processes.

Fig. 4. A sheet of amorphous matrix from a 72 h control culture. Fibrous components (arrows) are seen peripheral to the amorphous matrix.

Fig. 5. A higher magnification of a 60 h control culture. The transition from a fibrous and granular matrix (short arrow) to an amorphous matrix (long arrow) is apparent.

Fig. 6. Fibres and granules in the vicinity of a cell process in a 24 h PGA-stimulated somite explant. To be compared with Figs 1 and 2.

Enzymatic digestion of matrix

Although the SEM evidence indicated that cartilage matrix was being observed, it was necessary to obtain less equivocal evidence. Since cartilage matrix is composed primarily of collagen and proteoglycans, enzymatic digestion was used to determine whether the fibres and granules could be selectively removed. Selective digestion was performed using highly purified collagenase and chondroitinase ABC (methodology according to Daniel *et al.*, 1973). When the fibres and granules appearing in the young cultures were digested with chondroitinase the granules were removed (Figs 10 and 12). Subsequent digestion with collagenase removed the remaining fibres (Fig. 11). These results substantiate the interpretation that the fibres and granules seen in the SEM are collagen and proteoglycans, and that they represent cartilage matrix deposition. The selective removal of the matrix components is more striking when the amorphous matrix of the older explants is selectively digested with enzymes. The initial digestion with chondroitinase removes the proteoglycans, and exposes the fibrous collagenous network within the matrix (Fig. 12). Subsequent digestion with collagenase again removes the remaining fibrous material.

CONCLUSIONS AND DISCUSSION

As the result of a fortuitous observation of the under surface of Nuclepore filters containing somite explants, it was possible to study the deposition of cartilage matrix unencumbered by the tissue mass on the other side of the filter. A reconstruction of the events seen in cultures of various ages indicate that the collagen fibres appear first in the vicinity of the cell processes, and then deposition occurs at greater and greater distances from visible cell processes. The distance between fibres

Fig. 7. Fibres and granules secreted by a 48 h PGA-stimulated somite explant. The beaded appearance of the fibres is due to the proteoglycan granules.

Fig. 8. An abundance of fibres and granules are seen in a 60 h PGA-stimulated culture.

Fig. 9. A higher magnification of a 60 h PGA-stimulated culture. The proteoglycan granules almost mask the fibres.

Fig. 10. Chondroitinase digestion of a 60 h control explant. The fibrous components of the amorphous matrix become apparent after the removal of the proteoglycans.

Fig. 11. The subsequent digestion of a chondroitinase-treated culture (cf. Fig. 10) with collagenase removes practically all of the fibres and granules.

Fig. 12. A higher magnification of a somite explant digested with chondroitinase (cf. Fig. 10). The remaining fibrous elements are clearly seen.

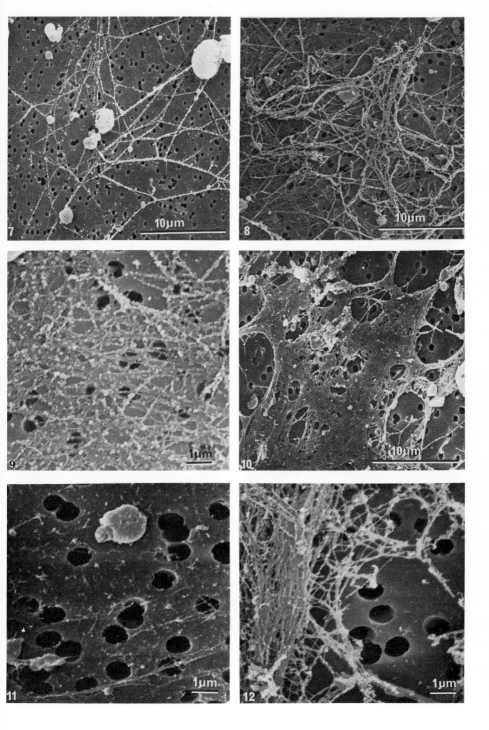

and visible processes may be misleading, since cell processes may be projecting into the nearby pores, but not emerging. The direct association of the proteoglycan granules with the cell processes is seen in cultures of all ages. The granules seem to appear first on, or at, the surface of the processes, and then on, or at, the surface of the fibres. The beaded appearance of the fibres is a constant feature of the developing matrix before it becomes amorphous. The proteoglycan granules are visibly more plentiful in the PGA stimulated somites, which is consonant with the fact that these explants are forming more cartilage.

Heretofore, assays on induced tissue differentiation have relied on rather indirect measurements of tissue-characteristic macromolecules which had been extracted from the induced tissue, or on more direct observations on tissue morphology, either at the light-microscope level or at the ultrastructural level. In the former instance, the heterogeneity of the tissue being assayed was a factor which was difficult to account for in the interpretation of some results, and in the latter instance, the area sampled was of necessity a very small portion of the tissue under study. With the SEM analyses of the under surface of Nuclepore filters containing somite explants, the observations show unequivocally some of the post-translational events postulated for cartilage matrix deposition. Both collagen and proteoglycans are known to be extruded from the cartilage-forming cells (Revel and Hay, 1963; Weinstock, 1975). It is also known that these two molecules become associated extracellularly, probably through electrostatic forces between the glycosaminoglycans of the proteoglycans and the collagen fibres (Oegema *et al.*, 1976). The fibres seen in the earliest cultures represent cross-linked fibres, as evidenced by their size.

Biochemical as well as histological observations have established conclusively that cartilage formation in somite explants is enhanced by association with notochordal tissue or molecular products which are similar to those produced by the notochord (Kosher *et al.*, 1973; Kosher and Lash, 1974). The fact that these molecular products alone can stimulate the somite cells strongly supports the contention that cell contact is not a prerequisite for an effective interaction between the inducing tissue (notochord) and the responding tissue (somite). This conclusion is supported by the ultrastructural analyses of Ruggeri (1972) and Minor (1973).

The sequence of events in the stimulation of chondrogenic cells can be established as approximating the following sequential events. The precartilaginous somite cells synthesize glycosaminoglycans at a low level (as are all other embryonic tissue, Abrahamsohn *et al.*, 1975; Solursh, 1976), and there is no accumulation of matrix products. As the somites develop, the precartilaginous cells of the sclerotomal region migrate medially towards the notochord. As the cells are migrating, the notochord is synthesizing collagen Type II and proteoglycans, and the cells become surrounded by a halo of these materials. As the result of events still unclear, the presence of the extracellular matrix products surrounding the sclerotomal cells stimulate them to synthesize matrix products at a faster rate, and to begin to accumulate these products as hyaline cartilage.

ACKNOWLEDGEMENT

This study was supported by NIH Research Grant HD-00380 and Training Grant GM-00281.

REFERENCES

Abrahamson, P. A., Lash, J. W., Kosher, R. A. and Minor, R. R. The ubiquitous occurrence of chondroitin sulfates in chick embryos. *J. exp. Zool.* **194**, 511–518 (1975).

Belsky, E., Lash, J. W. and Vasan, N. S. Scanning Electron Microscopy of Induced Cartilage Matrix Deposition by Embryonic Chick Somites. In preparation.

Daniel, J. C., Kosher, R. A., Lash, J. W. and Hertz, J. The synthesis of matrix components by chondrocytes *in vitro* in the presence of 5-Bromodeoxyuridine. *Cell Diff.* **2**, 285–298 (1973).

Galbraith, D. B. and Kollar, E. J. Procollagen enhancement of mouse tooth germ development *in vitro*. *Amer. Zool.* **16**, 183 (abst.) (1976).

Gordon, J. S. and Lash, J. W. *In vitro* chondrogenesis and differential cell viability. *Develop. Biol.* **36**, 88–104 (1974).

Grobstein, C. Trans-filter induction of tubules in mouse metanephrogenic mesenchyme. *Exp. Cell Res.* **10**, 424–440 (1956).

Grobstein, C. Comments in H. C. Slavkin and R. C. Gruelich (ed.), Extracellular Matrix Influences on Gene Expression, Academic Press, New York (1975).

Grobstein, C. and Dalton, A. J. Kidney tubule induction in mouse metanephrogenic mesenchyme without cytoplasmic contact. *J. exp. Zool* **135**, 57–73 (1957).

Hamburger, V. and Hamilton, H. L. A series of normal stages in the development of the chick embryo. *J. Morph.* **88**, 49–92 (1951).

Kosher, R. A. and Church, R. L. Stimulation of *in vitro* somite chondrogenesis by procollagen and collagen. *Nature* (*Lond.*) **258**, 327–330 (1975).

Kosher, R. A. and Lash, J. W. Notochordal stimulation of *in vitro* somite chondrogenesis before and after enzymatic removal of perinotochordal materials. *Develop. Biol.* **42**, 362–378 (1975).

Kosher, R. A., Lash. J. W. and Minor, R. R. Environmental enhancement of *in vitro* chondrogenesis. IV. Stimulation of somite chondrogenesis by exogenous chondromucoprotein. *Develop. Biol.* **35**, 210–220 (1973).

Lash, J. W. Tissue interaction and specific metabolic responses: Chondrogenic induction and differentiation, in M. Locke (ed.), Cytodifferentiation and Macromolecular Synthesis. pp. 235–260, Academic Press, New York (1963).

Lash, J. W. Somitic mesenchyme and its response to cartilage induction, in R. Fleisch-majer and R. F. Billingham (ed.), Epithelial-Mesenchymal Interactions, pp. 165–172, Williams & Wilkins Co., Baltimore, Md (1968).

Lash, J. W., Holtzer, S. and Holtzer, H. An experimental analysis of the development of the spinal column. IV. Aspects of cartilage induction. *Exp. Cell Res.* **13**, 292–303 (1957).

Lash, J. W. and Vasan, N. S. Tissue interactions and extracellular components, in J. W. Lash and M. Burger (ed.), Cell and Tissue Interactions, Raven Press, New York, in press (1977).

Lehtonen, E., Wartiovaara, J., Nordling, S. and Saxén, L. Demonstration of cytoplasmic processes in Millipore filters permitting kidney tubule induction. *J. Embryol. exp. Morph.* **33**, 187–203 (1975).

Linsenmayer, T. F., Trelstad, R. L. and Gross, J. The collagen of chick embryonic notochord. *Biochem. biophys. Res. Commun.* **53**, 39–45 (1973).

Marzullo, G. and Lash, J. W. Control of phenotypic expression in cultured chondrocytes: Investigations on the mechanism. *Develop. Biol.* **22**, 638–654 (1070).

Meier, S. and Hay, E. D. Stimulation of extracellular matrix synthesis in the developing cornea by glycosaminoglycans. *Proc. nat. Acad. Sci (Wash.)* **71**, 2310–2313 (1974).

Meier, S. and Hay, E. D. Stimulation of corneal differentiation by interaction between cell surface and extracellular matrix. I. Morphometric analysis of transfilter "induction". *J. Cell Biol.* **66**, 275–291 (1975).

Miller, E. J. and Mathews, M. B. Characterization of notochord collagen as a cartilage-type collagen. *Biochem. biophys. Res. Commun.* **60**, 424–430 (1974).

Minor, R. R. Somite chondrogenesis: A structural analysis. *J. Cell Biol.* **56**, 27–50 (1973).

Oegema, Jr, T. R., Laidlaw, J., Hascall, V. C. and Dziewiatkowski, D. D. The effect of proteoglycans on the formation of fibrils from collagen solution. *Arch. Biochem. Biophys.* **170**, 698–709 (1976).

Porter, M. C. A novel membrane filter for the laboratory. *Amer. Lab.* **6**, 63–76 (1974).

Revel, J. P. and Hay, E. D. An autoradiographic and electron microscopic study of collagen synthesis in differentiating cartilage. *Z. Zellforsch.* **61**, 110–144 (1963).

Ruggeri, A. Ultrastructural, histochemical, and autoradiographic studies on the developing chick notochord. *Z. Anat. Entwickl. Gesch,* **138**, 20–33 (1972).

Slavkin, H. C. and Bringas, Jr, P. Epithelial-mesenchyme interactions during odontogenesis. IV. Morphological evidence for direct heterotypic cell–cell contacts. *Develop. Biol.* **50**, 428–442 (1976).

Solursh, M. Glycosaminoglycan synthesis in the chick gastrula. *Develop. Biol.* **50**, 525–530 (1976).

Wartiovaara, J., Nordling, S., Lehtonen, E. and Saxén, L. Transfilter induction of kidney tubules: correlation with cytoplasmic penetration into Nucleopore filters. *J. Embryol. exp. Morph.* **31**, 667–687 (1974).

Weinstock, M. Elaboration of precursor collagen by osteoblasts as visualized by autoradiography after ^{3}H-proline administration, in H. C. Slavkin and R. C. Greulich (ed.), Extracellular Matrix Influences on Gene Expression, pp. 119–128, Academic Press, New York (1975).

IV
Molecular Mechanisms of Cell Contact Interactions

Molecular Mechanisms of Cell Contact Interactions: An Introduction

Leonard Weiss

Department of Experimental Pathology, Roswell Park Memorial Institute, Buffalo, New York 14263, USA

A variety of cell interactions leading to differentiation are documented in this volume. As discussed by Saxén (p. 148), some of these occur over thousands of nanometres, whereas others involve cell contact. In this section, some aspects of these interactions will be considered from the submolecular to the ultrastructural levels.

Intimacy of contact often implies cell adhesion. In practical terms, adhesion as recognized by biologists involves the formation of various adhesive bonds between interacting species which, from well-accepted physical principles, cannot be further apart than approximately 0.5 nm (Weiss and Harlos, 1971). Therefore, at first sight the physical definition of contact in a given cell system appears simple. It might be concluded that by means of electron microscopy, but not by light microscopy, examination of interacting cells would unambiguously reveal whether or not appropriate regions of their surfaces were separated by distances of approximately 0.5 nm or less, in which case they could be operationally defined as being in contact. If the distances were appreciably greater than 0.5 nm, then contact by this definition would not have taken place. Thus, according to this approach, the problem of associating differentiation or any other cellular reaction with contact, revolves around the accurate assessment of the shortest distance between the interacting cells.

In early electron micrographs, the measurement of intercellular distances was unambiguous, since the cell surface membranes were visualized as electron-dense, proteolipid, "75 Å units", with crisp, sharply defined surfaces. Since that time, electrokinetic and chemical studies, in addition to major advances in electron-microscopic techniques, have demonstrated that the use of the term "surface" in connection with cell contact phenomena is misleading in the sense that a "surface" is a two-dimensional, planar structure, whereas the *cell periphery* is a three-dimensional structure, the limits of which are perhaps best defined in terms of the interests of the observer. Thus, if one's interests lie in average interactions between charged groups in the cell periphery, then the appropriate distances are those between the plane of interacting charges of interacting cells. As many of these charges are often associated with carbohydrate components of proteoglycans projecting out from the lipid matrix of the membrane—in the human erythrocyte for example, some 70% of the surface charge is due to sialic acids—it will be appreciated that distances measured between these ionogenic moieties will be considerably less than those between the comparatively spatially discrete membrane matrices. Therefore, in terms of the critical distance of 0.5 nm with respect to contact, it can be seen if in the identical situation we were to measure the distance between the planes of the membrane matrices we might assume that contact had *not* occurred, whereas measurements taken between the planes of the charges might well indicate that it had.

From what has been said, it follows that any physical description of cell interactions involving parts of their peripheries requires not a knowledge of the gross physicochemical properties of their components, but also a description of their spatial location. In addition there are good indications that the nature of the cell periphery dynamically reflects both intracellular and environmental changes. To further complicate the issue, some cell interactions undoubtedly involve trigger-mechanisms in which quantitatively minor components of the cell peripheries might well have critical functional roles. Thus, macroscopic analyses, physical and chemical, are possibly not only unhelpful, but also misleading. At present, we can give with certainty neither the detailed physicochemical descriptions of the molecules involved in cell contact, nor their mode of action.

In the first part of the following section, I shall give a somewhat personal view of contact-making, contact-breaking and the relation of these phenomena to active cell movement. Grossberg will then consider the problem of specificity in physicochemical terms. Although the work described elsewhere in this volume by Moscona (p. 353) and Burger (pp. 357–374) indicates a very useful start on the chemical characterization of materials involved in the specific aggregation of cells, insufficient data are currently available to discuss the problem in physicochemical detail. Grossberg will therefore use as examples of specific interactions the well-documented work by himself and his colleagues on hapten-antibody interactions.

In tissues, the perplexing question of where the cell ends and the intercellular matrix begins is as difficult to answer satisfactorily as the question of original sin! Vaheri will give us some insight into the former question by discussing "fibronectin", a major cell surface-associated protein which is also a matrix component associated with basement membranes, which of course form natural boundaries. Fibronectin, interestingly enough, may act as an organizing protein during mouse and chicken development, although convincing evidence for this is lacking.

When cells make contact and consequentially differentiate, then information of some sort must pass between them. The primary information may initially be confined in its action to the cell surface. Alternatively but not exclusively, information may pass from the inside of one cell directly into the inside of another, via intercellular gap junctions. Gilula will discuss some of the remarkable progress made in this field in his analysis of junctional complexes from both the ultrastructural and functional viewpoints.

In vitro interactions taking place over distances of thousands of nanometres may result in contact-independent differentiation (Saxén, p. 148). *In vivo* differentiation is a specific event with respect to type, time and place. As diffusion over very great distances *in vivo* would result in large volumes of "non-target" tissues receiving potential morphogenetic signals, it follows that specificity in response would most readily be accomplished by specificity in either receptors on the "targets" or signal-destruction on the "non-targets". Either of these mechanisms would surely require pre-existing differentiation at the time the message was received. Wolpert (this volume, p. 85) discusses more sophisticated approaches to the diffusion problem involving double-gradients, which are more feasible and which are compatible with experimental data in hydra and certain insects. Pictet and Rutter will discuss cell-free, transmissible factors derived from mesenchymal cells, which control proliferation and cytodifferentiation in the developing pancreas.

REFERENCE

Weiss, L. and Harlos, J. P. Short-term interactions between cell surfaces. *Progr. Surface Sci.* **1**, 355–405 (1971).

Some Biophysical Aspects of Cell Contact, Detachment and Movement

Leonard Weiss

Department of Experimental Pathology, Roswell Park Memorial Institute,
Buffalo, New York 14263, USA

INTRODUCTION

When contact phenomena in general are analysed in terms of the structure of the cell periphery, emphasis often appears to be placed on somewhat static and average differences and/or similarities between cells of different types. However, when insight into mechanisms is required as distinct from anecdotal descriptions, or when extrapolation from experimental systems to a natural environment is sought, this former approach is lacking because it overlooks the essentially dynamic nature of the cell, and the possibility that at critical periods of time, the physicochemical and functional properties of macroscopic and microscopic regions of interacting cell peripheries may differ from or resemble each other. In the following discussion, I shall focus on some dynamic and microscopic aspects of cell contact and cell detachment and their relationship to movement, from a biophysical viewpoint.

CELL CONTACT AND ADHESION

In order to form adhesive bonds, cell surfaces or molecules attached to them must approach within approximately 0.5 nm of each other. Although in the case of "specific" adhesions the nature of the bonds formed may well be of major importance, in the case of adhesion in relation to cell movement, for example, such specificity seems an unlikely requirement, and if contact over an adequate area can be achieved, then any level of adhesive strength can be accomplished in biological context, by the formation of different numbers of a whole gamut of bonds of different types, ranging from comparatively weak hydrogen bonds to the high energy covalent bonds. Therefore, a major general problem is not only how cells form specific or non-specific ad- hesions, but also how they approach closely enough together to permit such adhesive bonds to be formed. I shall deal with this problem of close approach, and Grossberg (p. 291) has provided examples of physi- cochemical specificity.

All mammalian cells so far examined carry a net negative surface charge; contact between cells tends to be prevented by the mutual elec- trostatic repulsion between them. In colloid systems, whether or not contact between charged particles and subsequent flocculation occurs depends in general terms on the balance of the electrostatic forces of repulsion and the Van der Waals' forces of attraction, in relation to Brownian or other movements.

In computing average electrostatic interactions between cellular and other surfaces, the key experimental measurement is that of electro- phoretic mobility. It should be emphasized that this latter property applies to the cellular *elecktrokinetic surface*, which lies in the plane between the suspending fluid moving with the cell, and that remaining with the bulk phase of the solution through which the cell migrates in the electrophoresis apparatus, along a voltage gradient.

As the distance between surfaces gets less than approximately 2 nm, equations become progressively more complex and inaccurate for calculation of potential energies of repulsion, because of overlap of the compact parts of the ionic "double-layer" surrounding the surfaces, and partly because bulk-phase values for the dielectric constant of the medium become less relevant to molecules intimately related to these surfaces, such as water structures.

Van der Waals' attractive interactions occur between permanent dipoles, permanent and induced dipoles, and between induced dipoles. The latter are also known as London dispersion forces, and appear to be quantitatively the most important in biological materials. In the "microscopic" approach (Hamaker, 1937), the sum of the dispersion interactions between constituent molecules of the interacting bodies are taken into account, as are the molecular properties of the medium between them, in terms of their polarizibilities, densities and absorption frequencies. For distances of less than 10 nm, where orientation forces are small, this approach appears accurate enough for the present pur- poses (Nir, 1975).

A key question is whether use of the theories outlined above actually allow predictions to be made on the adhesiveness or non-adhesiveness of specified cells for specified surfaces. Some insight into this question was gained from studies of the adhesion of two cell types to protein-coated coverslips *in vitro*. D-cells derived from a human osteogenic sarcoma adhered to the coverslips within minutes. In contrast, M-cells derived from the murine P-815 mastocytoma did not adhere to the coverslips at all, over the course of days in stationary culture. In terms of adhesion, therefore, the differences between the D- and M-cells were gross.

The zeta-potentials for the D- and M-cells were −14.7 and −14.2 millivolts respectively. Making similar assumptions in enumerating the other parameters for both cell types, it was computed (Fig. 1) that insurmountable potential energy barriers of approximately 3×10^{-11} ergs prevent *both* types from making direct contact with the coverslips (Weiss and Harlos, 1971). It is well documented that, with recognized qualifications, inert particles do follow the predictions of colloid theory, and electrostatic and Van der Waals' interactions occur between all

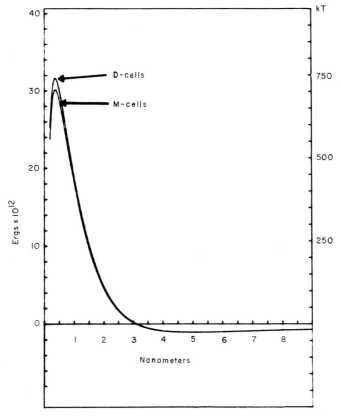

Fig. 1. The computed interaction energies between non-adhesive M-cells or adhesive D-cells, and protein-coated coverslips *in vitro*.

charged materials, living and non-living. It appears reasonable at present, therefore, to conclude that colloid theory more or less correctly presents the problem, which is how cells can make contact with net negatively charged surfaces, cellular or non-cellular, in spite of apparently insurmountable potential energy barriers. However, this particular direct approach does not provide the answers.

Computations of the type discussed above reflect the *average* situation since, from the viewpoint of electrostatic repulsion, the use of surface potentials derived from measurements of cell electrophoretic mobility reflect the *average* net electrical properties of the cell periphery near to the slip-plane generated as a cell moves through the electrophoresis apparatus. It was therefore suggested that if filopodial projections from one cell made contact with lower-than-average charge density regions of another, then the paradox of contact in spite of the high *average* potential energy would be solved, since the average barrier would be irrelevant at the very localized sites of contact which occupy only a small proportion of total cell surface area.

The possibility that negatively charged regions of one cell make contact with positively charged regions of another has also been explored by reacting cells and/or their substrata with non-lethal reagents for amino and other cationic cites. The results indicate that in both the case of cancer cells adhering to protein-coated dishes (Weiss, 1974a) and reaggregating chick neural retina cells (Maslow and Weiss, 1976), such treatment indeed reduces the rate of cell adhesion, but only by small amounts. Therefore although surface cationic sites play a demonstrable role in cell adhesion, it appears to be a minor one.

Attempts have been made by my colleagues and myself to study the patterns of negative charges at the cell periphery by marking them with electron-dense, positively charged colloidal iron hydroxide (CIH) and other particles, and determining their densities and distributions in sections of known thickness under the electron microscope. Parallel studies with cell electrophoresis have revealed that in some cells each CIH-particle, which has a diameter of approximately 10 nm, corresponds to a cluster of 20 to 30 negative charges, at the cellular electrokinetic surface. Estimates of this type are subject to serious reservations (e.g. Weiss and Subjeck, 1974a).

Computer reconstructions of the particle distribution patterns seen in thin sections reveal the presence of gaps which are large enough to accommodate the tips of contacting filopods (Weiss, 1973). Charge heterogeneity on this scale resembles the heterogeneity seen in antigenic determinant surface binding sites for example, although our procedures reveal no analogies to capping.

Sub-confluent cultures of EAT cells, which appeared to be making contact with each other by means of filopodial projections when viewed by phase-contrast microscopy, were also examined by electron microscopy after reaction with CIH. Potential contact situations were examined by projecting the axes of "microvilli" to the opposing surfaces of other cells, and enumerating the particle densities with respect to these intercepts. The particle densities in the immeidate region of the intercepts were significantly lower than in the surrounding regions (Weiss and Subjeck, 1974b). This alignment phenomenon has been observed when the apparent surfaces of filopods and the opposing low density regions are separated by more than 200 nm. Formal calculations of the range of electrostatic and Van der Waals' interactions ruled out direct surface interactions as a cause of alignment at these distances, and the frequency of the phenomenon ($>85\%$ of cases) ruled out an entirely random process. Therefore, by default, diffusion processes must be considered in causing alignment (Weiss *et al.*, 1975).

In considering the diffusion of materials from cells, distinction is often made between loosely bound peripheral constituents which may be readily released, and exudates of more immediate cellular origin. Material released from a low charge density region of one cell could guide or trap an exploring filopod from another; alternatively material released from the region of a filopod could in some way either seek out a pre-existing region of low charge density or make one. It is relevant to this hypothesis that a close correlation has been observed by Maslow and Weiss (1972) between the adhesion of cells to cellular monolayers and the release of proteinaceous materials from them. In addition, following reaction with ricin-ferritin conjugates which react with β-D-Gal and structurally similar residues, amorphous material was seen bridging the gap between aligned regions in approximately 30% of the cases examined. The failure to demonstrate this material in the majority of cases is taken to reflect the ease with which it is removed during washing and other processing for electron microscopy, and indicates its diffusible nature and/or its weak attachment to the cell surface. In the case of erythrocytes, a wide variety of charged and neutral, elongated macromolecules, generally having molecular weights in excess of 40 000 daltons, will form intercellular bridges and induce aggregation (Hardwicke and Squire, 1952; Chien and Jan, 1973). Moscona deals with this ligand problem elsewhere in this volume (p. 354); but this type of intercellular linkage is surely initial and does not constitute the more intimate type of junction recognized by the electron microscopists.

Computations have also established the feasibility of negatively charged macromolecules originating near to or from filopods, displacing mobile charge clusters on responding cells by electrostatic repulsion (Weiss *et al.*, 1975). The ability of antigenic-determinant sites to move in the liquid lipid membrane matrix is well recognized, and it would not be surprising if charged groups also possessed this dynamic property. Thus, material diffusing from one cell to another could modify the surface of the approached cell, making contact energetically more likely.

To sum up, it seems indisputable that potential energy and viscous barriers tend to prevent cells from making contact with, and adhering to, other cell surfaces. It is suggested that cells make contact in spite of these apparently insurmountable *average* barriers by making contact through comparatively small, non-average, low charge density, surface regions—in other words, *the essence of contact is microheterogeneity.*

CELL SEPARATION

Cell separation is not the reverse of cell adhesion. The demonstration that cells detached from substrata, such as protein-coated glass, leave material behind was taken as a clear indication that cell separation was not simply the reverse of cell adhesion, but that in this situation, non-lethal cohesive failure occurred within the cell periphery (Weiss, 1961). This residual material was a glycoprotein as evidenced by its species-specific antigenic nature and was left behind when cells broke contact with solid substrate either by passive detachment (Weiss and Coombs, 1963) or by virtue of their active movements (Weiss and Lachmann, 1964).

Any structure breaks at its weakest point. If, as argued earlier, cells make contact with surfaces to which they adhere by means of small adhesive "points", and by virtue of diffusible macromolecules, then we are simply stating that the adhesion forces are greater at the interface between the "points" and/or the diffusing molecules and the substratum to which the cell is attached, than the cohesive forces holding the "points" of the diffusing molecules together. If the attachment regions tend to spread out on their substrata, then the force per unit area applied at these adhesive interfaces will be less than that applied over the attachment process at its smallest cross-sectional area, and the stress concentration will favour breakages in this region. Other factors are discussed at length elsewhere (Weiss, 1967).

DYNAMIC ASPECTS OF CELL DETACHMENT

Increases in cell detachment may accompany growth and mitotic processes; thus, *in vitro*, increasing growth-rate facilitates cell detachment from solid surfaces and it is a common observation that cells grown attached to coverslips detach during the rounding-up process associated with metaphase. Other factors often associated with growth also facilitate the separation of cells from solid surfaces, including lysosome-activation, increased endocytosis (Weiss, 1967) and changes in membrane permeability (Weiss, 1974b). In addition, changes associated with degeneration such as lysosome-activation (Weiss, 1967) and decreased metabolism (Weiss, 1974b) also promote cell detachment. In this respect it should be remembered that lysosomal activation represents a final common pathway for many pathobiologic processes.

An attempt has been made to observe the effects of the growth processes *per se* in tissues, by studying parenchymal cell detachment from

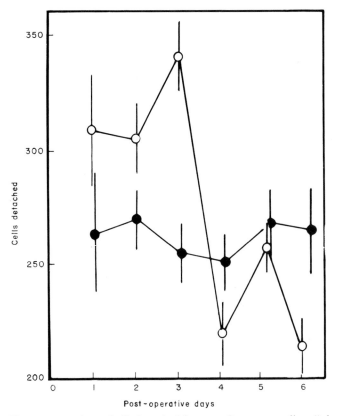

Fig. 2. The mean numbers of cells detached from two 2 mm mouse liver "plugs" ($\pm$S.E.) at indicated times after partial hepatectomy. Each point represents 30 observations on sham-operated (●——●) or partially hepatectomized (○——○) animals.

regenerating mouse liver following 50% hepatectomy, where regeneration was maximal over the second to fourth post-operative days. Very accurately sized cylinders of liver were removed at the time of the chemical determinations, mechanically shaken under very reproducible conditions, and the numbers of parenchymal cells released were determined. The detachment data summarized in Fig. 2 show that, compared with a sham operated group, significantly ($p < 0.001$) more parenchymal cells were released from partially hepatectomized animals on the second and third post-operative days.

An indication that regeneration is different from generation, with respect to facilitation of cell detachment, comes from studies on the liberation of parenchymal cells from the liver of early and late embryonic, neonatal and adult rats, where no significant differences in detachment were detected (Table I).

TABLE I

Release of cells from rat livers at different stages of development

Age	No. released × 1000	Volume function
Foetal	190 + 11 (113)	1.91
Neonatal	216 + 17 (78)	2.05
Adult	350 + 16 (84)	3.19

Mean numbers ($\pm$ standard error; (Numbers of observations)) of parenchymal cells and multicellular aggregates shaken free from standard-size cylinders of rat liver.
Volume function = (mean radius in μm)^3 × mean number × 10^{-9}.

A good example of effects of external factors on cell detachment comes from studies of Walker adenocarcinomata transplanted into the livers of rats. It was observed that significantly more parenchymal cells were shaken free of liver plugs taken adjacent to the tumour edge than 0.5 or 1 cm out from it. Presumably this effect was due to some product of the tumour diffusing out into the surrounding liver (Table II).

The conclusions that I draw from these data are firstly that cell detachment is not simply the reverse of cell adhesion; secondly, that the ease with which a cell detaches from a substratum is a dynamic property related to other aspects of cell activity and environment; and thirdly, that it is feasible to assume that as much complexity is associated with cell detachment as with adhesion.

TABLE II

Release of liver parenchymal cells as function of
distance from tumour edge

Distance from tumour edge	Volume ratio (cf. 0 cm)
0 cm	1.0
0.5 cm	0.8
1.0 cm	0.5

Ratios of volume functions of parenchymal cells
shaken free from standard cylinders obtained from
normal liver surrounding transplanted Walker 256
tumours, at various distances from the tumour
edge. Differences significant at 1% level.

CELL ATTACHMENT, DETACHMENT AND MOVEMENT

In order to move actively over or through tissues, or indeed any substratum, a cell must:

(1) Make contact with and adhere to the substratum, to obtain a "hold" against which it can "push", or "pull", or both.
(2) Be capable of generating the necessary locomotor energy to propel itself.
(3) Break contact with or detach from regions of adhesion with its substratum.

It is a common observation that on non-adhesive substrata, or in the presence of enzymes, e.g. trypsin, which prevent the establishment of cell contacts, active movements of cells across solid surfaces do not occur. However, the role of cell detachment in cell movements has not been clearly defined, and the mere observation of the absence of movement in normally motile cells does not it itself indicate which of the three components of movement are affected.

Attempts to define and quantify the detachment-dependent arm of locomotion in macrophages have been described by Weiss and Glaves (1974). Materials produced by lymphoid cells (MIF; Remold and David, 1971) under defined conditions inhibit the migration of macrophages *in vitro*. MIF was also added to 18 h cultures of macrophages which were adherent to glass; after 60 min exposure, the cells were exposed to shearing pressures of approximately $8\,\mathrm{N.m^{-2}}$, transmitted through the fluid covering them for 2 min, and the percentages detached were determined. Significant inhibition of detachment was observed following exposure to MIF, compared with appropriate controls (Table III). This macrophage-detachment test provides evidence for the detachment-dependent arm of locomotion, and an example of factors acting on it. It should perhaps be emphasized that these latter experiments yield no information on the effects of MIF on macrophage locomotor forces and/or adhesion, which are also expected to partially regulate migration.

TABLE III

Effects of supernatant fluids derived from lymphocytes exposed to PPD, from BCG-sensitized and control mice on guinea-pig macrophage migration and detachment

Lymphocyte supernatant source	% Migration	% Detachment
BCG-sensitized	78.5	59
Control	100	85
	$(0.05 > p > 0.01)$	$(p < 0.001)$

Effects of supernatant fluids derived from lymphocytes exposed to PPD, from BCG-sensitized and control mice on guinea-pig macrophage migration and detachment.

CONCLUSIONS

The evidence presented lends support to the initial suggestion that the ability of cells to adhere to, separate from and move over or through tissue is very variable within individual cell types, and further suggests that such cellular variations in response to intracellular and external events must be considered in addition to comparisons between different cell types.

ACKNOWLEDGEMENT

This study was partially supported by Grant PDT-14 from the American Cancer Society, Inc.

REFERENCES

Chien, S. and Jan, K. M. Ultrastructural basis of the mechanism of rouleaux formation. *Microvasc. Res.* **5**, 155–166 (1973).

Hamaker, H. C. The London–Van der Waals attraction between spherical particles. *Physica* **4**, 1058–1072 (1937).

Hardwicke, J. and Squire, J. R. The basis of the erythrocyte sedimentation rate. *Clin. Sci.* **11**, 333–355 (1952).

Maslow, D. E. and Weiss, L. Cell exudation and cell adhesion. *Exp. Cell Res.* **71**, 204–208 (1972).

Maslow, D. E. and Weiss, L. Some effects of positively charged surface groups on cell aggregation. *J. Cell Sci.* **21** (2), 219–225 (1976).

Nir, S. Long range intermolecular forces between macroscopic bodies. Macroscopic and microscopic approaches. *J. theor. Biol.* **53**, 83–100 (1975).

Remold, H. G. and David, J. R. Further studies on migration inhibitory factor (MIF): Evidence for its glycoprotein nature. *J. Immunol.* **107**, 1090–1098 (1971).

Weiss, L. Studies on cellular adhesion in tissue culture: IV. The alteration of substrata by cell surfaces. *Exp. Cell Res.* **25**, 504–517 (1961).

Weiss, L. The Cell Periphery, Metastasis, and Other Contact Phenomena, North-Holland Publishing Co., Amsterdam (1967).

Weiss, L. Biophysical aspects of metastasis: a personal viewpoint, in G. P. Murphy (ed.), Perspectives in Cancer Research and Treatment, pp. 387–398, Liss, New York (1973).

Weiss, L. Studies on cell adhesion in tissue culture. XIV. Positively charged surface groups and the rate of cell adhesion. *Exp. Cell Res.* **83**, 311–318 (1974a).

Weiss, L. Studies on cell adhesion in tissue culture. XV. Some effects of cycloheximide on cell detachment. *Exp. Cell Res.* **86**, 223–232 (1974b).

Weiss, L. and Coombs, R. R. A. The demonstration of rupture of cell surfaces by an immunological technique. *Exp. Cell Res.* **30**, 331–338 (1963).

Weiss, L. and Glaves, D. Effects of migration inhibiting factor(s) on the *in vitro* detachment of macrophages. *J. Immunol.* **115**, 1362–1365 (1975).

Weiss, L. and Harlos, J. P. Short-term interactions between cell surfaces. *Progr. Surface Sci.* **1**, 355–405 (1971).

Weiss, L. and Lachmann, R. J. The origin of an antigenic zone surrounding HeLa cells cultured on glass. *Exp. Cell Res.* **36**, 86–91 (1964).

Weiss, L., Nir, S., Harlos, J. P. and Subjeck, J. R. Long-distance interactions between Ehrlich ascites tumour cells. *J. Theor. Biol.* **51**, 439–454 (1975).

Weiss, L. and Subjeck, J. R. Electrical heterogeneity of the surfaces of Ehrlich ascites tumour cells. *Ann. N.Y. Acad. Sci.* **238**, 352–360 (1974a).

Weiss, L. and Subjeck, J. R. Interactions between the peripheries of Ehrlich ascites tumour cells as indicated by the binding of colloidal iron hydroxide particles. *Int. J. Cancer* **13**, 143–150 (1974b).

Some Chemical Bases for Biological Specificity as Exemplified by Hapten-Antibody Interactions

ALLAN L. GROSSBERG

Department of Immunology Research, Roswell Park Memorial Institute,[1] Buffalo, New York 14263, USA

INTRODUCTION

Specific biological interactions, including, for example, those specific interactions which are involved in cell–cell recognition or in interaction of cell surface receptors with macromolecules, hormones and drugs, all depend on a complementary fit of one structure to another. That is, one structure provides a binding region or "site" available for interaction with a region on the other structure—interaction occurring because there is close steric approach of the two regions and because the chemical groups making up the two regions are correctly aligned so that sufficient interaction energy is attained.

Investigations of the molecular basis of such specific interactions have been carried out, for example, with reactions involving enzymes and their substrates and inhibitors. However, there have been certain limitations in such studies. A different system which has proved to have wide applicability as a model for studying biological binding sites is that involving the interaction of antihapten antibodies with haptens. This system is particularly advantageous (1) antibodies can be made at will against an almost unlimited variety of structures; (2) haptens of known structure can be chosen which contain chemical groupings of interest; (3) interactions are easily followed and do not involve a second step of covalent bond formation (or splitting) as occurs in enzyme-substrate interactions; (4) antibodies demonstrate an exquisite

[1] A unit of the New York State Department of Health.

selectivity in their combination with antigens or haptens, i.e. they have very great specificity. This combination of circumstances makes antibodies uniquely useful in studying the chemical bases of specific biological interactions.

In the discussion to follow I will give selected examples of the experience in our laboratory and others, accumulated since the 1940s with a variety of antihapten-antibody systems (Pressman and Grossberg, 1968), leading to the view that the important considerations for specific biological interactions are the structural details of the interacting surfaces which lead to a close complementary fit. The closeness of the fit is required since the forces involved in the interactions are short-range ones. These forces (Table I) include the London dispersion forces providing van der Waals' attraction and hydrophobic forces resulting from

TABLE I

Forces involved in antigen-antibody interactions

1. London dispersion forces (van der Waals' interactions)
2. Electrostatic forces (charge interactions)
3. Hydrophobic forces (nonpolar interactions)
4. Hydrogen bond formation
5. Dipole interactions
6. Entropy increase (from release of bound water)

displacement of ordered water from nonpolar surfaces into the bulk water, with resultant increase in entropy. In addition, complementarity is required so that groups may be aligned to provide effective interactions. Thus charge interaction requires the approach of two groups of opposite charge; hydrogen bond formation requires the approach of the proton donor group and the proton acceptor group; dipole interactions require the approach of two atoms of different electronegativity.

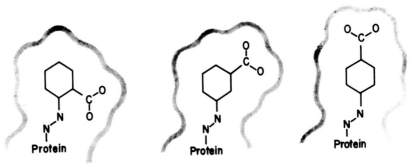

Fig. 1. Van der Waals' outlines of the o-, m- and p-azobenzoates.

CLOSENESS OF FIT

The degree of closeness of fit has been examined for a variety of anti-hapten-antibody systems. Examples illustrating the general principles involved are provided by the antibodies directed against the o-, m-. p-azobenzoate groups (Fig. 1). By the technique of hapten inhibition of specific precipitation one can obtain combining constants of haptens with antibody, K_{rel},[1] relative to a reference hapten (e.g. unsubstituted benzoate). These K_{rel} values can be converted to $\triangle F_{rel}$ values ($-\triangle F_{rel} = RT\ln K_{rel}$). A large positive value of $\triangle F_{rel}$ indicates poor combination of hapten with antibody and therefore a large degree of steric interaction. Figure 2 illustrates the effect of chlorine substituents on the combination of p'-hydroxyphenyl-azobenzoates with anti-o-, anti-m- and anti-p-azobenzoate antibodies (Pressman et $al.$, 1954). The van der Waals outline of the hapten is shown superimposed on the outline of the immunizing group. The results of the analysis of data for $\triangle F_{rel}$ in Fig. 2 are summarized in Fig. 3. There is a large degree of steric interaction between antibody surface and the positions ortho to the carboxyl group in the anti-m-azobenzoate and anti-p-azobenzoate systems, as shown by the large values in $\triangle F_{rel}$, indicating a close fit of the antibodies around these positions. There is much less steric interaction at the ortho position in the anti-o-azobenzoate system. This is probably due to the fact that in o-azobenzoate the carboxylate is twisted out of the plane of the benzene ring, whereas in the other two systems the carboxylate lies in the plane of the ring. Thus the former group is thicker and the antibody site directed against it cannot approach as close to the ortho position, so that an ortho substituent can be accommodated.

Additional information can be obtained from combinations in which two substitutents are placed on a benzoate. Representative data are

---

[1] K_{rel} values are calculated from the values of the concentrations of reference hapten (H_{ref}) and hapten of interest (H) which cause 50% inhibition of specific precipitation. Thus:

$$K_{rel} = \frac{K_H}{K_{ref\,hapten}}$$
$$= \frac{(HAb)_{50\%}}{(H)_{50\%}(Ab)_{50\%}} \Bigg/ \frac{(H_{ref}Ab)_{50\%}}{(H)_{50\%}(Ab)_{50\%}}$$
$$= \frac{(H_{ref})_{50\%}}{(H_{ref})_{50\%}}$$

Since the total antibody concentration is usually small compared to the total hapten concentrations, the concentration of free hapten is essentially that of total hapten. Hence:

$$K_{rel} = \frac{(\text{total concentration of reference hapten added to give 50\% inhibition})}{(\text{total concentration of hapten added to give 50\% inhibition})}$$

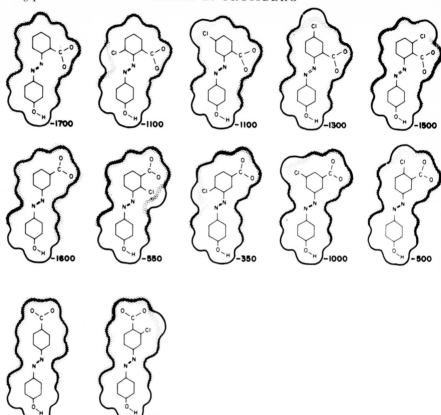

Fig. 2. Effect of chlorine substituents on the combination of p'-hydroxyphenylazobenzoates with anti-azobenzoate antibodies. The van der Waals outline of the inhibiting hapten is shown superimposed on the outline of the injected haptenic group. Values given are for $\triangle F_{rel}$ (Pressman et al., 1954).

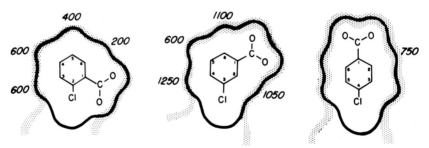

Fig. 3. Effect of chloro substituents on the combination of p'-hydroxyphenylazobenzoates with antibody. Values given are $\triangle F_{rel}$ for the chlorine in the indicated position (Pressman et al., 1954).

in Table II. In these examples, compounds having ortho, meta or para substituents were used to obtain K_{rel} values. Then the K_{rel} values were determined for compounds in which two substituents, either ortho and para or meta and meta, were present. If one assumes that $\triangle F_{rel}$ values for each substituent in a given position should be additive, one can calculate K_{rel} values to be expected as the product of individual substituent K_{rel} values. As seen in Table II, this holds for substituents in the ortho and para positions, but not for meta-meta substituted compounds, for which K_{rel} values observed are much smaller than the calculated K_{rel}. This is because the fit of the antibody around the benzene ring is quite close. A single substituent in the ortho position is accommodated with some shift in orientation of the hapten and this shift does not affect accommodation of a para substituent. However when a meta substituent is accommodated, there is no room for additional accommodation of a second meta substituent on the opposite side of the ring and thus the K_{rel} for these disubstituted haptens is very small.

TABLE II

The effect of two substituents on combination in the anti-p-azobenzoate system

Hapten	K_{rel} (observed)	K_{rel} (calculated)
o-Nitrobenzoate	0.006	
m-Nitrobenzoate	0.40	
p-Nitrobenzoate	11.5	
Isophthalate	0.83	
2,4-Dinitrobenzoate	0.06	0.07 (0.006×11.5)
3,5-Dinitrobenzoate	0.004	0.16 (0.4×0.4)
Trimesate	0.0004	0.64 (0.8×0.8)

Data from Pressman et al. (1954).

VAN DER WAALS' INTERACTIONS

The contribution of van der Waals' interactions has been quantified in many systems. An example is given in Table III for the effect of substituents in the attachment-homologous position on benzoates for the anti-o-, m- and p-azobenzoate systems. For the methyl, chloro, bromo and iodo substituents, there is increased interaction energy due to van der Waals' attraction when the substituents are in the position corresponding to the azo group in the immunizing hapten. The order of increased K_{rel} corresponds well to the order of increasing van der Waals' attraction for the groups listed.

TABLE III

Effect of substituents in the attachment-homologous position on combination in the anti-*o*-, -*m*- and -*p*-azobenzoate systems (X_o, X_m, X_p)

| Substituent | K_{rel} in the system: | | |
	X_o	X_m	X_p
None	1.0	1.0	1.0
H_3C-	1.7	1.9	2.6
$Cl-$	2.9	3.0	2.8
$Br-$	2.5	4.8	5.1
$I-$	4.2	6.0	6.5

Data from Pressman *et al.* (1954).

HYDROGEN BONDING

The importance of configuration relevant to hydrogen bond formation in an anti-sugar system is illustrated by the interaction of antibody against *p*-azophenyl-β-lactoside with lactose and cellobiose (Karush, 1957). The former sugar, glucosyl-β-galactoside, differs in configuration from the latter, glucosyl-β-glucoside, only in the configuration of the hydrogen and hydroxyl groups on the terminal ($4'$) carbon atom of the second ring. This difference is sufficient to decrease the K_{rel} value of cellobiose to 0.0025, relative to lactose. The decreased interaction is probably due in part to the necessity of the hydroxyl group in the $4'$ position to be in the proper orientation to participate in hydrogen bond formation with a group in the antibody site.

STERIC FIT AND HYDROPHOBIC INTERACTION

The role of optical configuration in antigen-antibody combination has been investigated with antibodies against the D- and L-forms of phenyl(*p*-azobenzoylamino)acetate. Antibody against each of the optical isomers combines well with its homologous isomer and very poorly ($K_{rel} < 0.01$) with the antipode (Table IV) (Karush, 1956).

In addition this system demonstrates the contribution of hydrophobic interaction to the combination. Thus benzoylglycine—the

TABLE IV

Combination of optical isomers with antibodies in the anti-D- and anti-L-phenyl(p-azobenzoylamino)acetate systems

Hapten		Anti-D-K_{rel}	Anti-L-K_{rel}
D-Phenyl-(p-nitrobenzoyl-amino)acetate	$O_2N-\phi-CO-NH-\overset{H}{\underset{\phi}{C}}-COO-$	1.00	0.0090
L-Phenyl-(p-nitrobenzoyl-amino)acetate	$O_2N-\phi-CO-NH-\overset{\phi}{\underset{H}{C}}-COO-$	0.006	1.00
D-Phenyl-(p-benzoyl-amino)acetate	$\phi-CO-NH-\overset{H}{\underset{\phi}{C}}-COO-$	0.45	
L-Phenyl-(p-benzoyl-amino)acetate	$\phi-CO-NH-\overset{\phi}{\underset{H}{C}}-COO-$		0.61
Phenylacetate	$H-\overset{H}{\underset{\phi}{C}}-COO-$	0.03	
Benzoylglycine	$\phi-CO-NH-CH_2COO$	0.0006	0.00064

Data from Karush (1956).

structure lacking one benzene ring of the immunizing hapten—combines with either antibody with a K_{rel} of 0.0006 (Table IV). This is equivalent to a $\triangle F_{rel}$ of 4 kcal per mole, which is the $\triangle F$ observed when benzene is transferred from liquid benzene to water. This suggests that the benzene ring of the hapten is in a benzene-like or hydrophobic region when it is combined with antibody.

Evidence for hydrophobic interaction is found in many other systems in which a benzene ring is part of the immunizing haptenic group. In all these cases, the ring provides a very appreciable part of the interaction, through both hydrophobic and van der Waals' interaction. Thus in the anti-p-azobenzoate system, acetate gives a $\triangle F_{rel} > 4000$ cal relative to benzoate (Pressman et al., 1954). In the anti-p-azophenyltrimethylammonium system, tetramethyl ammonium gives a $\triangle F_{rel}$ of 1700 cal relative to phenyltrimethylammonium (Pressman et al., 1946). In the o-, m- and p-azobenzene arsonate systems, methylarsonate gives a $\triangle F_{rel}$ value in each case > 4000 cal/mole relative to benzenearsonate (Pressman et al., 1945). In the anti-4-azophthalate system, maleate gives a $\triangle F_{rel} > 3000$ cal/mole relative to phthalate (Pressman and Pauling, 1949).

An interesting example showing the contribution of dipolar and steric interaction is the combination of cyclohexanecarboxylate with anti-*p*-azobenzoate antibody. The $\triangle F_{rel}$ is 3500 cal relative to benzoate (Pressman and Grossberg, 1968). The hydrophobic effect of the cyclohexane ring is the same as that of the benzene ring, but the former has much less polarizability and also is thicker, since the ring is puckered rather than planar.

STERIC FIT—CONFIGURATIONAL ASPECTS

The ability of a combining site structure to select one of the alternative conformations that a ligand may assume has important implications for many biological interactions. Such an ability has been demonstrated in the case of the antibody site against the flexible *p*-azosuccinanilate structure. This structure may exist in either the extended or coiled form (Fig. 4). The latter form can be stablized by hydrogen bonding, as shown. The antibody is apparently directed mainly against the coiled form (Pressman *et al.*, 1948) since the *cis*-maleanilate ion (Fig. 4) has a larger K_{rel} value (0.1 to 0.25) than does the fumaranilate ion ($K_{rel} = 0.01$) which can only assume the extended (*trans*) form.

That antibodies can easily distinguish between *cis* and *trans* forms is shown by antibodies against *p*-azomaleanilate and *p*-azofumaranilate (Siegel and Pressman, 1954). In each case $\triangle F_{rel}$ for combination with the non-homologous structure exceeds 4000 cal. Succinanilate combines rather poorly with anti-*cis*-maleanilate, even though it can assume

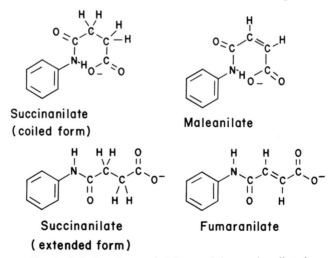

Fig. 4. Coiled and extended forms of the succinanilate ions.

the *cis* configuration. However, it is considerably thicker than the flat maleanilate. Antibody against *trans*-fumaranilate combines rather well with succinanilate ($K_{rel} = 0.14$) since the hapten can exist to some extent in the *trans* form. This observation reinforces the significance of the preferred combination of anti-succinanilate antibody with the *cis* form of succinanilate.

WATER OF HYDRATION AS A STRUCTURAL FEATURE

The occurrence of hydrated groups as components of biologically important structures is widespread. In an aqueous environment water of hydration forms an important and stable part of charged groups such as carboxylate and ammonium as well as uncharged proton donor and acceptor groups such as carbonyl, hydroxyl and amide nitrogen groups. In addition, nonpolar (hydrophobic) surfaces often have stable ordered water structure associated with them. Thus water molecules often are an integral part of a structure, and their presence must be taken into account in interpreting the results of experiments involving interaction of one structure with another.

One of the clearest examples of the importance of water of hydration in a hapten-antihapten antibody system is that involving the 3-azopyridine group (Fig. 5). The ring nitrogen of pyridine is hydrated and it

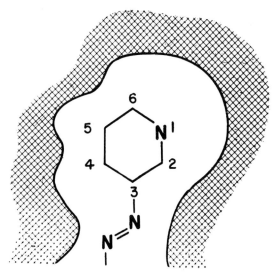

Fig. 5. Sketch illustrating an antibody site directed against the hydrated 3-azopyridine group.

is a very important part of the structure against which antibodies are made (Nisonoff and Pressman, 1957a). Anti-3-azopyridine antibodies combine 77 times stronger with N-(3-pyridyl)succinamate than they do with N-(phenyl)succinamate, indicating that the ring nitrogen contributes -2400 cal/mole to the $\triangle F$ of combination. The antibodies appear to have a large concavity to accommodate the hydrated ring nitrogen, large enough to accommodate substituents on the ring of phenylsuccinamate (Table V). The fact that substituents at the 3 or 4 position of phenylsuccinamate give K_{rel} values from about 2 to 50 times greater than the unsubstituted compound indicates that this region of the antibody site accommodates the substituents and that their presence increases the extent of interaction. Thus it is apparent that antibody was formed against the hydrated pyridine molecule and that the water of hydration is required for good interaction between the ligand and the antibody site.

TABLE V

Effect of substituents on the benzene ring of phenylsuccinamates on combination in the anti-3-azopyridine system

Substituent	K_{rel} for compound with substituent in the indicated position		
CH$_3$	0.22	1.4	1.4
Cl	0.9	1.75	7.2
Br	0.94	2.3	8.1
I	0.81	14	22
NO$_2$	1.3	3.8	10
CH$_3$CO	—	13	56
Ring N (pyridyl)	2.0	78	3.2

Su = succinamate; values are relative to unsubstituted phenylsuccinamate. Data from Nisonoff and Pressman (1957a).

CHARGE INTERACTIONS

Positively and negatively charged groups are present on most biologically important macromolecules and on many smaller metabolites, hormones and effector substances as well. The electrostatic forces involved often play a commanding role in the interaction between structures containing such groups. This fact is clearly demonstrated in the binding of charged haptens to antibodies against them. Thus antibody against the positively charged p-azophenyltrimethylammonium group combines effectively with this group but very ineffectively with the isosteric p-azophenyl-t-butyl group, which has the same size, shape and polarizability, but lacks the positive charge. The difference in observed K_{rel} values (Table VI) leads to a calculated value of $\triangle F_{rel}$ of -1150 cal in favour of the positively charged compound. This corresponds to the interaction of the positive charge with a negative charge in the antibody site at an estimated distance of about 8 Å, which is within 2 to 3 Å of the distance of closest approach of the hapten to a negatively charged carboxylate in the protein.

TABLE VI

Contribution of the positive charge to combination in the
anti-p-azophenyltrimethylammonium system

Hapten		K_{rel}	$\triangle F_{rel}$ (cal)
Phenyltrimethylammonium	$\phi-\overset{+}{N}(CH_3)_3$	1.00	0
"H–Acid"-p-azophenyl-trimethylammonium	$(H–Acid)—N=N-\phi-\overset{+}{N}(CH_3)_3$	5.2	−930
"H–Acid"-p-azo-t-butyl benzene	$(H–Acid)—N=N-\phi-C(CH_3)_3$	0.68	220
"H–Acid"-azobenzene	$(H–Acid)—N=N-\phi$	0.06	1570

Data from Pressman and Siegel (1953).

A similar observation has been made for antibody against the negatively charge p-(p-azophenylazo)benzoate group in its combination with the negative p-azobenzoate group or the uncharged isosteric p-azonitrobenzene group (Nisonoff and Pressman, 1957b). Here the difference in K_{rel} values leads to a calculated $\triangle F_{rel}$ of -4900 cal in favour of the charged group. The estimated distance of separation between the negative carboxylate and a positive group in the antibody site is less than 4.5 Å, which is within 1 Å of the distance of closest approach to an ammonium group in the antibody site.

CHARGED GROUPS IN ANTIBODY SITES

Direct evidence is available for the presence of charge groups in antibody sites against charged haptens. The evidence has been obtained by chemically modifying either the negatively charged carboxylate groups on antibody or the positively charged ammonium or guanidinium groups. When a charged group is present in the antibody site, its modification results in loss of binding activity. Furthermore, to establish that the loss of activity is not merely due to general denaturation of the protein, the same extent of modification is carried out with the site occupied by hapten, thereby protecting the charged group in the site from modification, with resultant retention of binding activity (Fig. 6). Moreover, when a given charged group is not present in the site, the modification procedure does not affect the binding activity.

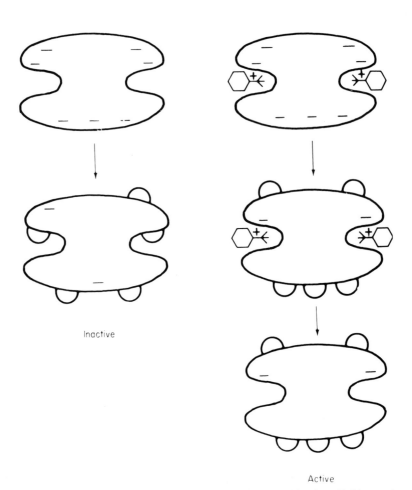

Fig. 6. Protection of antibody sites by hapten during esterification. Evidence that a carboxylate is in the site.

Evidence of the kind outlined above has been obtained for the presence of a negative carboxylate in the combining site of antibody against the positively charged p-azophenyltrimethylammonium (Ap) group (Grossberg and Pressman, 1960) by esterifying the protein carboxyl groups with diazoacetamide:

$$(\text{protein } COOH + N_2CHCONH_2 \rightarrow \text{protein } COOCH_2CONH_2 + N_2)$$

As shown in Fig. 7 anti-Ap antibody activity is lost with increasing extent of esterification. In a typical experiment involving protection of anti-Ap antibody sites with hapten, antibody esterified with no protection lost 58% of its binding activity. When esterified to the same extent (18% carboxyls esterified) in the presence of 0.005 M p-iodophenyltrimethylammonium, only 19% of anti-Ap binding activity was lost. Table VII shows that when a mixture of anti-Ap antibody and anti-p-azobenzoate (anti-Xp antibody is esterified only anti-Ap activity is lost and not activity against the negatively charged (Xp) group.

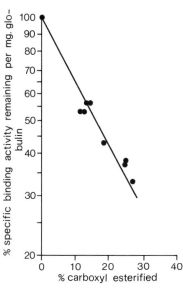

Fig. 7. The effect of esterification of anti-p-azophenyltrimethylammonium antibody (Grossberg and Pressmen, 1960).

A similar result has been obtained with antibody against the negatively charge p-azobenzenearsonate group, in that esterification of 20% of the carboxyls in this antibody causes no loss of sites and no change in their binding constant compared to unmodified antibody (Fig. 8) (Pressman and Grossberg, 1968).

TABLE VII

Effect of esterification by diazoacetamide on the hapten binding activities of mixtures of anti-Ap and anti-Xp antibodies

Sample	Carboxyl esterified (%)	Binding activity of Anti-A$_p$	Anti-X$_p$
		(% of original)	
Esterified mixture 1	24	36	105
Esterified mixture 2	27	33	106

Data from Grossberg and Pressman (1960).

When antibodies against negatively charged groups are treated with diacetyl or glyoxal, which modify the positively charged guanidinium group of arginine, antibody sites are lost. Hapten in the sites protects them against loss by treatment with diacetyl or glyoxal. Figure 9 shows this result for treatment with diacetyl of antibody against the negatively charged p-azophenylarsonate group (Grossberg and Pressman, 1968). A similar result has been obtained showing the presence of arginyl residues in the sites of antibodies against the negatively charged p-azo-benzoate (Grossberg and Pressman, 1968), 4-azophthalate (Mayers et al., 1973) and 5-azoisophthalate groups (Mayers et al., 1972). Notably no loss of sites or effect on the binding constant is found by glyoxal or diacetyl treatment of antibodies against the positively charged Ap group or the neutral 3-azopyridine group, indicating that the positively charged guanidinium is not present in the sites of these antibodies.

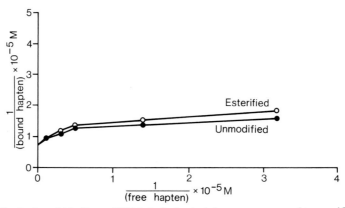

Fig. 8. Similarity of binding of ^{131}I-labelled p-iodobenzenearsonate by esterified and by untreated anti-p-azobenzenearsonate antibody, showing that esterification of this antibody results in no loss of sites (Pressman and Grossberg, 1968).

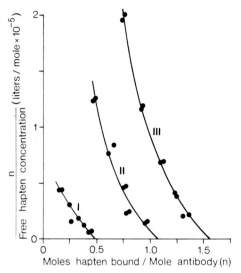

Fig. 9. Binding curves showing the extent of loss of sites due to modification by diacetyl of 54% of the arginyl residues in acetylated specifically purified anti-p-azobenzenearsonate antibody in the presence and in the absence of hapten. Fewer sites are lost when the modification is carried out in the presence of 0.1 M p-nitrobenzenearsonate (Grossberg and Pressman, 1968). I, Anti-Rp, 54%, arginine modified in absence of hapten; II, Anti-Rp, 54%, arginine modified in presence of hapten; III, Anti-Rp arginine not modified.

In the case of antibody sites against p-azophenylphosphorylcholine (Krausz et al., 1976) or for myeloma protein sites which bind phosphorylcholine (Grossberg et al., 1974), both a negative carboxylate and the positive guanidinium have been shown to be present by the above methods. These groups are thus complementary to, and involved in the binding of, the positively charged choline moiety and the negatively charged phosphoryl group, respectively.

We have examined all of the above-mentioned antibodies and others for the presence of the positively charged ammonium group in their sites (Chen et al., 1962; Pressman and Grossberg, 1968). Various reagents have been used to modify amino groups on proteins, including especially maleic anhydride (Freedman et al., 1968). The results, in summary, are that antibody sites against neutral and positively charged groups show no indication of the presence of amino groups. For antibody sites against negatively charged groups, only a variable, usually small, proportion appear to contain amino groups. Thus antibodies appear to prefer to interact with negative charges through the use of arginine as a binding site constituent rather than lysine. It would be interesting to determine if this is so for other proteins having binding sites with specificity for negative groups.

The above discussion has provided specific examples of the structural features important for specific interaction of antibody binding sites with ligands. A few additional comments of a general nature are in order:

(1) For many antihapten systems studied, the strength of combination shows a degree of regulation. That is, the value of $-\triangle F$ of combination usually is between 6 and 8 kcal/mole (Table VIII). When a group that is important for interaction is removed, there is usually an increase in $\triangle F$ of 2 kcal or more, as noted above in the discussion of hydrophobic interactions involving the benzene ring (Table IV and text). For a ligand with two charged groups that are important for combination, such as the carboxylates in the phthalate and isophthalate systems (Table VIII), each group contributes at least 2 kcal to the combination. Since the total strength of combination is limited, the other part of the ligand (the benzene ring) must contribute less than it does in the benzoate system where only one carboxyl is present. The observed values seem to bear this out—the benzene ring contribution in benzoate is >4 kcal; in phthalate it is ~ 3 kcal.

TABLE VIII

Limited strength of combination in some antihapten-antibody systems

Antibody system	Range of values of:		Ref.
	$K_H \times 10^{-5}$ (l/mole)	$-\triangle F$ (kcal/mole)	
Anti-p-azobenzoate	0.05–0.8	4.7–6.3	(1)
Anti-p-azophenyltrimethylammonium	4–11	7.1–7.8	(2)
Anti-3-azopyridine	0.4–1.6	5.9–6.7	(3)
Anti-p-azobenzenearsonate	1–10	6.4–7.7	(4)
Anti-4-azophthalate	8–60	7.5–8.7	(5)
Anti-5-azoisophthalate	6–30	7.4–8.3	(6)
Anti-p-azobenzenesulphonate	3–15	7.0–7.9	(7)
Anti-p-azophenylphosphorylcholine	8–12	7.5–7.8	(8)
Anti-D-phenyl(p-azobenzoylamino)acetate	4.4–6.7	7.3–7.5	(9)
Anti-p-azophenyl-β–lactoside	3–4.5	7.0–7.3	(10)
Anti-dinitrophenyl	0.4–8.7	5.9–7.6	(11)

Data from: (1) Nisonoff and Pressman, 1958; Kitagawa *et al.*, 1965. (2) Grossberg and Pressman, 1960; Grossberg and Pressman, 1963; Grossberg *et al.*, 1962. (3) Grossberg and Pressman, 1963; Grossberg *et al.*, 1962 (4) Grossberg *et al.*, 1962; Grossberg and Pressman, 1963; Kreiter and Pressman, 1964. (5) Mayers *et al.*, 1973. (6) Mayers *et al.*, 1972. (7) Grossberg, unpublished. (8) Krausz *et al.*, 1976. (9) Karush, 1956. (10) Karush, 1957. (11) Eisen and Siskind, 1964.

The limited strength of combination in antihapten-antibody systems suggests that in other interacting systems a similar constraint may exist. Thus cell contacts may involve forces limited in extent so that stronger interactions can compete successfully with weaker ones to allow cell recognition mechanisms to operate. If a weaker interaction were to have too great a $-\triangle F$, then any stronger interaction could not effectively replace it in a reasonable time interval (e.g. the lifetime of a cell).

(2) Antibodies against a given hapten are usually heterogeneous— i.e. their sites are not all of the same configuration so that an individual animal produces a population of antibodies with a range of binding constants. In addition it is often observed that, with increasing time after immunization, the average binding constant of such a population increases ("maturation" effect). The two effects mean that antibodies of varying degrees of cross reactivity are produced at various times during the immune response to a determinant. The fact that the expression of binding sites on antibodies changes or "matures" with time suggests that other biologically important binding sites may do the same. One could postulate a mechanism in cell differentiation in which binding sites relevant to functional activity at the cell surface change in this manner, as do the receptor sites on the lymphocytes which are involved in antigen recognition, so that more efficient interactions are promoted as it becomes important for the surface of a given cell to recognize a similar cell or a related cell.

ACKNOWLEDGEMENT

These investigations have been supported in part by United States Public Health Service Research Grants AI-03962, AI-10454 and CA-11656.

REFERENCES

Chen, C. C., Grossberg, A. L. and Pressman, D. Effect of cyanate on several anti-hapten antibodies: Evidence for the presence of an amino group in the site of anti-*p*-azobenzenearsonate antibody. *Biochemistry* **1**, 1025–1030 (1962).

Eisen, H. N. and Siskind, G. W. Variations in affinities of antibodies during the immune response. *Biochemistry* **3**, 996–1008 (1964).

Freedman, M. H., Grossberg, A. L. and Pressman, D. Evidence for ammonium and guanidinium groups in the combining sites of anti-*p*-azobenzenearsonate antibodies—separation of two different populations of antibody molecules. *J. biol. Chem.* **243**, 6186–6195 (1968).

Grossberg, A. L., Krausz, L. M., Rendina, L. and Pressman, D. The presence of arginyl residues and carboxylate groups in the phosphorylcholine-binding site of mouse myeloma protein, HOPC 8, *J. Immunol.* **113**, 1807–1814 (1974).

Grossberg, A. L. and Pressman, D. Nature of the combining site of antibody against a hapten bearing a positive charge. *J. Amer. chem. Soc.* **82**, 5478–5482 (1960).

Grossberg, A. L. and Pressman, D. Effect of acetylation on the active site of several antihapten antibodies: Further evidence for the presence of tyrosine in each site. *Biochemistry* **2**, 90–96 (1963).

Grossberg, A. L. and Pressman, D. Modification of arginine in the active sites of antibodies. *Biochemistry* **7**, 272–279 (1968).

Grossberg, A. L., Radzimski, G. and Pressman. D. Effect of iodination on the active site of several antihapten antibodies. *Biochemistry* **1**, 391–401 (1962).

Karush, F. The interaction of purified antibody with optically isomeric haptens. *J. Amer. chem. Soc.* **78**, 5519-5526 (1956).

Karush, F. The interaction of purified anti-β-lactoside antibody with haptens. *J. Amer. chem. Soc.* **79**, 3380-3384 (1957).

Kitagawa, M., Yagi, Y. and Pressman, D. The heterogeneity of combining sites of antibodies as determined by specific immunoadsorbents. III. Further characterization of purified antibody fractions obtained from anti-p-azobenzoate antibodies. *J. Immunol.* **95**, 991–1001 (1965).

Krausz, L. M., Grossberg, A. L. and Pressman, D. The chemical nature of the combining site of rabbit anti-p-azophenylphosphorylcholine antibody. *Immunochemistry* **13**, 51–57 (1976).

Kreiter, V. P. and Pressman, D. Fractionation of anti-p-azobenzenearsonate antibody by means of immunoadsorbents. *Immunochemistry* **1**, 91–108 (1964).

Mayers, G. L., Grossberg, A. L. and Pressman, D. Arginine and lysine in binding sites of anti-5-azoisophthalate antibodies. *Immunochemistry* **9**, 169–178 (1972).

Mayers, G. L., Grossberg, A. L. and Pressman, D. Arginine and lysine in binding sites of anti-4-azophthalate antibodies. *Immunochemistry* **10**, 37–41 (1973).

Nisonoff, A. and Pressman, D. The annular nitrogen of pyridine as a determinant of immunologic specificity. *J. Amer. chem. Soc.* **79**, 5565–5572 (1957a).

Nisonoff, A., and Pressman, D., Closeness of fit and forces involved in the reactions of antibody homologous to the p-(p'-azophenylazo)benzoate ion group. *J. Amer. chem. Soc.* **79**, 1616–1622 (1957b).

Nisonoff, A. and Pressman, D. Heterogeneity and average combining constants of antibodies from individual rabbits. *J. Immunol.* **80**, 417–428 (1958).

Pressman, D., Bryden, J. H. and Pauling, L. The reactions of antiserum homologous to the p-azosuccinanilate ion group. *J. Amer. chem. Soc.* **70**, 1352–1358 (1948).

Pressman, D. and Grossberg, A. L. The Structural Basis of Antibody Specificity, W. A. Benjamin, Inc., Reading, Mass. (1968).

Pressman, D., Grossberg, A. L., Pence, L. H. and Pauling, L. The reactions of antiserum homologous to the p-azophenyltrimethylammonium group. *J. Amer. chem. Soc.* **68**, 250–255 (1946).

Pressman, D., Paradee, A. B. and Pauling, L. The reactions of antisera homologous to various azophenylarsonic acid groups and the p-azophenylmethylarsinic acid group with some heterologous haptens. *J. Amer. chem. Soc.* **67**, 1602–1606 (1945).

Pressman, D. and Pauling, L. The reactions of antiserum homologous to the 4-azophthalate ion. *J. Amer. chem. Soc.* **71**, 2893–2899 (1949).

Pressman, D. and Siegel, M. The binding of simple substances to serum proteins and its effect on apparent antibody-hapten combination constants. *J. Amer. chem. Soc.* **75**, 686–693 (1953).

Pressman, D., Siegel, M. and Hall, L. A. R. The closeness of fit of antibenzoate antibodies about haptens and the orientation of the haptens in combination. *J. Amer. chem. Soc.* **76**, 6336–6341 (1954).

Siegel, M. and Pressman, D. The reactions of antiserum homologous to the *p*-azomaleanilate and *p*-azofumaranilate ion groups. *J. Amer. chem. Soc.* **76**, 2863–2866 (1954).

Interactions of Fibronectin, a Cell-Type Specific Surface-Associated Glycoprotein

A. Vaheri, D. Mosher, J. Wartiovaara, J. Keski-Oja,
M. Kurkinen and S. Stenman

*Department of Virology, and Third Department of Pathology, University of Helsinki,
SF-00290 Helsinki 29, Finland*

INTRODUCTION

Fibronectin is a major protein of cultured fibroblastic cells and has recently drawn attention because loss of surface-associated fibronectin represents a major quantitative difference between those normal and malignantly transformed cells so far examined ((Hynes, 1976; Vaheri *et al.*, 1976a). The significance of this loss can only be understood when the functions of fibronectin in normal cells are known. Proteins antigenically related to the fibronectin of cultured fibroblasts are present in embryonic mesenchyme and basement membranes, connective tissue, in the circulation, and at later stages of ontogenesis. Our purpose here is to describe the structure, metabolism and interactions of the various forms of fibronectin. The present information suggests that fibronectin is a structural protein that may play an important role in cell–cell and cell–matrix interactions.

HIGH MOLECULAR WEIGHT GLYCOPROTEIN

Fibronectin is a glycoprotein composed of high molecular weight subunits (Table I). The protein has been studied using various experimental approaches and has been given a number of names (Table II). Although the chemical identity of the immunologically cross-reactive forms of fibronectin has not been established, the information available indicates that, aside from minor differences, the protein is the same. The name fibronectin is suggested to emphasize the interactions of the protein (*fibra* (Lat.), fibre; *nectere* (Lat.), connect, link) to fibrillar structures such as fibrin (Ruoslahti and Vaheri, 1975; Mosher, 1976), collagen (Pearlstein, 1976), and the characteristic structures in which it is found on cultured fibroblasts (Wartiovaara *et al.*, 1974).

TABLE I

Characteristics of fibronectin

Property	Surface-associated in cultured fibroblasts	Soluble in plasma
M,	440 000, dimer of disulphide-bonded 220 000 subunits	440 000–450 000 (12–14S), dimer of disulphide-bonded 200 000 subunits
Carbohydrate	Present	5%
Electrophoretic mobility	β_1—α_2	β_1—α_2
Antigenicity	Indistinguishable within each vertebrate species Cross-reaction between avian and mammalian fibronectins	
Concentration	Major component	300–600 μg/ml in human plasma
Affinity for fibrin and fibrinogen	No?	Yes
Cross-linked by plasma transglutaminase	Yes	Yes
Affinity for collagen	Yes	Yes

For details and references see text. Other forms of fibronectin (intracellular fibronectin, fibronectin shed by normal or malignantly transformed cells, fibronectin of connective tissue matrix and basement membranes) have been less fully characterized.

The cellular and plasma (serum) fibronectins share several properties (Table I). Their mobility upon immunoelectrophoresis (Ruoslahti *et al.*, 1973; Ruoslahti and Vaheri, 1974) and subunit composition (dimers of disulphide-bonded 200 000–220 000 polypeptides; Mosesson *et al.*, 1975; Mosher, 1975a; Keski-Oja *et al.*, 1977) are similar. Both are substrates of plasma transglutaminase (fibrin-stabilizing factor, blood coagulation factor XIII: Mosher, 1975b; Keski-Oja *et al.*, 1976a), a thrombin-activated cross-linking enzyme. The subunit of plasma fibronectin seems to be slightly smaller than that of the cellular form (Mosher, 1975b; Keski-Oja *et al.*, 1977).

Fibronectin has been detected in fibroblasts and sera of various vertebrate species. Antigen cross-reactions between avian and human fibronectins (Kuusela *et al.*, 1976) suggest that fibronectin

TABLE II

Fibronectin: synonyms, pseudonyms, acronyms

Name	Reference
Cold-insoluble globulin (CIG)	Morrison et al., 1948
Surface fibroblast antigen (SF)	Ruoslahti et al., 1973
SF210 polypeptide (SF210)	Vaheri and Ruoslahti, 1975
Large external transformation sensitive protein (LETS)	Hynes and Bye, 1974
L1 band protein	Hogg, 1974
Cell surface protein (CSP)	Yamada and Weston, 1974
Galactoprotein a	Gahmberg et al., 1974
Zeta protein (Z)	Robbins et al., 1974
Band 1 protein	Pearlstein and Waterfield, 1974; Hunt et al., 1975
Cell adhesion factor (CAF)	Pearlstein, 1976
Large noncollagenous fibroblast protein	Lukens, 1976; Sear et al., 1976

has remained highly conserved in evolution. This invariance may have to do with its multiple interactions and structural functions as a cell surface-associated protein. Besides fibroblasts, fibronectin has been detected in cultured astroglial cells (Vaheri *et al.*, 1976b), myoblasts (Hynes, personal communication; our unpublished observations), and in smooth muscle cells (unpublished). Within one species the different forms of fibronectin (such as those produced by normal or transformed fibroblasts or glial cells, plasma fibronectin) are antigenically indistinguishable (Vaheri *et al.*, 1976b).

The cross-reactivity among the various forms of fibronectin in a given species and among fibronectins from different vertebrates make it possible to use antiserum produced against one fibronectin, e.g. purified from human plasma (Mosher, 1975a), as a general reagent to study fibronectin in a variety of species by double antibody techniques, e.g. indirect immunofluorescence, radioimmunoassay, and immunoprecipitation of metabolically labelled protein. Because of cross-reactivity, the rabbit antiserum should be absorbed with heterologous serum (usually calf serum) used in cell or organ culture medium.

FIBRONECTIN IN CELL CULTURES

In cell culture conditions fibronectin is present in four states (Vaheri *et al.*, 1976a): (1) intracellularly, seen as cytoplasmic patches in

314 A. VAHERI et al.

immunofluorescence (Fig. 1), (2) on cell surface in association with fibrillar (Fig. 1) or punctuate (Fig. 2) structures, (3) extracellularly on growth substrate and as an intercellular fibrillar network, particularly in dense cultures (Fig. 3), and (4) in a soluble form in medium (as a result of cellular synthesis). In addition 10% calf serum used to supplement growth medium probably contains about 10–50 $\mu g/ml$ of bovine fibronectin. The amount of fibronectin produced to the medium is considerable. Growing human embryonic WI-38 cells produced about 45μg and corresponding SV40 transformed cells about 30 μg fibronectin per mg cell-associated total protein in 24 hours (Vaheri et al., 1976b).

Surface iodination (Hynes, 1973; Hogg, 1974) and galactose oxidase tritiation (Gahmberg and Hakomori, 1973) label surface-associated fibronectin (or a selective fraction of it) when present on the cell surface, on the growth substrate, or in the intercellular fibrillar matrix. Fibronectin clearly is not a conventional membrane protein: when surface-associated it is located to fibrillar structures (Wartiovaara et al., 1974; Vaheri et al., 1976b) and in a dense (G=1.25–1.26) particular fraction upon cell fractionation (Graham et al., 1975). Moreover, both urea and nonionic detergents are needed to fully extract the protein (Vaheri and Ruoslahti, 1974), although urea (Yamada and Weston, 1974) and Triton X-100 separately will partially extract cellular fibronectin (Mosher and Vaheri, in preparation).

Surface iodination of normal fibroblasts prominently labels two components: monomeric 220 000 fibronectin (or its dimer in nonreducing conditions) and very high m.w. material (VHMW) that does not enter 3% polyacrylamide gels (Keski-Oja et al., 1976b). In the presence of physiological concentration of plasma transglutaminase, ^{125}I–220 000 fibronectin is cross-linked to the VHMW form (Keski-Oja et al., 1976b). Conversion of 220 000 to VHMW also seems to take place slowly in

Figs 1–3. Visualization of fibronectin by indirect immunofluorescence in cultured adherent cells from human embryos.

Fig. 1. Fibroblastic cells (embryo HE6) grown in Eagle's basal medium supplemented with 10% foetal calf serum, 100 units/ml penicillin and 50 units/ml streptomycin, from skin, studied at third serial passage one day after seeding. Note patchy intracellular and fibrillar surface-associated fibronectin immunofluorescence. Fixation by formaldehyde and acetone.
Fig. 2. Large flattened cells (embryo HE6) grown from kidney, studied at third passage one day after seeding. Note punctate surface-associated and patchy intracellular fluorescence. Fixation by formaldehyde and acetone.
Fig. 3. Dense culture of fibroblastic cells grown from embryonic lung (MRC-5 line), studied at 14th passage 3 days after seeding. Note dense fibrillar network of fibronectin fluorescence. Fixation by formaldehyde and acetone.

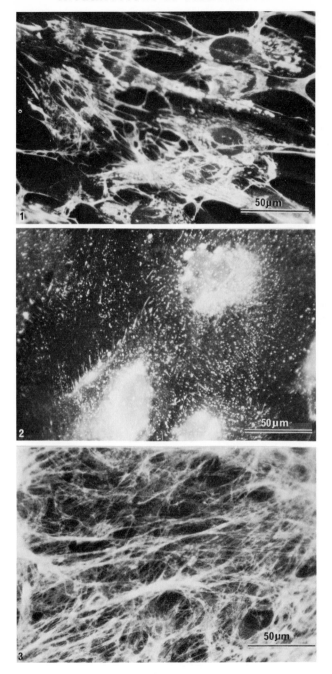

conventional cell culture conditions (Keski-Oja, 1976). Like the 220 000 form, the VHMW fibronectin is trypsin-sensitive and missing from virus-transformed fibroblasts.

Expression of surface-associated fibronectin is cell cycle dependent. We have examined exponentially growing unsynchronized human fibroblasts by a combination of ^{3}H-thymidine incorporation autoradiography, immunofluorescence and immuno-scanning electron microscopy (Stenman *et al.*, 1977). Meta- and anaphase cells had little surface fibronectin; when present it was localized to mitotic retraction fibres and filopedia-like structures. In telophase cells fibronectin reappeared at the leading cytoplasmic edge of the spreading cells. Cells in G and S phases had large amounts of surface fibronectin. Intracellular fibronectin was present throughout the cell cycle. Decreased levels of surface fibronectin in mitotic cells have also been detected using surface labelling techniques by Hynes and Bye (1974) and Pearlstein and Waterfield (1974).

A characteristic feature of surface-associated fibronectin is its extreme susceptibility to proteases such as papain and trypsin (Ruoslahti *et al.*, 1973; Hynes, 1973). Bacterial collagenase has been reported to release the protein (Blumberg and Robbins, 1975) although other workers have concluded that this effect was due to contamination of collagenase preparations with another protease (Yamada and Weston, 1974; Hynes and Humphryes, 1974). To date, little work has been directed at studying the interactions of cultured fibroblasts with homologous proteases that the cells would be expected to encounter *in vivo*.

The microfilament disrupting agent, cytochalasin B, will release surface-associated fibronectin from cultured fibroblasts whereas microtubule poisons have no such effect (Kurkinen *et al.*, in preparation). It is noteworthy that loss of surface-associated fibronectin in transformed, mitotic, protease-treated and cytochalasin B-treated cells is paralleled by loss of thick microfilament bundles (actin cables) under these cellular conditions (Spooner *et al.*, 1971; Spudich, 1973; Pollack *et al.*, 1975; Pollack and Rifkin, 1975). This indicates that complex interactions may exist among surface-associated fibronectin, microfilament proteins, and proteases.

FIBRONECTIN IN DEVELOPING EMBRYOS

In fully differentiated tissues fibronectin is found ubiquitously in basement membranes, vascular walls and surface of smooth muscle cells (Fig. 4). The onset of expression of fibronectin was studied in early embryos and in later stages of development. Presence of fibronectin in early embryos has been studied using inbred mice (CBA/Balb C cross) and indirect antifibronectin immunofluorescence (unpublished results). Evidence so far indicates that fibronectin is not found in early precompaction stage (2–16 cells, 1–3 days old) embryos.

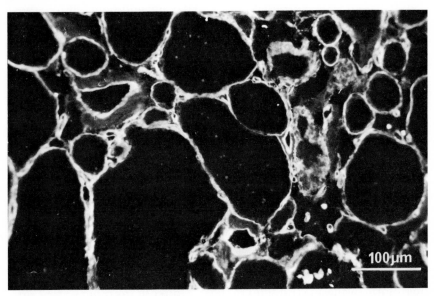

Fig. 4. Distribution of fibronectin in adult human thyroid gland. Paraffin section stained by indirect immunofluorescence for visualization of fibronectin. Basement membranes of thyroid follicles and of blood vessel endothelium are strongly positive. Interstitial connective tissue between follicles is negative.

The distribution of fibronectin in later developmental stages has been studied both in chicken and mouse using cryostat sections and immunofluorescence. Chick fibronectin (Linder *et al.*, 1975) is present (1) widely in loose connective tissues and in the primitive mesenchymal tissues of embryos incubated for 2–3 days, (2) limiting tissue membranes throughout the organism, such as the glomerular and tubular basement membranes of the kidney, the boundary membranes of the notochord, yolk sac and liver sinusoids, and (3) in soluble form in plasma and allantoic fluid (unpublished observations).

The differentiation of primitive mesenchymal cells into parenchymal cells is involved in a number of organogenetic phenomena. In kidney development mesenchymal cells differentiate into tubular epithelium. Immunofluorescent studies indicated that during chicken mesonephros development *in vivo* (Linder *et al.*, 1975) and mouse metanephros formation in organ cultures (Wartiovaara *et al.*, 1976) fibronectin is lost from the developing kidney tubular epithial cells and becomes localized to basement membrane structures surrounding the epithelium (Figs 5 and 6). Observations on fibronectin distribution in developing chick embryos (Linder *et al.*, 1975) indicate that, like kidney epithelial cells, parenchymal cells of striated muscle, cartilage and bone, also derived from the mesenchyme, had no detectable fibronectin. Thus, expression of fibronectin seems to be a property of primitive mesenchymal cells acquired early in ontogenesis. Its expression appears to be decreased during the differentiation of primitive mesenchymal cells into parenchymal cells.

PROPOSED STRUCTURAL AND ORGANIZING ROLES OF FIBRONECTIN

Based on our observations, we should like to consider two questions. First, does fibronectin, by itself or in association with collagen and other matrix components, contribute to the organization of tissues? Second, does fibronectin present on primitive cells play a role in embryonic differentiation?

Surface-associated fibronectin in cell culture has the properties of a structural element. It has a tendency to form disulphide-bonded multimers (Keski-Oja *et al.*, 1977) and is covalenty cross-linked into polymeric form by plasma transglutaminase. Unlike many other surface proteins, it has a highly nonrandom fibrillar or punctuated distribution. In dense cultures of normal fibroblastic cells a large part of external fibronectin is in form of a fibrous intercellular network. There are indications that fibronectin in this network may interact with collagen (Pearlstein, 1976) and elastin (Muir *et al.*, 1976). As discussed above, immunofluorescence studies demonstrate large amounts of fibronectin in basement membranes, indicating that fibronectin may be a major noncollagenous glycoprotein (Kefalides, 1971) of these structures. Vracko (1974) has pointed out the importance of intact basal lamina in the maintenance of orderly tissue structure.

Figs 5–6. Fibronectin in kidney rudiment cultures from 11-day-old mouse embryos. Fixation with paraformaldehyde, cryostat-sectioning and indirect fibronectin immunofluorescence.

Fig. 5. Primitive kidney mesenchymal cells at onset of cultivation. Note widespread fibronectin immunofluorescence.
Fig. 6. At 24 h of culture tubular epithelium has developed from mesenchymal cells. Note brightly stained basement membrane-like structure around tubular epithelial cells. The cells themselves show relative absence of fluorescence.

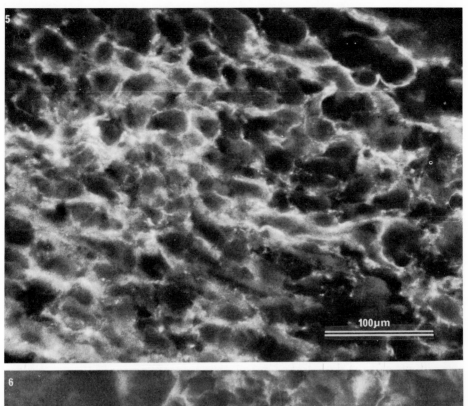

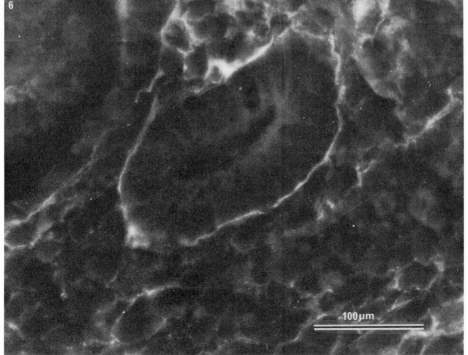

Fibronectin in connective tissue matrix does not necessarily represent a stable structure. Any fibronectin that does become fixed is likely to be a substrate for plasminogen activator–plasmin system thought to be important in embryonic cell migrations (Strickland *et al.*, 1976). Finally, the cell culture experiments indicate that surface-associated fibronectin would be released during the cell cycle (Stenman *et al.*, 1977).

Thus fibronectin could well participate in transient and variable interactions occuring during cell migration and proliferation. Once fixed, it could provide a matrix for differentiated cells. The different patterns of surface fibronectin in adherent cells derived from different human embryonic tissues (Figs 1 and 2) may represent a stable differentiated function and be a means for these different adherent cells to influence epithelial cells in their respective tissues. Changeux (1969) has pointed out that the genome is not large enough to code for specific products for all possible cell–cell interactions and proposed that differences in distribution of common surface components may provide the variability needed to account for diverse interactions. This idea has been developed further by Oosawa *et al.* (1972) and Wallach (1975). Fibronectin is a good candidate for such a surface component.

There are are two other points that may be important. First, plasma fibronectin is cross-linked by plasma transglutaminase (factor XIII) to fibrin; thus, approximately 50% of fibronectin is bound to fibrin when normal plasma is clotted at $+37°C$ while no fibronectin is bound when plasma transglutaminase deficient plasma is clotted (Mosher, 1976). Approximately 20–25% of factor XIII-deficient patients suffer from poor wound healing (Duckert, 1972), suggesting that fibronectin bound to fibrin may serve as a temporary matrix supportive of cell proliferation and migration. To date, there are no other functions for the circulating form. Second, transformed fibroblastic cells, as mentioned above, synthesize fibronectin but do not retain it bound to their surfaces (Vaheri and Ruoslahti, 1975). The inability to recognize and "use" fibronectin may contribute to the disorderly growth of malignant cells.

ACKNOWLEDGEMENTS

This work was supported by grants from the Sigrid Jusélius Foundation, the Finnish Medical Research Council, the Finnish Cancer Foundation and a grant from the National Cancer Institute, DHEW.

REFERENCES

Blumberg, P. M. and Robbins, P. W. Effect of proteases on activation of resting chick embryo fibroblasts and on cell surface proteins. *Cell* **6**, 137–147 (1975).

Changeux, J.-P. Symmetry and function of biological systems at the macromolecular level, in A. Engström and B. Strandberg (ed.), The 11th Nobel Symposium, p. 235, Wiley Interscience, John Wiley, New York (1969).

Duckert, F. Documentation of plasma factor XIII deficiency in man. *Ann. N.Y. Acad. Sci.* **202**, 190–199 (1972).

Gahmberg, C. G. and Hakomori, S. Altered growth behaviour of malignant cells associated with changes in externally labelled glycoprotein and glycolipid. *Proc. nat. Acad. Sci. (Wash.)* **70**, 3329–3333 (1973).

Gahmberg, C. G., Kiehn, D. and Hakomori, S. Changes in a surface-labelled galacto-protein and in glycolipid concentrations in cell transformed by a temperature-sensitive polyoma virus mutant. *Nature (Lond.)* **248**, 413–415 (1974).

Graham, J. M., Hynes, R. O., Davidson, E. A. and Bainton, D. F. The location of proteins labeled by the ¹²⁵I-lactoperoxidase system in the NIL 8 hamster fibroblast. *Cell* **4**, 353–365 (1975).

Hogg, N. M. A comparison of membrane proteins of normal and transformed cells by lactoperoxidase labelling. *Proc. nat. Acad. Sci. (Wash.)* **71**, 489–492 (1974).

Hunt, R. C., Gold, E. and Brown, J. C. Cell cycle dependent exposure of a high molecular weight protein on the surface of mouse L cells. *Biochim. biophys. Acta* **413**, 453–458 (1975).

Hynes, R. O. Alteration of cell-surface proteins by viral transformation and proteolysis. *Proc. nat. Acad. Sci. (Wash.)* **70**, 3170–3174 (1973).

Hynes, R. O. Cell surface proteins and malignant transformation. *Biochim. biophys. Acta* **458**, 73–107 (1976).

Hynes, R. O. and Bye, J. M. Density and cell cycle dependence of cell surface proteins in hamster fibroblasts. *Cell* **3**, 113–120 (1974).

Hynes, R. O. and Humphryes, K. C. Characterization of the external proteins of hamster fibroblasts. *J. Cell Biol.* **62**, 438–448 (1974).

Kefalides, N. A. Chemical properties of basement membranes. *Int. Rev. exp. Path.* **10**, 1–39 (1971).

Keski-Oja, J. Polymerization of a major surface-associated glycoprotein, fibronectin, in cultured fibroblasts. *FEBS Letters* **71**, 325–329 (1976).

Keski-Oja, J., Mosher, D. F. and Vaheri, A. Cross-linking of a major fibroblast surface-associated protein (fibronectin) by blood coagulation factor XIII. *Cell* **9**, 29–35 (1976a).

Keski-Oja, J., Vaheri, A. and Ruoslahti, E. Fibroblast surface antigen (SF): The external glycoprotein lost in proteolytic stimulation and malignant transformation. *Int. J. Cancer* **17**, 261–269 (1976b).

Keski-Oja, J., Mosher, D. and Vaheri, A. Dimeric character of fibronectin, a major cell surface-associated glycoprotein. *Biochem. biophys. Res. Commun.* **74**, 699–706 (1977).

Kuusela, P., Ruoslahti, E., Engvall, E. and Vaheri, A. Immunological interspecies cross-reactions of fibroblast surface antigen (fibronectin). *Immunochemistry* **13**, 639–642 (1976).

Linder, E., Vaheri, A., Ruoslahti, E. and Wartiovaara, J. Distribution of fibroblast surface antigen in the developing chick embryo. *J. exp. Med.* **142**, 41–49 (1975).

Lukens, L. N. Time of occurrence of disulfide linking between procollagen chains. *J. biol. Chem.* **251**, 3530–3538 (1976).

Morrison, P., Edsall, R. and Miller, S. G. Preparation and properties of serum and plasma proteins. XVIII. Separation of purified fibrinogen from fraction I of human plasma. *J. Amer. chem. Soc.* **70**, 3103–3108 (1948).

Mosesson,.M. W., Chen, A. B. and Huseby, R. M. The cold-insoluble globulin of human plasma: studies of its essential structural features. *Biochim. biophys. Acta (Amst.)* **386**, 509–524 (1975).

Mosher, D. F. Cross-linking of cold-insoluble globulin by fibrin-stabilizing factor. *J. biol. Chem.* **250**, 6614–6621 (1975a).

Mosher, D. F. Labeling of cold-insoluble globulin and a fibroblast protein by fibrin-stablizing factor. *Fed. Proc.* **34**, 498 (1975b).

Mosher, D. F. Action of fibrin-stabilizing factor on cold-insoluble globulin and α_2-macroglobulin in clotting plasma. *J. biol. Chem.* **251**, 1639–1645 (1976).

Muir, L. W., Bornstein, P. and Ross, R. A presumptive subunit of elastic fibre microfibrils secreted by arterial smooth muscle cells in culture. *Europ. J. Biochem.* **64**, 105–114 (1976).

Oosawa, F., Maruyama, M. and Fugima, S. Orientation distribution of globular protein molecules in a two dimensional lattice: computer stimulation. *J. theor. Biol.* **36**, 203–221 (1972).

Pearlstein, E. Plasma membrane glycoprotein which mediates adhesion of fibroblasts to collagen. *Nature (Lond.)* **262**, 497–500 (1976).

Pearlstein, E. and Waterfield, M. D. Metabolic studies on ^{125}I-labeled baby hamster kidney cell plasma membranes. *Biochim. biophys. Acta (Amst.)* **362**, 1–12 (1974).

Pollack, R., Osborn, M. and Weber, K. Patterns of organization of actin and myosin in normal and transformed cultured cells. *Proc. nat. Acid. Sci (Wash.)* **72**, 994–998 (1975).

Pollack, R. and Rifkin, D. Actin-containing cables within anchorage-dependent rat embryo cells are dissociated by plasmin and trypsin. *Cell* **6**, 495–506 (1975).

Robbins, P. W., Wickus, G. G., Branton, P. E., Gaffney, B. J., Hirschberg, C. B., Fuchs, P. and Blumberg, P. M. The chick fibroblast cell surface after transformation by Rous sarcoma virus. *Cold Spr. Harb. Symp. quant. Biol.* **39**, 1173–1180 (1974).

Ruoslahti, E. and Vaheri, A. A novel type of human serum protein from the plasma membrane of fibroblasts. *Nature (Lond.)* **248**, 789–791 (1974).

Ruoslahti, E. and Vaheri, A. Interaction of soluble fibroblast surface antigen with fibrinogen and fibrin. Identity with cold insoluble globulin of human plasma. *J. exp. Med.* **141**, 497–501 (1975).

Ruoslahti, E., Vaheri, A., Kuusela, P. and Linder, E. Fibroblast surface antigen: a new serum protein. *Biochim. biophys. Acta (Amst.)* **322**, 352–358 (1973).

Sear, C. H. J., Grant, M. E. and Jackson, D. S. Identification of a major extracellular non-collageneous glycoprotein synthesized by human skin fibroblasts in culture. *Biochem. biophys. Res. Commun.* **7**, 379–384 (1976).

Spooner, B. S., Yamada, K. M. and Wessels, N. K. Microfilaments and cell locomotion. *J. Cell Biol.* **49**, 595–613 (1971).

Spudich, J. A. Effects of cytochalasin B on actin filaments. *Cold Spr. Harb. Symp. quant. Biol.* **37**, 585–593 (1973).

Stenman, S., Wartiovaara, J. and Vaheri, A. Changes in the distribution of a major fibroblast protein, fibronectin, during mitosis and interphase. *J. Cell Biol.* In press.

Strickland, S., Sherman, M. and Reich, E. Plasminogen activator in early embryogenesis: Enzyme production by trophoblast and parietal endoderm. *Cell* **9**, 231–240 (1976).

Vaheri, A. and Ruoslahti, E. Disappearance of a major cell-type specific surface antigen

(SF) after transformation of fibroblasts by Rous sarcoma virus. *Int. J. Cancer* **13**, 579–586 (1974).

Vaheri, A. and Ruoslahti, E. Fibroblast surface antigen produced but not retained by virus-transformed human cells. *J. exp. Med.* **142**, 530–535 (1975).

Vaheri, A., Ruoslahti, E., Linder, E., Wartiovaara, J., Keski-Oja, J., Kuusela, P. and Saksela, O. Fibroblast surface antigen (SF): Molecular properties, distribution *in vitro* and *in vivo*, and altered expression in transformed cells. *J. supramolec. Struc.* **4**, 63–70 (1976a).

Vaheri, A., Ruoslahti, E., Westermark, B. and Pontén, J. A common cell-type specific surface antigen (SF) in cultured human glial cells and fibroblasts: loss in malignant cells. *J. exp. Med.* **143**, 64–72 (1976b).

Wallach, D. F. H. Membrane Molecular Biology of Malignant Cells, pp. 392, Elsevier, Amsterdam, Oxford, New York (1975).

Wartiovaara, J., Linder, E., Ruoslahti, E. and Vaheri, A. Distribution of fibroblast surface antigen. Association with fibrillar structures of normal cells and loss upon viral transformation. *J. exp. Med.* **140**, 1522–1533 (1974).

Wartiovaara, J., Stenman, S. and Vaheri, A. Changes in expression of fibroblast surface antigen (SFA) in induced cytodifferentiation and in heterokaryon formation. *Differentation* **5**, 85–89 (1976).

Vracko, R. Basal lamina scaffold-anatomy and significance for maintenance of orderly tissue structure. *Amer. J. Path.* **77**, 314–346 (1974).

Yamada, K. M. and Weston, J. A. Isolation of a major cell surface glycoprotein from fibroblasts. *Proc. nat. Acad. Sci. (Wash.)* **71**, 3492–3496 (1974).

Gap Junctions and Cell Contacts

NORTON B. GILULA

The Rockefeller University, New York, New York 10021, USA

INTRODUCTION

Metazoan cells can interact via diffusible substances or direct physical contact. These two types of interaction have been associated with a variety of biological phenomena that are clearly related to the process of differentiation in both embryonic as well as mature organisms. Diffusible substances, such as hormones, cyclic nucleotides, proteases, etc., are frequently transmitted between cells that are not in physical contact with each other. In fact, the most extensive of these "long-range" interactions would include the hormonally mediated interaction between the pituitary gland and a specific target cell in a different organ. Perhaps the most common method for cells to interact in solid tissue masses is by direct physical contact (short-range interactions). Many different types of direct physical interactions have been described between cells, and they are generally referred to as cell contacts. These contacts include cell surface specializations that are involved in cell-to-cell adhesion, communication, permeability regulation (occlusion), synaptic transmission, etc.

The purpose of this discussion is: (1) to briefly describe some of the contacts that are present between cells; (2) to examine some of the recent biochemistry of a specific cell contact, the gap junction; and (3) to discuss some of the relevant observations that have been made on the potential role of gap junctions (cell communication) in certain differentiation processes.

TYPES OF CELL CONTACTS

In this format, I will briefly describe a few of the cell contacts that have been characterized between cells. Several extensive reviews are

available that can provide additional information about these structures (Gilula, 1974a; Staehelin, 1974; McNutt and Weinstein, 1973;
Overton, 1974; Weinstein *et al.*, 1976).

Desmosomes

These structures are very abundant in many tissues, particularly in epithelia and myocardium (Farquhar and Palade, 1963; Kelley, 1966;
McNutt and Weinstein, 1973; Overton, 1974). There are many types
of desmosomes that are present between cells; however, they are all
considered to have a relevant role in adhesive interactions between
cells, cells and substrates, and they may provide insertion sites for cytoplasmic microfilaments. The intercellular organization of many tissues
is dependent on the integrity of the desmosomes. In general, desmosomes may be disrupted by protease treatment or by the removal of
divalent cations, such as calcium or magnesium (Muir, 1967; Berry
and Friend, 1969). In thin-section electron microscopy, the desmosome
(macula adhaerens) is characterized by a bipartite arrangement of
material on the cytoplasmic surfaces of the desmosomal membranes.
A dense plaque, that frequently serves as the insertion site for filaments,
is closely associated with the cytoplasmic surface of desmosomal membranes. The intervening intercellular space (about 20–30 nm wide)
contains a dense crystalline material that can be penetrated by extracellular tracers. In freeze-fracture replicas, some desmosomal membranes have a nonpolygonal arrangement of closely packed particles
or granules that can be detected on both the outer membrane half (E
fracture face) and the inner (cytoplasmic) membrane half (P fracture
face) (McNutt *et al.*, 1971; Breathnach *et al.*, 1972). In some tissues,
the desmosomal membrane components are not readily detectable
(Friend and Gilula, 1972).

Tight junctions

The tight junction or zonula occludens has been implicated as the cell
contact that is involved in transepithelial permeability regulation. This
structure clearly provides an occluding barrier for the intercellular diffusion of large molecules between cells (Farquhar and Palade, 1963;
Friend and Gilula, 1972). From recent studies, the tight junction has
been considered as a possible pathway for the transepithelial movements of small ionic molecules (Erlij and Martinez-Palomo, 1972;
Wade *et al.*, 1973; Di Bona and Civan, 1973). Ultrastructurally, the
tight junction can be resolved as a series of focal fusions between the

Fig. 1. Freeze-fracture appearance of the apical junctional complex between mammary gland epithelial cells (guinea pig). The region is characterized by microvilli (m), the tight junction, and desmosomes. The tight junctional region contains complementary ridges and furrows, while the desmosomal membranes (d) contain a plaque of heterogeneous intramembrane particles.

membranes of adjacent cells. The junctional membranes have a combined diameter of 14–15 nm at the points of fusion. The junction may exist as a belt-like structure (zonula) or as an isolated band of membrane fusion (fascia). With freeze-fracturing, the tight junction is characterized by two internal membrane fracture face components that are complementary (Fig. 1). A mesh-like arrangement of anastomosing ridges (8–9 nm in diameter) are present on the P fracture face (inner membrane half), while a complementary arrangement of grooves is present on the E fracture face (outer membrane half) (Chalcroft and Bullivant, 1970; Wade and Karnovsky, 1974; Staehelin, 1973). The tight junctions are generally insensitive to proteases; however, they can be "unzipped" or opened by treatment with hypertonic solutions (Goodenough and Gilula, 1974).

Gap junctions

The first electrotonic or low-resistance synapse was described in the crayfish central nervous system (Furshpan and Potter, 1959). Since that time, low-resistance pathways have been described between cells in both excitable and non-excitable tissues (Loewenstein, 1966; Furshpan and Potter, 1968; Bennett, 1973). Subsequently, the cell-to-cell transfer of small metabolites was described by Subak-Sharpe et al., (1969). In recent studies, it has been demonstrated that both metabolic and

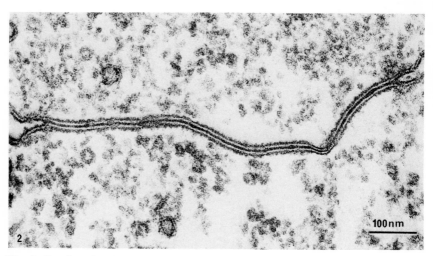

Fig. 2. Gap junction in this section between granulosa cells in a rat ovarian follicle. The septilaminar junction has been treated with tannic acid to emphasize the extracellular space or gap between the junctional membranes.

electrotonic transmission can occur between cells (Gilula *et al.*, 1972; Azarnia *et al.*, 1972), and a specific cell contact, the gap junction, serves as the structural pathway for these two types of cell-to-cell transfer (Gilula *et al.*, 1972). This type of cellular interaction where molecules are transmitted via a low-resistance pathway between cells is generally referred to as intercellular communication or cell-to-cell communication. This type of interaction is clearly responsible for the synchroniza-

Fig. 3. Typical freeze-fracture appearance of a gap junction between granulosa cells in a rat ovarian follicle. The junctional membrane is present as a plaque that contains a polygonal arrangement of complementary particles and pits.

tion of activities in certain regions of the nervous system, the myocardium, and smooth muscle. The gap junction in its present form was
first described by Revel and Karnovsky in 1967; however, this is identical to the structure that was previously described as the nexus (Dewey
and Barr, 1962). In thin sections, the gap junction is recognizable as
a septilaminar structure comprised of two 7.5 nm unit membranes
separated by a 2–4 nm space or "gap" (Fig. 2). This space can be penetrated by extracellular dense tracers such as lanthanum hydroxide. A
polygonal lattice of 7–8 nm subunits can be resolved in *en face* views
of lanthanum-impregnated gap junctions (Revel and Karnovsky, 1967;
McNutt and Weinstein, 1970). In freeze-fracture replicas, the inner
membrane half (P face) contains a specialized plaque of polygonally
arranged particles (9–10 nm centre-to-centre spacing) at the site of a
gap junction (Fig. 3). The outer membrane half (E face) at this site
contains a similar arrangement of complementary pits or depressions.
Gap junctions can occur between practically all types of metazoan cells
in a variety of pleiomorphic forms ·(Gilula, 1974a). At present, gap
junctions have been detected in a large number of systems that are elec-
trotonically coupled (Bennett, 1973; Gilula. 1974).

Other cell contacts

Many other types of cell contacts have been extensively characterized.
These include endothelial junctions (Yee and Revel, 1975; Simionescu
et al., 1975), Sertoli junctions in the testis (Gilula *et al.*, 1976), inverte-
brate septate junctions (Gilula *et al.*, 1970), and chemical synapses.
In all of these cases the cell contacts have been defined on the basis
of detectable features in the electron microscope. In many instances,
cells interact at very close range in regions where there are no detectable
specializations of the plasma membrane or intercellular space. These
regions cannot be adequately examined with respect to cell contact involvement until additional probes are developed.

RECENT OBSERVATIONS ON GAP JUNCTIONS

Physiological

Permeability properties of gap junctions

On the basis of electrotonic measurements, the junctional pathway
must contain channels that can permit the flow of hydrated inorganic
ions, such as sodium, potassium and chloride. Thus, the channels must
be in the order of 1–2 nm in diameter. In addition to inorganic ions,

injection studies have extended the permeability of coupling channels to include molecules such as fluorescein (330 m.w.), procion yellow (625 m.w.) and Chicago Sky Blue (993 m.w.), (Loewenstein, 1966; Furshpan and Potter, 1968; Bennett, 1973). Recent studies by Pitts and others (Pitts, unpublished observations; Tsien and Weingart, 1974; Rieske *et al.*, 1975) indicate that a variety of molecules, such as amino acids, sugars, phosphorylated sugars, nucleotides and cyclic nucleotides, may be transferred between cells. Thus, it appears from available data that the junctional channels are permeable to small metabolites and inorganic ions, but not to large macromolecules such as proteins and nucleic acids.

Regulation of junctional permeability

It has been firmly established in a recent study on the Dipteran salivary gland that junctional permeability can be controlled by intracellular calcium concentrations (Rose and Loewenstein, 1975). Furthermore, the data from this study can be interpreted to suggest that the calcium concentration must be elevated in the region of the cell contacts in order to affect the junctional permeability. In the studies on Dipteran salivary glands, the junctional permeability can be reduced by either increasing or decreasing the intracellular calcium concentration from the normal cytoplasmic level. Studies on two other systems, canine Purkinjie fibres (De Mello, 1974) and human lymphocytes (Oliveira-Castro and Barcinski, 1974), indicate a role for calcium in regulating junctional permeability. Studies on other systems have failed to demonstrate the same type of calcium involvement (Gilula and Epstein, 1975).

Structural

Gap junctions have now been identified in practically all multicellular systems that have been examined, with the notable exceptions of circulating lymphoid and erythroid cells, and mature skeletal muscle. In all cases, the identified gap junctions are basically similar, with the exception of those junctions in Arthropod tissues (Gilula, 1974a). Recent studies on isolated gap junctions have produced additional information on the junctional particles and the potential location of junctional channels (Zampighi and Robertson, 1973; Gilula, 1974b; Peracchia, 1975; Goodenough, 1976). In addition, some X-ray diffraction patterns have been obtained from isolated mouse liver gap junctions (Goodenough and Stoeckenius, 1972; Goodenough *et al.*, 1974). Gap junctional formation or assembly has recently been studied between cells in culture (Johnson *et al.*, 1974) and in developing tissues

(Decker and Friend, 1974; Revel *et al.*, 1974; Benedetti *et al.*, 1974; Decker, 1976). The formation process is characterized in freeze-fracture replicas by: (1) the appearance of formation plaques; (2) clustering or aggregation of large precursor particles in these regions; (3) appearance and polygonal packing of smaller gap junctional particles (Fig. 4); and (4) enlargement of particle plaques. In at least one of these systems, the formation process is not inhibited by inhibiting protein synthesis or metabolic energy production (Epstein and Sheridan, 1974).

Fig. 4. Gap junction formation between granulosa cells in the rat ovary. This intermediate stage in the formation process is characterized by the presence of large particles that exist in a loose arrangement in close association with smaller particles that are assembled into polygonal aggregates. The polygonal arrangement of small particles is the characteristic intramembrane specialization of "mature" gap junctions. In this image there are two regions of "mature" gap junctions that are associated with the larger "precursor" particles.

Biochemical

Gap junctions have been isolated as enriched subcellular fractions, primarily from rat and mouse liver (Benedetti and Emmelot, 1968; Goodenough and Stoeckenius, 1972; Evans and Gurd, 1972; Gilula, 1974b; Goodenough, 1974; Duguid and Revel, 1975). At present, no endogenous activity has been detected in these fractions, so the only criterion for purity is ultrastructural analysis. In all of these fractions that have been analysed biochemically, the junctions have been affected by a proteolytic treatment that is used to reduce the collagen contamination. Nonetheless, it appears that a 25 000 daltons polypeptide can be detected, together with other polypeptides, as a prominent component in both mouse and rat liver preparations. Thus far, no carbohydrate has been reported in any of the fractions. Recently, Goodenough (1976) has utilized an additional protease treatment with trypsin to produce some gap junctional vesicles in the mouse liver fraction.

ROLE OF GAP JUNCTIONS IN DIFFERENTIATION

Since gap junctional communication exists in a variety of developing tissues, it has been attractive to consider the potential regulatory role of this type of cellular interaction during differentiation. Unfortunately, it has been very difficult to demonstrate that gap junctional communication is indeed directly or indirectly involved in any differentiation process. It is clear from a study on the differentiation of epidermis (concerning intersegmental boundaries) that junctional communication can exist across apparent differentiation boundaries (Warner and Lawrence, 1973; Lawrence and Green, 1975). Recently, there have been several encouraging studies that may provide us with a good opportunity to carefully explore the role of gap junctions in differentiation. Some of these recent observations are discussed below.

In most early developing embryos that have been examined, gap junctional communication is well established. This communication, in the form of inorganic ions, may be present already at the 2-cell stage. However, several studies have indicated that while ions can be transferred between the embryonic cells, larger molecular weight dyes can not cross the junctions (Slack and Palmer, 1969; Bennett et al., 1972; Tupper and Saunders, 1972). In at least one case, the *Fundulus* embryo, the embryonic gap junctions are morphologically indistinguishable from the junctions in adult tissues (Bennett and Gilula, 1974). This restrictive junctional permeability in early embryos is particularly intriguing because it suggests that a variety of metabolities, such as nucleotides, sugars, etc., cannot be transmitted between the early embryonic cells, but they can cross the junctions at some later stage of development. Thus far, this is the only case where a qualitative selectivity has been associated with communication.

In a recent study, Blackshaw and Warner (1976) have found that during somite formation in the amphibian embryos *Bombina* and *Xenopus* the myotome cells in the segmented regions are coupled (communicating), the cells in the segmenting somite are poorly coupled to the cells in the segmented region, and the mesodermal cells in the unsegmented region are completely uncoupled from the cells in the segmented region. Thus, Blackshaw and Warner have suggested that this coupling pattern potentially reflects the number of cells that have completed their morphogenetic movements. In these embryos, coupling is re-established between cells of adjacent somites after segmentation is completed. Thus, during this differentiation process communication is initially present, lost and then re-established after the morphogenetic movements have ceased.

There have been some indications that gap junctional formation or proliferation can be induced during certain differentiation processes. In a study by Elias and Friend (1976), it has been reported that vitamin A-induced mucous metaplasia in 14-day chick embryo shank skin (organ culture) is accompanied by a significant proliferation of gap junctions after one day of treatment. Merk *et al.* (1972) have observed an increased number of gap junctions between granulosa cells in rat ovarian follicles that have been stimulated to proliferate by treatment with exogenous hormones. In hypophysectomized amphibian larvae, Decker (1976) has demonstrated that gap junctional proliferation between differentiating ependymoglial cells can be stimulated by treatment with thyroid hormone. The gap junctions that are formed in response to the hormone treatment appear 20–40 h after hormone treatment, and protein synthesis is required for the generation of these gap junctions.

A proliferation of gap junctions has been reported during apical ectodermal ridge formation in human limb morphogenesis (Kelley and Fallon, 1976). During this morphogenetic process, few gap junctions are present between the two cell layered epithelium. However, gap junctions dramatically increase in number during the subsequent proliferation of the ectoderm. At a later stage, the gap junctions are very sparse and the ectoderm returns to a two cell layer structure. In this system, there is a striking proliferation of gap junctions during the time when epithelial-mesenchymal interactions are associated with the morphogenetic process.

We have recently found that gap junctional communication is intimately associated with the follicular differentiation process in the mammalian ovary (Epstein *et al.*, 1976). Gap junctional communication, including fluorescent dye transfer, is present between the oocyte and the cumulus oophorus (granulosa) cells in immature follicles. This communication is established by granulosa cell processes that penetrate the zona pellucida and form bouton-like gap junctional contacts on the surface of the oocyte in the rat ovary. During follicular development, the communication gradually decreases as the time of ovulation approaches. After ovulation, the communication is not detectable between the cumulus cells and oocytes that are obtained from the oviduct. Hence, this pattern of communication between the cumulus and oocyte is perhaps closely related to the pattern of oocyte maturation that occurs during the same stages of development.

In the future, a clear definition of the relationship of cell communication to differentiation will depend on: (1) selecting a differentiation system that has a well-defined pattern of communication; and (2) developing procedures that will either facilitate the experimental manipulation of communication before it has been established or disrupt communication that has been previously established.

ACKNOWLEDGEMENTS

These studies have been supported by the Irma T. Hirschl Trust, the Andrew Mellon Fund, USPHS Grant HL-16507, and NIH Career Development Award HL-00110.

REFERENCES

Azarnia, R., Michalke, W. and Loewenstein, W. R. Intercellular communication and tissue growth. VI. Failure of exchange of endogenous molecules between cancer cells with defective junctions and noncancerous cells. *J. Membrane Biol.* **10**, 247–258 (1972).

Benedetti, E. L., Dunia, I. and Bloemendal, H. Development of junctions during differentiation of lens fiber. *Proc. nat. Acad Sci. (Wash.)* **71**, 5073–5077 (1974).

Benedetti, E. L. and Emmelot, P. Hexagonal array of subunits in tight junctions separated from isolated rat liver plasma membranes. *J. Cell Biol.* **38**, 15–24 (1968).

Bennett, M. V. L. Function of electrotonic junctions in embryonic and adult tissues. *Fed. Proc.* **32**, 65–75 (1973).

Bennett, M. V. L. and Gilula, N. B. Membranes and junctions in developing *Fundulus* embryos, freeze fracture and electrophysiology. *J. Cell Biol.* **63**, 21 (1974).

Bennett, M. V. L., Spira, M. E. and Pappas, G. D. Properties of electrotonic junctions between embryonic cells of *Fundulus. Develop. Biol.* **29**, 419–435 (1972).

Berry, M. N. and Friend, D. S. High yield preparation of isolated rat liver parenchymal cells: a biochemical and fine structure study. *J. Cell Biol.* **43**, 506–520 (1969).

Blackshaw, S. E. and Warner, A. E. Low resistance junctions between mesoderm cells during development of trunk muscles. *J. Physiol. (Lond.)* **255**, 209–230 (1976).

Breathnach, A. S., Stolinski, C. and Gross, M. Ultrastructure of fetal and post-natal human skin as revealed by the freeze-fracture replica technique. *Micron* **3**, 287–304 (1972).

Chalcroft, J. P. and Bullivant, S. An interpretation of liver cell membrane and junction structure based on observations of freeze-fracture replicas of both sides of the fracture. *J. Cell Biol.* **47**, 49–60 (1970).

Decker, R. S. Hormonal regulation of gap junction differentiation. *J. Cell Biol.* **69**, 669–685 (1976).

Decker, R. S. and Friend, D. S. Assembly of gap junctions during amphibian neurulation. *J. Cell Biol.* **62**, 32–47 (1974).

De Mello, W. Intracellular Ca injection and cell communication in heart. *Fed. Proc.* **33**, 445 (1974).

Dewey, M. M. and Barr, L. Intercellular connection between smooth muscle cells: the nexus. *Science* **137**, 670–672 (1962).

DiBona, D. R. and Civan, M. M. Pathways for movement of ions and water across toad urinary bladder. I. Anatomic site of transepithelial shunt pathways. *J. Membrane Biol.* **12**, 101–128 (1973).

Duguid, J. R. and Revel, J. P. The protein components of the gap junction. *Cold Spr. Harb. Symp. quant. Biol.* **40**, 45–47 (1975).

Elias, P. M. and Friend, D. S. Vitamin-A-induced mucous metaplasia. An *in vitro* system for modulating tight and gap junction differentiation. *J. Cell Biol.* **68**, 173–188 (1976).

Epstein, M. L., Beers, W. H. and Gilula, N. B. Cell communication between the rat cumulus oophorus and the oocyte. *J. Cell Biol.* **70**, 302 (1976).

Epstein, M. and Sheridan, J. Formation of low-resistance junctions in the absence of protein synthesis and metabolic energy production. *J. Cell Biol.* **63**, 95 (1974).

Erlij, D. and Martinez-Palomo, A. Opening of tight junctions in frog skin by hypertonic urea solutions. *J. Membrane Biol.* **9**, 229–240 (1972).

Evans, W. H. and Gurd, J. W. Preparation and properties of nexuses and lipid-enriched vesicles from mouse liver plasma membranes. *Biochem. J.* **128**, 691–700 (1972).

Farquhar, M. G. and Palade, G. E. Junctional complexes in various epithelia. *J. Cell Biol.* **17**, 375–412 (1963).

Friend, D. S. and Gilula, N. B. Variations in tight and gap junctions in mammalian tissues. *J. Cell Biol.* **53**, 758–776 (1972).

Furshpan, E. J. and Potter, D. D. Transmission at giant motor synapses of the crayfish. *J. Physiol.* (*Lond.*) **143**, 289–325 (1959).

Furshpan, E. J. and Potter, D. D. Low-resistance junctions between cells in embryos and tissue culture. *Curr. Top. Develop. Biol.* **3**, 95–127 (1968).

Gilula, N. B. Junctions between cells, in R. P. Cox (ed.), Cell Communication, pp. 1–29, John Wiley and Sons, Inc., New York (1974a).

Gilula, N. B. Isolation of rat liver gap junctions and characterization of the polypeptides. *Abstract. J. Cell Biol.* **63**, 111 (1974b).

Gilula, N. B., Branton, D., and Satir, P. The septate junction: a structural basis for intercellular coupling. *Proc. nat. Acad. Sci.* (*Wash.*) **67**, 213–220 (1970).

Gilula, N. B. and Epstein, M. L. Cell-to-cell communication gap junctions and calcium. *Soc. exp. Biol. Symp.* **30**, 257–272 (1975).

Gilula, N. B., Fawcett, D. W. and Aoki, A. The Sertoli cell occluding junctions and gap junctions in mature and developing testis. *Develop. Biol.* **50**, 142–168 (1976).

Gilula, N. B., Reeves, O. R. and Steinbach, A. Metabolic coupling, ionic coupling, and cell contacts. *Nature* (*Lond.*) **235**, 262–265 (1972).

Goodenough, D. A. Bulk isolation of mouse hepatocyte gap junctions. *J. Cell Biol.* **61**, 557–563 (1974).

Goodenough, D. A. *In vitro* formation of gap junction vesicles. *J. Cell Biol.* **68**, 220–231 (1976).

Goodenough, D. A., Caspar, D. L. D., Makowski, L. and Phillips, W. C. X-ray diffraction of isolated gap junctions. *J. Cell Biol.* **63**, 115 (1974).

Goodenough, D. A. and Gilula, N. B. The splitting of hepatocyte gap junctions and zonulae occludentes with hypertonic disaccharides. *J. Cell Biol.* **61**, 575–590 (1974).

Goodenough, D. A. and Stoeckenius, W. The isolation of mouse hepatocyte gap junctions. Preliminary chemical characterization and X-ray diffraction. *J. Cell Biol.* **54**, 646–656 (1972).

Johnson, R. G., Hammer, M., Sheridan, J. and Revel, J. P. Gap junction formation between reaggregated Novikoff hepatoma cells. *Proc. nat. Acad. Sci. (Wash.)* **71**, 4536–4540 (1974).

Kelley, D. E. Fine structure of desmosomes, hemidesmosomes, and an adepidermal globular layer in developing newt epidermis. *J. Cell Biol.* **28**, 51–72 (1966).

Kelley, R. O. and Fallon, J. F. Ultrastructural analysis of the apical ectodermal ridge during vertebrate limb morphogenesis. *Develop. Biol.* **51**, 241–256 (1976).

Lawrence, P. A. and Green, S. M. The anatomy of a compartment border. The intersegmental boundary in *Oncopeltus. J. Cell Biol.* **65**, 373–382 (1975).

Loewenstein, W. R. Permeability of membrane junctions. *Ann. N.Y. Acad. Sci.* **137**, 441–472 (1966).

McNutt, N. S., Hershberg, R. A. and Weinstein, R. S. Further observations on the occurrence of nexuses in benign and malignant human cervical epithelium. *J. Cell Biol.* **51**, 805–825 (1971).

McNutt, N. S. and Weinstein, R. S. The ultrastructure of the nexus. A correlated thin-section and freeze-cleave study. *J. Cell Biol.* **47**, 666–687 (1970).

McNutt, N. S. and Weinstein, R. S. Membrane ultrastructure at mammalian intercellular junctions. *Progr. Biophys.* **26**, 45–101 (1973).

Merk, F. B., Boticelli, C. R. and Albright, J. T. An intercellular response to estrogen by granulosa cells in the rat ovary; an electron microscope study. *Endocrinology* **90**, 992–1007 (1972).

Muir, A. R. The effect of divalent cations on the ultrastructure of the perfused rat heart. *J. Anat.* **101**, 239–262 (1967).

Oliveira-Castro, G. M. and Barcinski, M. A. Calcium-induced uncoupling in coupling in communicating human lymphocytes. *Biochem. biophys. Acta (Amst.)* **352**, 338–343 (1974).

Overton, J. Cell junctions and their development. *Progr. Surface Membrane Sci.* **8**, 161–208 (1974).

Peracchia, C. and Fernandez-Jaimovich, M. E. Isolation of intramembrane particles from gap junctions. *J. Cell Biol.* **67**, 330 (1975).

Revel, J. P., Chang, L. and Yip, P. Cell junctions in the early chick embryo: a freeze-etch study. *Develop. Biol.* **35**, 302–317 (1973).

Revel, J. P. and Karnovsky, M. J. Hexagonal array of subunits in intercellular junctions of the mouse heart and liver. *J. Cell Biol.* **33**, C7-C12 (1967).

Rieske, E., Schubert, P. and Kreutzberg, G. W. Transfer of radioactive material between electrically coupled neurons of the leech central nervous system. *Brain Res.* **84**, 365–382 (1975).

Rose, B. and Loewenstein, W. R. Permeability of cell junction depends on local cytoplasmic calcium activity. *Nature (Lond.)* **254**, 250–252 (1975).

Simionescu, M., Simionescu, N. and Palade, G. E. Segmental differentiations of cell junctions in the vascular endothelium. The microvasculature. *J. Cell Biol.* **67**, 863–886 (1975).

Slack, C. and Palmer, J. F. The permeability of intercellular junctions in the early embryo of *Xenopus laevis*, studied with a fluorescent tracer. *Exp. Cell Res.* **55**, 416–419 (1969).

Staehelin, L. A. Further observations on the fine structure of freeze-cleaved tight junctions. *J. Cell Sci.* **13**, 763–786 (1973).

Staehelin, L. A. Structure and function of intercellular junctions. *Int. Rev. Cytol.* **39**, 191–283 (1974).

Subak-Sharpe, J. H., Burk, R. R. and Pitts, J. D. Metabolic cooperation between biochemically marked mammalian cells in tissue culture. *J. Cell Sci.* **4**, 353–367 (1969).

Tsien, R. W. and Weingart, R. Cyclic AMP: cell-to-cell movement and ionotropic effect in ventricular muscle, studied by a cut-end method. *J. Physiol. (Lond.)* **242**, 95–96P (1974).

Tupper, J. T. and Saunders, J. W., Jr. Intercellular permeability in the early *Asterias* embryo. *Develop. Biol.* **27**, 546–554 (1972).

Wade, J. B. and Karnovsky, M. J. The structure of the zonula occludens. A single fibril model based on freeze-fracture. *J. Cell Biol.* **60**, 168–180 (1974).

Wade, J. B., Revel, J. P. and DiScala, V. A. Effect of osmotic gradients on intercellular junctions of the toad bladder. *Amer. J. Physiol.* **224**, 407–415 (1973).

Warner, A. E. and Lawrence, P. A. Electrical coupling across developmental boundaries in insect epidermis. *Nature (Lond.)* **245**, 47–48 (1973).

Weinstein, R. S., Merk, F. B. and Alroy, J. The structure and function of intercellular junctions in cancer. *Advanc. Cancer Res.* **23**, 23–89 (1976).

Yee, A. G. and Revel, J. P. Endothelial cell junctions. *J. Cell Biol.* **66**, 200–204 (1975).

Zampighi, G. and Robertson, J. D. Fine structure of the synaptic discs separated from the goldfish medulla oblongata. *J. Cell Biol.* **56**, 92–105 (1973).

The Molecular Basis of the Mesenchymal-Epithelial Interactions in Pancreatic Development

RAYMOND L. PICTET and WILLIAM J. RUTTER

Department of Biochemistry and Biophysics, University of California, San Francisco, California 94143, USA

INTRODUCTION

The regulation of the function of cells by other cells, e.g. the nervous system, or through the agency of hormones is relatively well perceived at the physiological and biochemical level. A molecular signal secreted by the effector cell interacts with a receptor of the target cell to produce the consequent physiological effect. It is becoming evident that hormone-like molecules play an important role in the terminal differentiation of many cells. Many of these molecules have been detected and several, such as nerve growth factor, epithelial growth factor and fibroblast growth factor, have been isolated and reasonably well characterized. Whether hormone-like signals are the basis of the cell interactions involved in so-called embryonic induction in which effector cells or tissues alter the growth and differentiation of target cells still remains a matter of conjecture. Tiedemann and his associates (1966) have reported the isolation of molecules (neuralizing and mesodermal factors) which exert inductive effects, but a number of non-specific chemicals and treatments are also known to produce these effects *in situ*. At a later stage of development it is well known that mesenchyme and epithelial interactions are involved in the development of most organ systems. The interaction of the mesenchymal component with its epithelial counterpart has been defined primarily by the work of Wolff and Grobstein and their respective colleagues (for review see Wolff, 1968; Grobstein, 1967). The results of the classical transfilter culture experiments suggested that close proximity but not direct contact between the cells was required; thus a chemical intermediary was implicated. Recently, however, Saxén and his group (Wartiovaara *et al.*, 1974; Lehtonen *et al.*, 1975) have shown that in kidney development processes of the cells can indeed pass through the filter and that inducing (spinal cord) and target (kidney mesenchyme) cells make contact; the extent of such contact appears roughly proportional to the degree of the observed effect. These results thus call into question the older experimental observations of Grobstein and co-workers since Kallman and Grobstein (1964) didn't detect interactions between the mesenchymal and epithelial cells by observations with the electron microscope of thin sections made through the Millipore filter.

MESENCHYMAL FACTOR

Our experiments have employed the pancreas as a system to study epithelial-mesenchymal interactions. Unlike many epithelial-mesenchymal systems the pancreatic epithelium has a non-specific requirement for mesenchyme for normal growth and differentiation. Early experiments showed that a factor mimicking the mesenchymal tissue was present in the insoluble membrane-rich fractions of whole embryos. This material was demonstrated to be present in mesenchymal but not epithelial tissues (Rutter *et al.*, 1964; Ronzio and Rutter,

Figs 1 and 2. Only the basal part of the epithelial cells interacts with MF-Sepharose, in analogy with the spatial relationship between the epithelial and mesenchymal cells in the intact tissue.

Fig. 1. Epithelial cell is *in vivo* separated from mesodermal cells by basal lamina.
Fig. 2. The explanted epithelium (12 days) has been deprived of mesenchyme and basal lamina by trypsin digestion and cell surface is in contact with the surface of the Sepharose beads. The polarization is more clearly seen when the acinar cells have differentiated (Levine *et al.*, 1973).

Fig. 3. Some of the cells on the surface of the beads show an unusually large amount of microfilaments along the plasma membrane and particularly where it is in contact with the bead surface. These filaments may be arranged perpendicular to the membrane as well as longitudinally as shown here. In some cases one can distinguish two layers of filaments (indicated as F_1 and F_2) which are oriented in the same plan but perpendicularly to one another. A variable amount of microtubules (arrows) are also generally present.

Fig. 4. Cytocholasin B (10μg/ml) does not prevent the attachment of the epithelial cells to the MF found to Sepharose beads (SB). The microfilaments are clearly altered and form an amorphous mass of material (arrows). Cytochalasin inhibits the cell motility and the cells not touching the beads are not brought into contact with them by the movement of already attached cells. This concentration of cytocholasin B is somewhat toxic and some necrotic cells are seen.

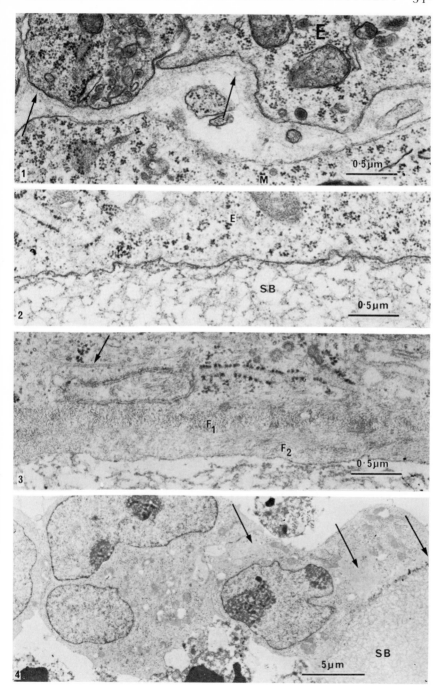

1973). This factor, termed mesenchymal factor (MF), is both necessary and sufficient to support the cell proliferation and normal differentiation of the pancreas. In particular, it elicits rapid DNA synthesis and the formation of a predominantly acinar tissue. No other known factors or hormones exert similar effects. The fact that the mesenchymal cells can be replaced by soluble material (derived therefrom) demonstrates that a physical interaction of the cells which may occur *in vivo* is not an absolute requirement; thus the transfer of biological information need not occur through direct exchange of intracellular components.

Chemical characteristics of MF

MF is partially purified; it can be extracted with high salt, and selectively precipitated with calcium ions at low ionic strength. Its activity is trypsin-sensitive, and insensitive to the action of RNAse and DNAse. It is, however, sensitive to treatment with periodate, under mild conditions in which there is selective oxidation of carbohydrate moieties. We therefore presume that MF is/are glycoprotein(s). More recently, we have shown that mesenchymal factor is fully active when convalently bound to Sepharose beads that are much too large to be engulfed by the cells (Levine *et al.*, 1973; Pictet *et al.*, 1975a) Pancreatic epithelia at around 35 somites isolated from their surrounding mesenchyme, and incubated with such MF-Sepharose, strongly attach and spread over the surface of the beads. The binding is restricted to the basal surface of the cells in an analogous fashion to the *in vivo* situation in which the basal surface of the cells faces the mesenchymal cells (Figs 1 and 2). After 24 h many of the cells attached to the MF-Sepharose beads show abundant filamentous structures (Fig. 3). These filaments seem to be related to the motility of the cells on the beads and are probably not required for the binding itself: cytochalasin B, which is known to destroy the filamentous structures, does not appear to block the binding of the cells, but does inhibit their spreading on the surface of the beads (Fig. 4). Since cytochalasin B has also been shown to inhibit glucose transport, these results suggest but do not prove that binding of the cells can occur in the absence of energy derived from the glycolytic pathway.

The very strong interaction of the epithelial cells with the MF-Sepharose beads suggests that mesenchymal factor interacts with receptors on the epithelial cell surface. By analogy with the known mode of action of hormones, we tested whether cAMP or cGMP analogues which penetrate cells could replace or affect the action of MF (Filosa *et al.*, 1975; Pictet *et al.*, 1975a). Neither cAMP nor cGMP analogues alone or in combination could replace MF, nor did they inhibit or potentiate its action (Table I). However, MF inactivated by periodate oxidation (IO_4-MF) is fully active in the presence of dibutyryl cAMP or 8-OHcAMP added to the medium. The activity (measured in terms of the incorporation of ^{3}H-thymidine into DNA) is dependent both on

TABLE I

MF inactivated by IO_4Na treatment stimulates DNA synthesis in the presence of cAMP but not cGMP. The pancreatic epithelia were isolated from rat embryos on the 12th day of gestation (about 35 somites) and the mesenchyme removed. They were incubated for 24 h in the presence of 600 μg/ml of IO_4-treated MF and 8-OHcAMP and/or 8-BrcGMP at the indicated concentrations. 10 μCi of ^{3}H-thymidine was added to the medium for the last 6 h of the culture and its incorporation into DNA was measured. The values are the average of 6–10 experimental points. (Modified from Filosa et al., 1975)

	8-OHcAMP (10^{-3}M)	8-BrcGMP (10^{-1}M)	^{3}H-Thymidine incorporated into DNA (cpm/epithelia/6 h)
MF	—	—	7330
	+	—	8680
	—	+	7495
IO_4-MF	—	—	290
	+	—	6500
	—	+	264

the concentration of the IO_4-MF and the concentration of the cAMP derivative (Fig. 5). This suggests that MF is a complex composed of at least two activities, one stimulating the adenylcyclase, and thereby increasing the intracellular level of cAMP, the second component acting through some other mechanism. Whether or not both activities correlate with one or several molecules is yet unknown. Our further attempts to purify the factor have been facilitated by the possibility of measuring the DNA synthetic activity in the presence of 8-OHcAMP since the moiety replaced by the cAMP derivative is very labile. It has been recently found that MF binds tightly to a macromolecule which is precipitated by calcium and is probably actin. The apparent specificity of this interaction is of some interest because of the role of actin and myosin in cell motility and the maintenance of cell shape and the morphogenetic activities of MF.

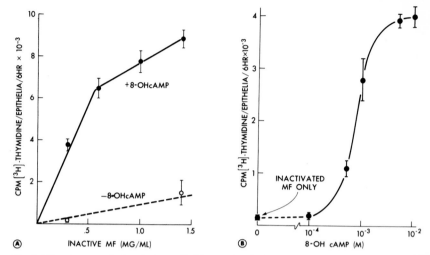

Fig. 5. The degree of stimulation of DNA synthesis is dependent on the concentration of both inactivated MF and the cAMP derivative. In isolated pancreatic epithelia DNA synthesis was measured by incorporation of ^{3}H-thymidine during the last 6 h of the 24 h incubation. (A) Sets of nine epithelia were cultured in the presence of various concentrations of inactivated factor in the presence ($\bullet$) or absence ($\circ$) of 10^{-3}M-OHcAMP. (B) Sets of 9 epithelia were cultured in the presence of 300 µg/ml of inactivated MF and various concentrations of 8-OHcAMP. The bars represent the SEM of the tritium counts. (From Filosa et al.)

Mode of action of MF

MF appears not to act as a trigger required at a single period of development and setting off a chain of developmental events. It is rather acting from the time of the appearance of the pancreatic diverticulum at 20–25 somites (11th day of embryonic development) until overt differentiation of the acinar cells (Fig. 6 and Table II). In the early developmental stages, the factor is absolutely required for DNA synthesis. If day 12 pancreatic epithelia are grown for 24 h in the presence of MF and kept for an additional day in the absence of it, incorporation of thymidine into DNA is reduced to background. In later periods, however, removal of MF decreases but does not suppress the DNA synthesis (Fig. 6) Thus in later stages of development the regulation of cell proliferation is no longer completely dependent upon the presence of mesenchymal factor. These results are consistent with earlier studies on the role of the mesenchyme on the stimulation of DNA synthesis and cell proliferation in epithelial tissues.

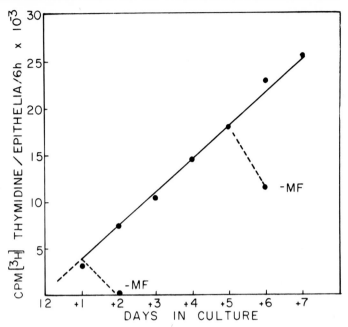

Fig. 6. Pancreatic epithelial buds were cultured for 7 days in the presence of mesen-chymal factor (MF). Every day 6 epithelia were given a 6 h pulse of H³-thymidine (to evaluate DNA synthesis). In addition 2 sets of 6 epithelia were pulsed on days 2 and 6 after 24 h of culture in albumin (instead of MF).

TABLE II

Cyclic AMP dependency during pancreatic development. Day 12 pancreatic epithelia prepared as described in Table I were cultured from 1–5 days in the presence of perio-date-inactivated factor and in the absence or presence of 10^{-3}M 8-OHcAMP. Ten μCi of ³H-thymidine were added to the medium for the last 6 h of culture and incorporation into DNA was measured. On both days there is a clear stimulation of DNA synthesis by the presence of cAMP (Pictet, Filosa and Rutter, unpublished results)

Days in culture	³H-Thymidine incorporation into DNA (cpm/epithelia/6 h)	
	− 8-OHcAMP	+ 8-OHcAMP
1	420	6260
5	1820	11 480

The differentiation of the pancreas is also qualitatively affected by the presence of the mesenchymal factor. The presence of MF in the culture medium for the 7 days it takes for the cells to differentiate *in vivo* leads to a pattern of the differentiation similar to *in vivo*. The large majority (about 80%) of the differentiated epithelial cells are acinar with about 6% endocrine B and 1% A. The remaining undifferentiated or duct cells correspond to approximately 13% of the total population. In contrast, pancreatic epithelia cultured in the absence of mesenchyme or MF in a defined medium containing albumin or in a medium containing serum developed very few acinar cells. The low proportion of acinar cells in the absence of MF is not due to a delay in the expression of differentiation since prolongation of the culture for another 7 days does not alter the appearance of the tissue (Fig. 7). An analysis of the tissue with the electron microscope revealed that most of the cells of the epithelium cultivated in the absence of MF are endocrine cells. The relative preponderance of the endocrine in relation to the acinar cells is especially high in albumin medium. Under these conditions, however, some of the cells of the rudiment slough off during incubation. This could be related to a nutritional insufficiency or the inability of the tissue to recovery after trypsinization or dissection. The very low proportion of acinar cells is not, however, due to the selective loss of acinar cell precursors since tissues cultured for a week in albumin-containing medium and an additional week in the presence of MF show numerous acinar cells (Fig. 8). Thus the tissues can still respond normally to MF. The change in the differentiation pattern may be linked to the proliferative effect of MF. Whether any endocrine A or B cells can undergo transdifferentiation to acinar cells is yet an unsettled question. The pancreatic epithelia can be cultivated more successfully in the presence of IO_4-MF. In these instances, the activity of IO_4-MF is no more than 5% of that of the original MF but is still significant. However, at the end of the 7–8 day cultivation period, the tissue still

Figs 7 and 8. The change in the pattern of differentiation occurring in the absence of MF is reversible. Fig. 7 is a section of a 12 day pancreatic epithelial bud grown in albumin-containing medium for 14 days. The large majority of the cells are light and only occasionally acinar cells containing zymogen granules are found. If after 7 days in such a culture medium, corresponding to the differentiation period *in vivo* or in epithelia grown in the presence of factor, MF is added to the medium, many acinar cells containing zymogen granules are present 7 days later (Fig. 8).

Fig. 9. Endocrine A cells in the islet of an *Acomys* on the 23rd day of embryonic development. At this time the pancreas contains only low levels of insulin and no B cells are detected. These numerous A cells correspond to a very high concentration in glucagon as shown on Fig. 11.

Fig. 10. Pancreas of an *Acomys* on the 23rd day of embryonic development. Individual epithelial digitation are dispersed in a loose mesenchyme (M).

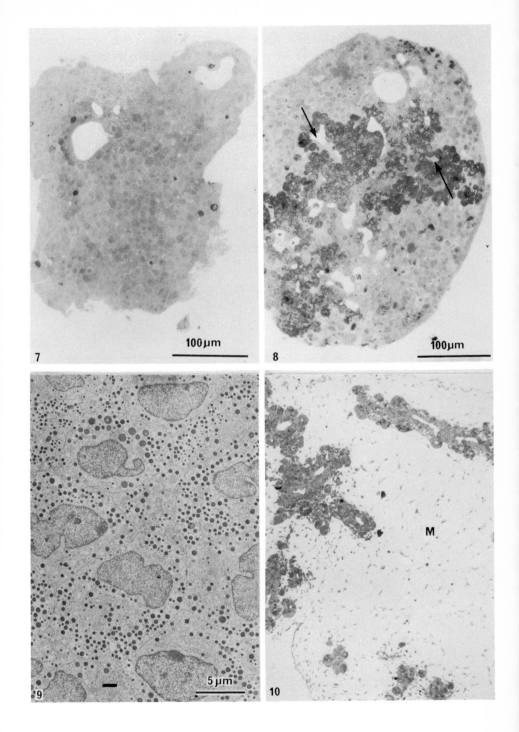

shows the inverted acinar/endocrine cell ratio typical of the explant grown in the absence of MF with over 60% of the differentiated cells endocrine and only about 20% acinar cells. The preponderant endocrine cells are A cells; about 15% of the total cells remain undifferentiated, corresponding to duct cells (Pictet *et al.*, 1975b).

Studies on *Acomys cahirinus*

These experimental results suggest that the mesenchymal factor plays a specific role in determination of the proportion of endocrine and acinar cells in the pancreas, either by a direct induction of the expression of the phenotype or by indirectly regulating the cell proliferation of the various precursors of the differentiated cell types. Thus, characteristics of the mesenchymal-epithelial interactions might qualitatively

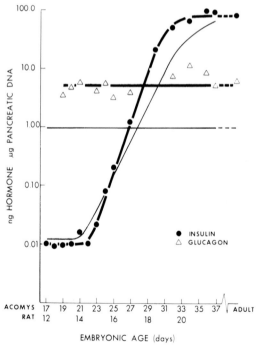

Fig. 11. Pattern of insulin and glucagon accumulation in *Acomys* (thick line) embryonic pancreas. The gestational period and the pancreatic differentiation are longer but the patterns of accumulation of the two hormones are similar to those in the rat (thin line). However, in *Acomys* there is a 5-fold increase in the glucagon to insulin ratio, possibly related to a deficiency in the development of the mesenchyme (see Fig. 10). (Pictet, R., DeGasparo, M., Rall, L. and Rutter, W. J., unpublished results.)

alter the cellular composition of the pancreas *in vivo*. Preliminary experiments supporting this thesis have been carried out with *Acomys cahirinus*, a small rodent from the eastern Mediterranean. The *Acomys* pancreas is characterized by a large proportion of endocrine relative to acinar tissue: endocrine cells are approximately 15% of the pancreas in adult and 25% of the pancreas at birth (Gonet *et al.*, 1965) (as compared with <2% in other species). In spite of the preponderance of endocrine cells, this species is known to develop a spontaneous diabetes. We have thus investigated the development of the pancreas during the 37 day gestational period (the gestational period of the mouse and rat is 19 and 22 days respectively). In contrast to rat and mouse pancreas, the development of the *Acomys* pancreas *in vitro* is poor. Histological observations show few mesenchymal cells in the pancreas of *Acomys* as compared to rat or mouse. Furthermore, there is a five-fold higher concentration of glucagon relative to insulin in the *Acomys* pancreas (Fig. 11). Thus the peculiar differentiation of the pancreas of *Acomys* may represent an example of a partial mesenchymal deficiency.

CONCLUDING REMARKS

The present experiments suggest that the mesenchymal-epithelial interactions in the pancreas are mediated by a protein or proteins which react with the cell surface. Whether this component diffuses from the mesenchymal to the epithelial cells or is presented by direct contact of the cell membrane remains to be demonstrated. Furthermore, the precise mechanism by which the components of MF exert their effect on cell proliferation and differentiation also remains to be elucidated. We presume, however, that the general mode of interaction is similar to that of other hormones with the important distinction that the action must occur over a very short range. We also presume that the mechanism is similar for all mesenchymal and epithelial interactions and indeed may be general for all cellular inductive reactions in embryological development.

ACKNOWLEDGEMENTS

This research was supported by grants from the National Foundation—March of Dimes (to Raymond L. Pictet) and the National Science Foundation, BMS72-02222 (to William J. Rutter). Raymond L. Pictet is a recipient of a Career Development Award of the National Institutes of Health.

REFERENCES

Filosa, S., Pictet, R. and Rutter, W. J. Positive control of cyclic AMP on mesenchymal factor controlled DNA synthesis in embryonic pancreas. *Nature (Lond.)* **257**, 702–705 (1975).

Gonet, A. E., Stauffacher, W., Pictet, R. and Renold, A. E. Obesity and diabetes mellitus with striking congenital hyperplasia of the islets of Langerhans in spiny mice (*Acomys cahirinus*). I. Histological findings and preliminary metabolic observations. *Diabetologia* **1**, 162–170 (1965).

Grobstein, C. Mechanism of organogenetic tissue interactions. *Nat. Cancer Inst. Monogr.* **26**, 279–299 (1967).

Kallman, F. and Grobstein, C. Fine structure of differentiating mouse pancreatic exocrine cells in transfilter culture. *J. Cell Biol.* **20**, 399–413 (1964).

Lehtonen, E., Wartiovaara, J., Nordling, S. and Saxén, L. Demonstration of cytoplasmic processes in Millipore filters permitting kidney tubule induction. *J. Embryol. exp. Morph.* **33**, 187–203 (1975).

Levine, S., Pictet, R. and Rutter, W. J. Control of cell proliferation and cytodifferentiation by a factor reacting with the cell surface. *Nature New Biol. (Lond.)* **246**, 49–52 (1973).

Pictet, R., Filosa, S., Phelps, P. and Rutter, W. J. Control of DNA synthesis in the embryonic pancreas: Interaction of the mesenchymal factor and cyclic AMP, in H. C. Slavkin and R. C. Greulich (ed.), Extracellular Matrix Influences on Gene Expression, pp. 531–540, Academic Press, New York (1975a).

Pictet, R., Rall, L. de Gasparo, M and Rutter, W. J. Regulation of differentiation of endocrine cells during pancreatic development *in vitro*, in R. A. Camerini-Davalos and H. S. Cole (ed.), Early Diabetes in Early Life, pp. 25–39, Academic Press, New York (1975b).

Ronzio, R. A. and Rutter, W. J. Effects of a partially-purified factor from chick embryos on macromolecular synthesis of embryonic pancreatic epithelia. *Develop. Biol.* **30**, 307–320 (1973).

Rutter, W. J., Wessells, N. K. and Grobstein, C. Control of specific synthesis in the developing pancreas. *Nat. Cancer Inst. Monogr.* **13**, 51–65 (1964).

Tiedemann, H. The molecular basis of differentiation in early development. *Curr. Top. develop. Biol.* **1**, 85–112 (1966).

Wartiovaara, J., Nordling, S., Lehtonen, E. and Saxén, L. Transfilter induction of kidney tubules: Correlation with cytoplasmic penetration into Nucleopore filters. *J. Embryol. exp. Morph.* **31**, 667–682 (1974).

Wolff, E. Specific interactions between tissues during organogenesis. *Curr. Top. develop. Biol.* **3**, 65-94 (1968).

V
Cell Recognition

Cell Recognition: An Introduction

A. A. Moscona

Department of Biology, University of Chicago,
Chicago, Ill., USA

The problems of "social interactions" among embryonic cells—i.e. how cells in the embryo associate into the diverse multicellular systems (tissues and organs) and cooperate in the construction of multiunit complexes—these problems are still rather poorly understood, and some of the major questions are only now beginning to come into sharp focus. Only few of these questions can, as yet, be phrased in precise biochemical and biophysical terms, and even fewer have been subjected to systematic analysis at the molecular-genetic levels.

The little that is known about these problems has drawn attention to the crucial role of the cell surface. The cell surface participates critically in two fundamental aspects of embryonic development. First, it relays signals from the outside to the inside of the cell and thereby participates in the control of cell differentiation and cell replication. It is now generally recognized that various factors extrinsic to the cell—including contact with other cells and with various extracellular components—generate changes in the cell surface which can alter intracellular processes; such internal alterations, in turn, can modify the cell surface, and thereby cell interactions; and it is such reciprocal sequences of inside-outside reactions, transacted through the cell surface, that drive embryonic cells along the course of morphogenesis and differentiation.

The other crucial role of the cell surface in embryogenesis derives from the fact that the surface is the site of mechanisms for *cell–cell recognition* and *selective cell adhesion*. These mechanisms are, of course, absolutely essential for embryonic morphogenesis because they enable cells in the embryo to identify one another, to adhere selectively into various characteristic configurations, and to become organized into the multicellular patterns which give rise to tissues and organs. Clarification of

the nature of these cell-recognition mechanisms is important for understanding not only normal embryonic development, but also the ontogeny of various congenital malformations which arise from defects in
morphogenetic processes. Furthermore, malfunctions in these mechanisms may also be an important aspect in the disruption of tissue architecture, such as is typical of many neoplasias.

The following four chapters present some of the most important conceptual and methodological approaches to the problems of cell–cell
recognition and the integration of cells into multicellular systems. Dr
Burger will discuss studies on the biochemical basis of species-specific
cell recognition and cell reaggregation of marine sponge cells. Dr
Gerisch will review the role of molecular signals in the assembly of
slime-mould cells into multicellular structures. Drs Bennett and Artzt
will discuss the role of the T-locus in the early morphogenesis of mouse
embryos. Dr Mäkelä will introduce the important problem of cell interactions in the genesis of immune response.

Studies in my laboratory on cell interactions in embryonic morphogenesis have led to the hypothesis that cell recognition and selective
cell adhesion are mediated by specific cell–surface constituents which
function as intercellular ligands or cell–cell receptors. We propose that
the specificity of cell recognition and of intercellular adhesion is determined by the chemical characteristics of such ligands as well as by the
topographic arrangement on the ligands on the cell surface, i.e. by a
combinatorial "code" which depends on both the qualitative and the
structural configurations of the units of which it consists. We further propose that cells belonging to different tissues or to different cell classes
may be surface-specified by different ligands conducive to *tissue-specific*
cell recognition; while *cell-type-specific* cell recognition (i.e. histological
organization of cells within a given tissue) may depend predominantly
on the topographic-temporal characteristics of ligands, i.e. on their display pattern on the cell surface (Moscona, 1974, 1975).

Consistent with the cell-ligand hypothesis have been (1) the finding
by immunological methods of tissue-specific antigens on embryonic cell
surfaces (Goldschneider and Moscona, 1972), and (2) the isolation from
embryonic cells and from cell membranes of embryonic cells of tissue-
specific cell-aggregating factors (Hausman and Moscona, 1975, 1976).
These factors have been purified and characterized as glycoproteins.

Thus, our results suggest that there exists on embryonic cell surfaces
a class of tissue-specific proteins which, under the experimental conditions employed by us, appear to function as morphogenetic cell-cell
ligands and to determine embryonic cell recognition. For operational
convenience I have referred to such proteins as "cognins", because

of their postulated role in cell recognition. Our present studies are concerned with the detailed biochemical characterization of these proteins, their organization on the cell surface, their biosynthesis and genetic regulation. In addition, their relationship to other "recognition-mechanisms" present on the cell surface is being investigated, such as immunoglobulins, histocompatibility antigens, T-locus products and lectin-binding sites.

The chapters in this section will demonstrate that studies on the morphogenetic cell interactions have made great strides in recent years; moreover, the results justify the expectation of further rapid progress towards elucidation of this complex problem in the near future.

REFERENCES

Goldschneider, I. and Moscona, A. A. Tissue-specific cell-surface antigens in embryonic cells. *J. Cell Biol.* **53**, 435–449 (1972).

Hausman, R. E. and Moscona, A. A. Purification and characterization of the retina-specific cell-aggregating factor. *Proc. nat. Acad. Sci. (Wash.)* **72**, 916–920 (1975).

Hausman, R. E. and Moscona, A. A. Isolation of retina-specific cell-aggregating factor from membranes of embryonic neural retina tissue. *Proc. nat. Acad. Sci. (Wash.)* **73**, 3594–3598 (1976).

Moscona, A. A. Cell surface specification in embryonic differentiation, in A. A. Moscona (ed.), The Cell Surface in Development, pp. 67–99, John Wiley and Sons, New York (1974).

Moscona, A. A. Embryonic cell surfaces: mechanism of cell recognition and morphogenetic cell adhesion, in O. McMahon and C. F. Fox (ed.), Developmental Biology—Pattern Formation, Gene Regulation, pp. 19–39, W. H. Benjamin, Inc., Reading, Mass. (1975).

Mechanisms of Cell–Cell Recognition: Some Comparisons between Lower Organisms and Vertebrates

Max M. Burger

*Department of Biochemistry, Biocenter, University of Basel,
Klingelbergstrasse 70, CH 4056 Basel, Switzerland*

INTRODUCTION

The early organ-specific cell-sorting experiments by the Mosconas (1952) and by Townes and Holtfreter (1955) have met with so much success that they have given rise to an outlook and overemphasis of a single cell surface phenomenon in development which seems to become too restrictive in two ways. It had hardly been the intention of these early workers to single out cell sorting as the most important event in embryogenesis since it is but one of the cell surface phenomena. Neither had it been their intention to focus exclusively on adhesion via complementary chemical receptors among the many possible mechanisms that can be considered to explain cell-cell recognition.

Despite the overwhelming emphasis that has been given to such cell-sorting systems one should keep in mind that there are other embryonal surface reactions. As to chemospecific adhesion, which is the most fashionable explanation given to cell sorting, we have to be aware that there are several other interpretations of cell sorting which eventually might turn out to be at least as if not more important for the *in vivo* processes as they occur.

The following phenomena can all be counted among the cell surface mediated developmental cell interactions:

(1) Inductions may turn out in part to be mediated via cell surfaces regardless whether they are of the instructive type or whether they simply activate precommitted cells (mesenchymal-epithelial inductions, formation of the neural plate, etc.) as discussed by Pictet and Rutter (this volume, p. 344).

(2) Restrictions of pluripotential cell groups by neighbouring cells into a specialized differentiation programme. This may eventually overlap with the phenomenon above but may, in some cell groups, lead to the ultimate type of restriction which is cell death—still initiated by neighbouring cells.

(3) Surface interactions can also be involved in the formation of the final adult configuration of two different cell types as, for example, the adeno- and the neurohypophysis, the adrenal medulla and the cortex, or the functionally important contact between nerve and target organ, giving rise to stable synapses. Such interactions can, of course, also be counted among the sorting-out processes, but they do not belong to the classical homotypic or self recognitions, and they have to be based on a heterotypic type of recognition in the contact zone.

(4) Since organ sorting out in the strict sense only aggregates cells of the same organ into an amorphous configuration, morphogenesis of tissue architecture (tubes, lobules) has to be considered separately since it requires the introduction of the concept of either differential adhesion of various parts of a given cell surface (Beug *et al.*, 1973) or cytoplasmic compartmentalization of tensile networks which deform the cells into characteristic groups of cells.

Considerations of hormone and other biological factor interactions with cells during embryogenesis concern most certainly also the surface membrane but are, as many others, beyond the scope of this short survey.

Since the most convincing results in the field of cell–cell recognition are based—as always in biochemical cell biology—on experiments with isolated and characterized molecules, the chemospecific recognition mechanisms may have been unduly overemphasized in the last decade. It is therefore appropriate in such a survey to consider other mechanisms that can explain morphogenesis. Thus, cells that migrate to their ultimate destination may simply end up at the target site by chemotaxis alone, and the high concentration of chemotactic agents can keep them at the appropriate location long enough, until they are physically trapped by gap and other junctions or extracellular matrix. It is interesting to note that the slime moulds, the model system *par excellence* for developmental chemotaxis, seem to require, besides chemotaxis, a second device to entrap such cells that have migrated to their final target. Such aggregative forces seem to be based on species-specific cell–cell adhesion (Beug *et al.*, 1970) for which lectin-like sugar-specific macromolecules are considered responsible (Barondes *et al.*, 1976).

Another mode of reaching the target can be via contact guidance as has, for example, been suggested for the migration of neurons in the foetal neocortex (Rakic, 1974). However, such a mechanism would in essence beg the question how the guiding element obtained its own orientation.

Once a cell has reached its final destination, be that by random walk, chemotaxis or contact guidance (Fig. 1), recognition of its proper location can be mediated by other than the most quoted cell surface

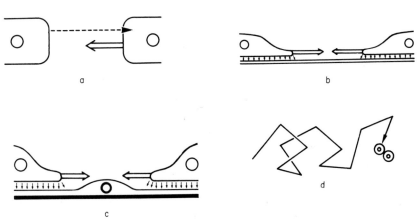

Fig. 1. Mechanisms by which specific cell–cell contacts can be established: (a) chemotaxis, (b) attraction-mediated and guided by an extracellular matrix or (c) by another cell, (d) random movement and entrapment due to specific cell–cell interaction. (From Burger, 1974.)

recognition mechanisms. Thus, recognition can come about by the exchange of cytoplasmatic components via gap or other junctions or via pino- or phagocytosis of elements between neighbouring cells (Fig. 2).

Wolpert's field position hypothesis is an elegant argument against each cell requiring its own specific surface molecules since the specific location of a given cell and general environmental effects can define its destiny (Wolpert, 1969). Weston's recent results support the notion that some migration patterns may simply be guided by physical barriers and low specificity environmental effects as exemplified by neural crest cell migration (Weston, 1972). Steinberg was one of the first to stress that physical interactions, as for example charge patterns, may be sufficient to explain the hierarchy of strengths of adhesion that lead to characteristic equilibrium distribution of different tissue cells at the end of the sorting-out process (Steinberg, 1970). Curtis (1972), on the other

hand, was one of the first to stress the possibility that timing differences
could determine morphogenetic processes. Thus, nerve cells might

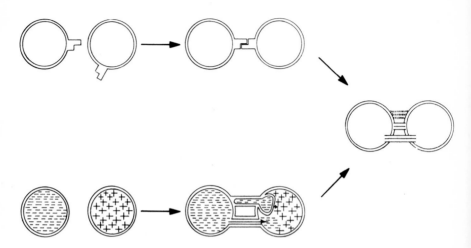

Fig. 2. Two posssible recognition mechanisms. In the upper portion of this figure a
mechanism is depicted where recognition occurs via surface macromolecules. In the
lower portion an alternative mechanism is considered where information from within
the cell is passed on to the neighbouring cell, either via vesicular exchange (upper
part) or via gap or other junctions (lower part). Such connections are suggested to
have a short life span and can be considered as a mutual probing, seen often at the
ruffling edge of a moving cell. In both cases such cell interactions will on the one hand
either lead to secondary stable linkages (intercytoplasmic, intermembranous or extracellular linkages, i.e. lower, middle and upper bridge in the last two adhering cells)
or on the other hand to the disengagement of the two partner cells (not shown). (From
Burger et al., 1975.)

mature at different times, send their processes out in a specific sequence
and arrive in the target region at different times. It is obvious that the
incoming fibres will go to different target cells if those also follow a
given timetable of division, since incoming fibres can only contact cells
that are available but not those that are only born later.

Any novice to the field of cell–cell recognition, particularly if he has
a biochemical background, will do well to keep such alternate explanations in mind while embarking on a biased study of the biochemically
more obvious and experimentally more available mechanisms.

SPONGE CELL RECOGNITION: A TWO-COMPONENT SYSTEM?

Marine sponges can be dissociated by squeezing them through a cheese cloth. If permitted to migrate they will reaggregate at the bottom of a watch glass. Such mechanically dissociated cells will not collect in a mixed aggregate if two differently pigmented species are mixed, but form two separate species-specific aggregates (Wilson, 1907 and 1911; Huxley, 1911; Galtsoff, 1925). Humphreys (1963) dissociated such sponges with calcium–magnesium-free sea water and could wash off large amounts of an aggregation-promoting factor (AF). Such chemically dissociated cells did not aggregate any more in the cold unless the aggregation factor was added, together with calcium, and the cells were slowly mixed on a rotating platform (Moscona and Moscona, 1952). At room temperature aggregation occurred in the absence of AF, presumably since the cells resynthesized or extruded AF again.

Isolation and characterization have been attempted earlier from different sponge species (Margiolash et al., 1965) and have recently been carried out on one particular sponge (Microciona parthena) in detail (Henkart et al., 1973; Cauldwell et al., 1973). It seems to be a huge "molecule" of 20×10^6 Daltons consisting of about 50% protein and 50% carbohydrate, the latter composed of one half galactose and 20% each hexosamines and uronic acid. Homogeneity has not been established in biochemically rigorous terms, since only sizing procedures have been used, because ion chromatography has not given any satisfactory results.

Species specificity of mechanically dissociated cells is good in some pairs of cells while not as convincing in others (McLennan, 1969; Spiegel 1954, 1955). Most of these graded differences can also be observed with antisponge antisera and find a plausible explanation in the closeness of taxonomic relationships among the species. Grafting experiments between a few sponges support the concept that only homotypic grafts will take and that aggregation specificity follows successful graft specificity (Moscona, 1968). Since Microciona prolifera AF is capable of aggregating Haliclona occulata cells that had been dissociated with $Ca^{++}-Mg^{++}$-free sea water, but not as well as Microciona cells that were dissociated the same way, we concluded that the available and probably not yet pure AF preparations do not show a qualitative absolute species specificity but a quantitative effect only (Turner and Burger, 1973). This conclusion, although compatible with the results of other aggregation promotion specificities, as for example of retina cells (Balsamo and Lilien, 1974), awaits confirmation with a homogeneous and fully pure AF, should such a preparation become available in the foreseeable future.

Many types of models can be drawn to explain how AF can bring about aggregation of the homotypic cells. We found inhibition of aggregation, particularly in the cold, by relatively high concentrations of glucuronic acid (Turner and Burger, 1973). Since AF contains uronic acid and could be inactivated by a β-glucuronidase preparation, we suggested that the cell surface should carry a protein molecule which would act as receptor for AF. Since the sugar-protein interaction is not yet fully established we suggested the noncommittal name of baseplate. Such a material was found after carefully shocking *Microciona prolifera* cells with low salt (Weinbaum and Burger, 1973). Other procedures tested included proteolytic enzymes, guanidinium chloride, butanol,

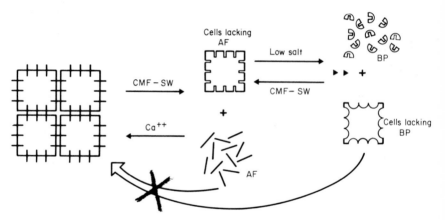

Fig. 3. Requirement of a membrane-bound baseplate for the specific sorting-out process. Sponge cells dissociated in $Ca^2 - Mg^2$-free sea water (CMF-SW) will release aggregation factor (AF) that is required in the presence of calcium (Ca^2) for reaggregation of the cells lacking the aggregation factor. Low salt treatment of such aggregation factor-less cells (hypotonic swelling) will release a baseplate or receptor (BP) for the aggregation factor. (From Burger, 1974.)

octanol, Triton X-100 and urea, but to no avail. Hypotonically shocked cells were much less or not at all aggregated with AF, and the shock medium contained a component which was capable of inhibiting AF (Fig. 3) as well as mechanically dissociated cells. This baseplate preparation could also restore aggregatability by AF to shocked cells if such cells were preincubated with shock medium and the excess baseplate subsequently removed by centrifugation (Fig. 3). Restoration experiments of this kind are not always successful and depend quite dramatically on the degree of damage inflicted to the cells by hypotonic swelling.

To demonstrate direct interaction between baseplate and AF, the macromolecules released by hypotonic shock were coupled to agarose beads. AF aggregated such beads very rapidly as much as did AF-coated beads in a mixed bead incubation (Weinbaum and Burger, 1973). The baseplate-coated beads were also capable of adhering to

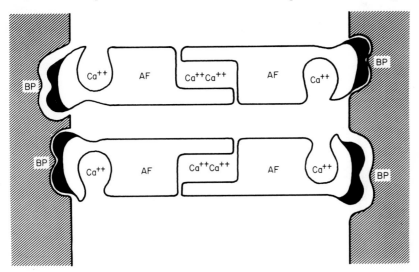

Fig. 4. Tentative model for the recognition observed in the sponge *Microciona prolifera*. Two macromolecular aggregation factors (AF) are illustrated each consisting of at least two subunits. The black termini at each pole carry the carbohydrates that are recognized by the baseplate (BP) anchored in adjoining cell surfaces. The calcium which keeps the subunits together in this model is required for function of the aggregation factor and is removed by EDTA or EGTA. The calcium at the periphery of the aggregation factor is removed by calcium–magnesium-free sea water and is meant to help in stabilizing the aggregation factor at the cell surface. It does not have to be bound to the aggregation factor though. It can fulfil such a stabilization function also by being bound to other molecules in the neighbourhood of the aggregation factor. (From Weinbaum and Burger, 1973.)

live, mechanically dissociated cells, i.e. cells that still carried AF, but not to cells that were devoid of a functionally sufficient amount of AF after treatment with calcium–magnesium-free sea water.

Since the baseplate preparation was not able to aggregate mechanically dissociated cells, but rather inhibited them, and furthermore since it did not precipitate AF and did not aggregate AF coated beads, we conclude that it must be a monovalent macromolecule, at least on a functional basis.

Baseplate cannot be sedimented at 105 000 × g for 90 min and it is non-dialysable and heat sensitive (60°C for 10 min). Unlike AF, baseplate appears to be stable towards EDTA (5mM), repeated freezing and thawing, as well as to lyophilization and a pH range of 3–12. After a 10-fold purification of the shock medium with the use of Sephadex G-200 and G-100, baseplate activity was eluted in the region of 40 000–60 000 daltons, using calibrating proteins (Jumblatt *et al.*, 1976).

Our present minimal model for the mechanism of *Microciona prolifera* aggregation is shown in Fig 4. It assumes a multivalent large glycosa-

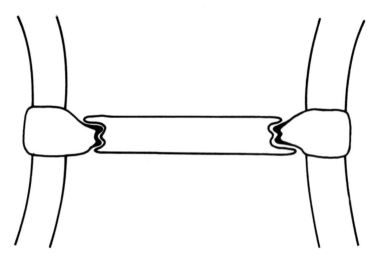

Fig. 5. Inversion of the polarity: the recognizing component lies between the cells and the antigenic or recognized component in the surface. (From Burger *et al.*, 1975.)

minoglycan-like AF made out of subunits which are held together by a tightly bound calcium pool that can only be released by EDTA. Another pool of calcium stabilizes the configuration which permits AF-cell interaction, in this case with the baseplate. This second type of calcium pool may either be a part of AF, as shown, or it can also be a part of the cell membrane. This second pool is easily removed when the cells are incubated in calcium–magnesium-free sea water, thereby also releasing the active AF. Each AF molecule has some loose surface attachment sites which are shown here at a baseplate, but could theoretically also be somewhere else. At these "weak" sites rupture occurs under mechanical stress. Although we do not know how the baseplate is built into the membrane we must assume that baseplates do not reach deep into the hydrophobic compartment of the surface membrane since shocking should release peripheral proteins only, and since the baseplate is fully water-soluble and does not form any micelles. If the baseplate is indeed a peripheral protein, we have to make a further assumption for the model, namely that it is attached to yet another, i.e. a third component. Whether this linkage is covalent or noncovalent, and whether it is to a lipid or a protein, is under investigation.

Recognition leading to adhesion can, of course, be achieved with a similar two-component system where the aggregation factor takes over

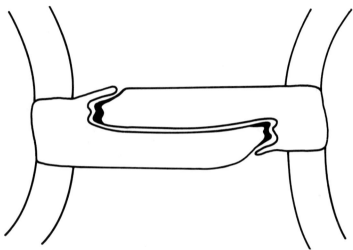

Fig. 6. Recognizing and recognized component can be built into the same cell surface membrane in homotypic cell–cell recognition. The two components can be separately embedded in the membrane or they can be part of the same molecule as shown here. (From Burger et al., 1975.)

the function of the recognizing unit and the baseplate is the component that is recognized (Fig. 5).

Yet another variant of the above two models would be a situation ideally adapted to homotypic or self recognition, where both recognizing and recognized sites are built into the same cell membrane as two molecules, or possibly even a single molecule (Fig. 6).

As was pointed out earlier, all these possibilities should be considered not only for species-specific sponge aggregation but also for cell–cell recognition in developmental cell–cell interactions in higher metazoans, since not enough is known at the present time about surface membrane molecules that are instrumental in cell recognition at that level of evolution.

PHENOMENA WHERE CARBOHYDRATES HAVE BEEN IMPLIED OR HAVE BEEN FOUND TO BE FUNCTIONAL IN CELL–CELL RECOGNITION

There is not enough and rigorous evidence that carbohydrates are involved in species-specific recognition of sponge cells in general. Even in *Microciona prolifera* aggregation can be inhibited with glucuronic acid, even though it should be pointed out that rather high doses of glucuronic acid have to be used to observe total inhibition. Similar as well as other problems apply to most other systems where carbohydrates are thought to be functionally involved in recognition. However, as will be explained below, one might be misled by standards for criteria that have been established in antigen-antibody reactions where bivalent high affinity and relative low molecular weight antibodies are involved. Many cell–cell recognition phenomena may be based on oligo- or polyvalent ligands, and furthermore one has to be aware of the fact that the receptors do not only consist of the monosaccharide that is used to inhibit the experimentally induced cell adhesion, but may be di- and oligosaccharides with considerably higher K_M's than the monosaccharide used in the test.

By far the best evidence that carbohydrates can be involved in cell–cell recognition comes from phage and viral interactions with cells. Thus mutants altered in specific sugars of teichoic acids (Glaser *et al.*, 1966) or lipopolysaccharides (Lindberg, 1973) lose their sensitivity to a particular phage infection. That is, the particular phage does not bind anymore. Moreover the isolated carbohydrate complex from the wild type bacterium does bind and inactivate the particular phage, while the mutant complex does not. Sialic acid was shown to be the receptor for influenza virus a long time ago although it was never clear whether the surface sites are exclusively sialic acid carrying glycoproteins (Schulze, 1973), or whether sialic acid carrying gangliosides could also function as receptors, as was recently suggested for Sendai virus (Haywood, 1974).

Not only infective agents, but also toxic agents, seem to interact with the cell surface via carbohydrate groups and to enter thereafter in some cases. Among the bacterial toxins, like those from diphtheria, tetanus or cholera, for which animal cell surface receptors can be defined, it turns out that at least cholera toxin binds carbohydrates specifically. It seems that the GM_1 ganglioside or its terminal galactose or *N*-acetylgalactosamine groups (Cuatrecasas, 1973) bind the toxin and convey it to the membrane-bound adenylcyclase (Bennett *et al.*, 1975) which is thereby activated and starts the chain of deleterious effects for the cell.

A plant toxin from *Ricinus communis*, which belongs to the class of compounds called lectins, has a very similar mechanism of action. It binds to plasma membrane-attached galactose and thereafter delivers to the ribosomes inside the cell a protein synthesis inhibitor which is part of the toxin complex. Lectins which are defined as carboydrate binding proteins have furthermore been implied in biological cell–cell recognition systems as diverse as plant root cells and symbiotic nitrogen-fixing bacteria (Hamblin and Kent, 1973) as liver cells, as aggregating slime moulds (Barondes *et al.*, 1976) and as fusing muscle cells (Stockert *et al.*, 1974; Teichberg *et al.*, 1975; Gartner and Podleski, 1975), the last being the least convincing example. In most of these studies the lectins are tested on erythrocytes and the sugar-binding capability is assessed, based on the fact that these lectins are isolated by affinity chromatography on carbohydrate containing column beds, and furthermore that the erythrocyte agglutination can be inhibited with some characteristic carbohydrates. Direct and unambiguous evidence for the carbohydrate as well as the lectin involvement in cell–cell recognition between the cells from which these lectins were isolated will still be a task for the future, although Barondes' group (1976), from where most convincing data come so far, is quite close to that goal.

Yeast cell mating types have yet a different sort of agglutinin on their surface which can have molecular weights up to 10^6 daltons. They are proteomannans, and although they lose their activity for agglutination of the opposite mating type cells when treated with exomannanase (Yen and Ballou, 1973), no results have been reported so far that mannan oligosaccharides could inhibit the aggregation reaction. Furthermore, since activity is protease-sensitive, the data available so far still have to be considered compatible with the activity of the agglutinin residing in the protein part; the exomannanase result would then have to be considered unusual and would require an explanation.

The best studied cell–cell recognition system in the development of higher metazoans is the neuroretinal cell of the chick embryo. Moscona introduced the notion of a chemoreceptive ligand. He, as well as Glaser's group, refrain from pinpointing the isolated factors which promote (McClay and Moscona, 1974) or inhibit aggregation (Glaser *et al.*, 1976) as components in a protein-carbohydrate recognizing system. In view of the fact that both groups use assays which require one of the recognizing cell partners at least to be alive, it may be wise to refrain from any overinterpretation of the available data.

Roth (personal communication) implicates carbohydrate-protein interactions for the retino-tectal recognition during development and so do Balsamo and Lilien (1974) for the retino-retinal recognition.

Roseman (1970) had suggested that an oligosaccharide receptor on one cell would be recognized by precisely that carbohydrate transferase on the neighbouring cell which should attach the next sugar in the oligo-saccharide chain. Roth's recent results imply that opposite gradients of enzyme and substrate occur on retina and in the tectum permitting an error-proof hook-up when the retina ganglion processes arrive at the tectum. Balsamo and Lilien (1974), on the other hand, found that a pronase-treated aggregation-promoting molecule, which after the treatment consisted essentially of carbohydrates only, still bound specifically to retina cells. Their model necessitates beside the aggrega-tion-promoting molecule yet another component which presumably ties together the aggregation-promoting molecules on opposite cells. Since these aggregation-promoting molecules require a baseplate-like sub-stance in the membrane, they were forced to the conclusion that they needed a three-component system to explain retina cell aggregation.

To prove beyond any doubt that carbohydrates are directly involved in the type of cell–cell interactions presented here will not be easy if rigorous biochemical criteria are applied. Moreover many, if not most, cell–cell recognitions may turn out to be based on protein–protein recognition. It would, however, be astonishing if nature would have gone through the trouble of evolving an amazing variability of oligosac-charides attached to surface lipids and surface proteins instead of keep-ing these surface carbohydrate varieties down to a few polysaccharides of the cellulose, glucan or mannan type without developing functions for these unique surface or extracellular products. The future will tell us what task or tasks these oligosaccharides have to fulfil.

A COMPARISON BETWEEN SPECIES-SPECIFIC SPONGE CELL RECOGNITION AND TISSUE-SPECIFIC CELL RECOGNITION IN DEVELOPING VERTEBRATE CELLS

Size of the promoting factor

The *Microciona prolifera* aggregation factor (Couldwell *et al.*, 1973) is with its "molecular weight" of 20×10^6 daltons the largest in this family of factors. Very similar, i.e. also a multisite factor, is the mating type 5 factor from the yeast *Hansenula wingei* (Ballou, 1976). It has an approximate molecular weight of 10^6 and comprises 6 binding sites of 12 000 m.w., each attached by bisulphide bridges to the core of about 9×10^5 daltons. It is interesting to note that also the counterparts, to which these factors are attached, are quite similar. The baseplate counterpart to the sponge factor is univalent and has a molecular weight between 40 000 and 60 000 daltons. Type 21 factor, the counter-part of the type 5 mating factor, is also univalent and with its sedimenta-tion coefficient of 2.9 $S_{20.W}$ in the same size class. The activity of both small molecular components resides in their protein moieties although both of them are glycoproteins.

A large molecular factor may also be instrumental in mating of the alga *Chlamydomonas* although only flagellar membrane pieces could be isolated so far in an effort to solubilize the activity (Snell, 1976).

The most specific factors that have been isolated so far in higher ani-mals, among which the neuroretinal cell aggregation promoting factor is certainly the prominent example, seem to be of small dimensions, according to Moscona's latest results about 50 000 daltons (McClay and Moscona, 1974).

We believe that a large factor may not require a high K_M for its inter-action with the site which it recognizes, simply because adhesion is guaranteed by the multiplicity of bonds it can establish based on its polyvalent nature. A small factor, which presumably has only few bind-ing sites, as for instance a bivalent aggregation factor, will however require a very high K_M as evidenced by another class of well-known "bivalent factors", the antibodies.

Such considerations might be important to explain why large factors may display less specificity (Turner and Burger, 1973) and require more hapten inhibitor *in vitro*, since they have lower K_M's. We do not believe, however, that this reasoning applies to the *in vivo* situation, since a cell itself becomes polyvalent if it has only a few monovalent factor molecules on its surface. Thus a cell can vary its polyvalency simply by setting either many or few factor molecules into its surface, be they small or large molecules with many or with few binding sites. The size distribution incidentally does not, as a general rule, drop with evolution. In other words, protozoa do not necessarily have large factors and higher metazoa small factors. The slime mould lectins, e.g. those found by Barondes' group, seem to be 250 000 and 100 000 daltons (Barondes, 1976). Furthermore a German group reported recently that even among the marine sponges a factor exists which is still active at about 20 000 daltons (Müller *et al.*, 1974). On the other hand, un-specific (Pessac and Defendi, 1972) and quite specific (Oppenheimer, 1975) aggregation factors can be found in vertebrates and mammalians which have molecular weights of several million daltons, and glucos-aminoglycan character.

The requirement for protein synthesis and live cells

Chick neuroretinal cell aggregation requires protein synthesis (Moscona, 1974). To what degree this reflects a requirement to repair the surface damage inflicted upon the cells when removing them from the tissue, be that with protease, EDTA or even with mechanical dissociation, is not established with certainty. One of the two cells that recognize each other has to be alive in most assay systems used so far. That again can reflect a requirement for repair. It can be interpreted, however, also as a requirement for microvillar formation and retraction energy-dependent receptor patch formation, subcellular fibrillar networks which might help zip up the interaction sites, or other processes which promote the specific recognition or promote the subsequent secondary phenomena which give rise to irreversible binding between the two plasma membranes. Even in a retinal recognition system, where one of the two partners is an isolated plasma membrane, the other partner, i.e. the cell, has to remain alive, or at least temperatures higher than 4°C are required (Glaser *et al.*, 1976).

It might be pointed out in this context that most adhesion and aggregation assays, particularly however those using live cells, suffer under the uncertainty that not recognition *per se* is measured but that recognition is only one among many steps that lead to the aggregated state, and that only the latter is scored. This criticism holds particularly for rate, but to some degree also for equilibrium measurements, since we do not know which the rate-limiting step is in rate measurements nor what the influences are of secondary interactions which can come before the recognition step (allowing the recognizing sites to come in proximity) or after (fixing the cells that have recognized each other). It might be pointed out that no specific inhibitor has been found which prevents specific cell aggregation and is similarly capable of dissociating cells again, and this is the best evidence that secondary steps must be occurring.

Mechanically dissociated sponge cells on the other hand can aggregate after fixation with 4% neutral formaldehyde (Moscona, 1968). Although there is a slight reduction in the rate of aggregation, as well as the ultimate size of the aggregate reached, we found chemically dissociated cells that were fixed with 1% glutaraldehyde to aggregate quite well with aggregation factor even after years of storage at 4°C. Many considerations which have to be brought into an analysis of retina cell aggregation, like those mentioned above (microvilli, energy-requiring systems, etc.), can therefore be discarded, which makes the sponge aggregation system a much simpler one, bringing it closer to the ideal bimolecular ligand-receptor system best suited for biochemical investigation and less prone to unforeseen biological pitfalls. On the other hand it is of course also, as all simple and evolutionary early systems, more remote from the real *in vivo* processes that occur in developmental morphogenesis of tissues.

Temperature

Retina cell aggregation is very temperature-dependent. At $28°C$ it is slowed down considerably and does not take place at $0°C$. Sponge cells—as can be deduced from the fact that glutaraldehyde fixed cells still display factor-dependent aggregation—do still aggregate specifically at $4°C$. The aggregation is much slower, but can reach almost but not quite as large aggregates as at room temperature. Thus aggregates of 0.5 mm diameter can be reached with the same cell density and the same concentration of factor in 5 min at room temperature, while it takes 210 min at $4°C$.

The high temperature sensitivity of retina cell aggregation may reflect a metabolic event. It can, however, as easily be explained as an absolute requirement for a fluid state of certain radial or tangential compartments of the plasma membrane. Before any detailed temperature profile has been established one cannot even begin to distinguish between these possibilities or even perhaps a dependence on the degree of association of cytoplasmatic elements like microtubuli.

Other differences

Under most assay conditions described so far only the retina factor has been found to promote aggregation. In other words, aggregation occurs at a slow rate and the factor speeds up the rate at which the equilibrium state of aggregation is achieved at a given speed of rotation. Sponge aggregation can be suppressed completely at $4°C$ if the factor has been carefully removed with enough calcium–magnesium-free sea water washes. Only factor addition can bring about aggregation for such a sponge cell preparation. The same holds for glutaraldehyde fixed cells.

Retina cell aggregation is very much dependent on the stage of development at which the cells were harvested (Moscona, 1962; Gottlieb *et al.*, 1974). Sponge aggregation has not yet been studied in embryonal development, but there is no necessity for such a temporal dependence since sponge aggregation is species specific and so far has not been shown to be tissue specific. Most sponge cell types seem to enter the final aggregate, although some differences in aggregatability cannot be ruled out and may be found in a careful analysis of that question.

POSSIBLE FUNCTIONS FOR THE PHENOMENON OF SPONGE CELL RECOGNITION

Since the revival of Wilson's (1907, 1911) earlier interest for sponge cell aggregation most of us have primarily been interested in how this cell recognition works as a possible model for morphogenesis in higher metazoans. The questions why it is there at all, and what the sponge uses it for, have hardly been raised at all, and if anybody began to consider these questions he has not done it publicly, or only in a peripheral context.

It seems somewhat unlikely that this aggregation, which is species specific, has any relevant function in sponge development. First, the factor material can still be isolated in large quantities from adult sponge cell surfaces, and second, it does not seem that only one specific type of sponge cell would be found in the final aggregate. It may be that some cells are less aggregated by AF, but that may be due to partial damage, or if it is a specific cell type, due to a higher susceptibility to damage. Usually with sufficient AF over 90% of the dissociated cells have entered the aggregate by the end of the assay.

Since AF is only homogeneous in size, but biochemical homogeneity has not yet been shown in a rigorous test, one can still maintain that several types of classes of AF molecules occur, and that our present AF preparations have histiotype-specific affinities for different sponge cells which may be relevant during embryogenesis. It would, however, not be obvious why AF should be species specific. Species specificity for histiotype-specific molecules becomes an even less likely postulate in view of the fact that those histiotype sorting-out processes, which have been tested in vertebrates and mammals, are almost as a rule not species specific. Thus while bird liver cells aggregate away from bird cartilage or heart cells, the same organ from bird, i.e. the liver, recognizes liver from mammalian origin as identical and aggregates into a mixed bird–mammalian liver clump.

If this species-specific recognition system is not used for embryogenesis in sponges, what is it then used for? Similar aggregation systems among organisms at a primitive evolutionary stage may furnish us with a possible hint provided their function is known. Species-specific aggregation of single slime mould amoebal cells into a slug and eventually a fruiting body has the function of preventing slime mould cells from other species from entering to form a heterospecific slug and forming a mixed spore. Sponge cells do not leave the organism except in the form of gametes. Those, however, may need a species-specific recognition mechanism.

Other more primitive eukaryotic cells have also evolved a gamete recognition or mating system. As mentioned earlier, yeast mating types insert into their surfaces macromolecules that are responsible for mating type adhesion or aggregation. It should be pointed out that the similarity of the components involved in yeast mating and in sponge aggregation is striking. Thus yeast mating type-5-factor is, as AF from *Microciona*, also a proteoglycan with a high carbohydrate content (85%). It has also, like AF, a high molecular weight (1×10^6) and its counterpart (type 21 factor) is not only a much smaller molecule, as is the baseplate, but it has also a much smaller sugar content, as does the baseplate.

Enough differences can be pointed out and questions raised to make that comparison just a superficial one. Thus, in the mating type interaction of yeast the two components are found on two different cells, while in the sponge system both AF and baseplate would possibly reside on each of the two gametes although a segregation of the two is theoretically possible and could still fulfil the proposed function. Furthermore the question arises of course why the somatic cells would still maintain a sizeable population of these molecules and not replace them by species-unspecific ones since there seems to be no obvious selection pressure to keep them after fertilization. On the other hand one can ask the question why the developing embryo would have to lose the species-specific characteristic of its cell surface coating, as long as it fulfilled the requirement for an intercellular ground substance similar to the proteoglycans in higher animals.

The advantage of a species-specific recognition for sponge gametes in their fluid marine environment, which can carry them anywhere including the neighbourhood of a different species of sponges, is self-evident. Sponge research is one of the more esoteric branches of biology, and our search for a complete description of gamete isolation has not borne any fruit so far, even when journals of the end of the last century were included, a time at which the development of such techniques was frontier research.

CONCLUSIONS

Sponge cell recognition is probably mediated by two components, the aggregation factor and the baseplate. Since the baseplate is not an integral membrane protein it must be attached to yet a third component of the system which would lie in the membrane and would not necessarily have to carry any specificity. Carbohydrate specificity is suspected for one sponge species but not absolutely proven. A baseplate-like carbohydrate-recognizing molecule has recently been found in a Mediterranean sponge species (Bretting and Kabat, 1976).

Several basic differences between sponge cell recognition and histiotypic sorting out of embryonal vertebrate tissue are discussed. They indicate that cell recognition, as it occurs in sponges, may become a useful model system to study the molecular components and mechanisms involved in aggregative cell recognition. Some of these may be preserved in higher organisms, but as one can already conclude at this stage, they are hardly in the exact form as they occur in sponges. The fact that sponges are found to still aggregate in a species-specific manner at 4°C, and that they do that even after glutaraldehyde fixation, makes them a good study object for the molecular components in cell recognition. Higher metazoan cell recognition seems to require, however, that at least one of the interacting cells be alive, and that would clearly indicate that besides recognition at the molecular level other processes are necessary to permit successful recognition and morphogenetic sorting out. We are, however, still at the stage where simple systems, such as sponge cell recognition, may contribute to the concepts which have to be worked out for the more elaborate multistep processes in vertebrate morphogenesis.

ACKNOWLEDGEMENTS

The author's work described in the text was carried out in collaboration with Drs George Weinbaum (Albert Einstein Hospital, Philadelphia), Robert Turner and J. Jumblatt, at the Marine Biological Laboratory, Woods Hole, Massachusetts, USA, and was entirely supported by the Swiss National Science Foundation, grant number 3.1330.73. I thank Dr G. Weinbaum for reading and commenting on the manuscript.

REFERENCES

Balou, C. in R. A. Bradshaw (ed.), Proceedings NATO Institute on Surface Receptors, Plenum Press (1976).

Balsamo, J. and Lilien, J. Functional identification of three components which mediate tissue specific embryonic cell adhesion. *Nature (Lond.)* **251**, 522–524 (1974).

Barondes, S. H. and Rosen, S. D. Cellular recognition in slime molds: Evidence for its mediation by cell surface species-specific lectins and complementary oligosaccharides, in R. A. Bradshaw (ed.), Proceedings NATO Institute on Surface Receptors, Plenum Press (1976).

Bennett, V., O'Keefe, E. and Cuatrecasas, P. Mechanism of action of cholera toxin and the mobile receptor theory of hormone receptor-adenylate cyclase interactions. *Proc. nat. Acad. Sci. (Wash.)* **72**, 33–37 (1975).

Beug, H., Gerisch, G., Kempff, S., Riedel, V. and Cremer, G. Specific inhibition of cell contact formation in Dictyostelium by univalent antibodies. *Exp. Cell Res.* **63**, 147–158 (1970).

Beug, H., Katz, F. E., Stein, A. and Gerisch, G. Quantitation of membrane sites in aggregating *Dictyostelium* cells using tritiated univalent antibody. *Proc. nat. Acad. Sci (Wash.)* **70**, 3150–3154 (1973).

Bretting, H. and Kabat, E. A. Purification and characterization of the agglutinins from the sponge *Axinella polypoides* and a study of their combining sites. *Biochemistry* **15**, 3228–3236 (1976).

Burger, M. M. The surface membrane and cell–cell interaction, in F. O. Schmitt (ed.), The Neurosciences: third study program, pp. 773–782, MIT Press, Cambridge (1974).

Burger, M. M., Turner R. S., Kuhns, W. J. and Weinbaum, G. A possible model for cell–cell recognition via surface macromolecules. *Phil Trans. B* **271**, 379–393 (1975).

Couldwell, C., Henkart, P. and Humphreys, T. Physical properties of sponge aggregation factor: a unique proteoglycan complex. *Biochemistry* **12**, 3051–3055 (1973).

Cuatrecasas, P. Interaction of *Vibrio cholerae* enterotoxin with cell membranes. *Biochemistry* **12**, 3547–3558 (1973).

Curtis, A. S. G. Adhesive interactions between organisms, in Functional Aspects of Parasite Surfaces, pp. 1–22, Blackwell's Scientific Publications, Oxford (1972).

Galtsoff, P. S. Regeneration after dissociation (an experimental study on sponges) *J. exp. Zool.* **42**, 183–251 (1925).

Gartner, T. K. and Podleski, T. R. Evidence that a membrane bound lectin mediates fusion of L_6 myoblasts. *Biochem. biophys. Res. Commun.* **67**, 972–978 (1975).

Glaser, L., Ionesco, H. and Schaeffer, P. Teichoic acids as components of a specific phage receptor in *Bacillus subtilis*. *Biochim. biophys. Acta (Amst.)* **124**, 415–417 (1966).

Glaser, L. *et al.*, in R. A. Bradshaw (ed.), Proceedings NATO Institute on Surface Receptors, Plenum Press (1976).

Gottlieb, D. I., Merrell, R. and Glaser, L. Temporal changes in embryonal cell surface recognition. *Proc. nat. Acad. Sci. (Wash.)* **71**, 1800–1802 (1974).

Hamblin, J. and Kent, S. P. Possible role of phytohaemagglutinin in *Phaseolus vulgaris* L. *Nature New Biol. (Lond.)* **245**, 28–30 (1973).

Haywood, A. M. J. Characteristics of Sendai virus receptors in a model membrane. *J. mol. Biol.* **83**, 427–436 (1974).

Henkart, P. S., Humphreys, S. and Humphreys, T. Characterization of sponge aggregation factor: A unique proteoglycan complex. *Biochemistry* **12**, 3045–3050 (1973).

Humphreys, T. Chemical dissolution and *in vitro* reconstruction of sponge cell adhesions. I. Isolation and functional demonstration of the components involved. *Develop. Biol.* **8**, 27–47 (1963).

Huxley, J. Some phenomena of regeneration in sycon with a note on the structure of its collar cells. *Phil. Trans. B* **202**, 165–189 (1912).

Jumblatt, J. E., Weinbaum, G., Turner, R. S., Ballmer, K. and Burger, M. M. Cell surface components mediating the reaggregation of sponge cells, in R. A. Bradshaw (ed.) Proceedings NATO Institute on Surface Receptors, Plenum Press (1976).

Lindberg, A. A. Bacteriophage receptors. *Ann. Rev. Microbiol.* **27**, 205–241 (1973).

Margiolash, E., Schenck, J. R., Hargie, N. P., Burokas, S., Richter, W. R., Barlow, G. H. and Moscona, A. A. Characterization of specific cell aggregating materials from sponge cells. *Biochem. biophys. Res. Commun.* **20**, 383–388 (1965).

McClay, D. R. and Moscona, A. A. Purification of the specific cell aggregating factor from embryonic neural retina cells. *Exp. Cell Res.* **87**, 438–443 (1974).

McLennan, A. P. Polysaccharides from sponges and their possible significance in cellular aggregation. *Symp. Zool. Soc. (Lond.)* **25**, 299–324 (1970).

Moscona, A. A. Cellular interactions in experimental histiogenesis. *Int. Rev. exp. Path.* **1**, 371–529 (1962).

Moscona, A. A. Cell aggregation: properties of specific cell-ligands and their role in the formation of multicellular systems. *Develop. Biol.* **18**, 250–277 (1968).

Moscona, A. A. The surface of embryonic cells, in A. A. Moscona (ed.), The Cell Surface in Development, pp. 67–99, Wiley, New York (1974).

Moscona, A. A. and Moscona, M. H. The dissociation and aggregation of cells from organ rudiments of the early chick embryo. *J. Anat.* **86**, 287–301 (1952).

Müller, W. E. G., Müller, I. and Zahn, R. K. Two different aggregation principles in reaggregation process of dissociated sponge cells (*Geodia cydonium*). *Experientia* **30**, 899–902, (1974).

Oppenheimer, S. B. Functional involvement of specific carbohydrates in teratoma cell adhesion factor. *Exp. Cell Res.* **92**, 122–126 (1975).

Pessac, B. and Defendi, V. Cell aggregation: Role of acid mucopolysaccharides. *Science* **175**, 898–900 (1972).

Rakic, P. Mode of cell migration to the superficial layers of fetal monkey neocortex. *J. comp. Neurol.* **145**, 61–83 (1974).

Roseman, S. The synthesis of complex carbohydrates by multiglycosyl transferase systems and their potential function in intercellular adhesion. *Chem. Phys. Lipids* **5**, 270–297 (1970).

Schulze, I. T. Structure of the influenza virion. *Advanc. Virus Res.* **18**. 1–55 (1973).

Snell, W. J. Mating in *Chlamydomonas*. A system for the study of specific cell adhesion. I. Ultrastructural and electrophoretic analyses of flagellar surface components involved in adhesion. *J. Cell Biol.* **68**, 48–49 (1976).

Spiegel, M. The reaggregation of dissociated sponge cells. *Ann. N.Y. Acad. Sci.* **60**, 1056–1078 (1954/55).

Steinberg, M. S. Does differential adhesion govern self-assembly processes? Equilibrium configurations and the emergence of a hierarchy among populations of embryonic cells. *J. exp. Zool.* **173**, 395–434 (1970).

Stockert, R. J., Morell, A. G. and Scheinberg, I. H. Mammalian hepatic lectin. *Science* **186**, 365–366 (1974).

Teichberg, V. I., Silman, I., Betisch, D. D. and Resheff, G. A β-D-Galactoside binding protein from electric organ tissue of *Electrophorus electricus*. *Proc. nat. Acad. Sci.* (*Wash.*) **72**, 1383–1387 (1975).

Townes, P. L. and Holtfreter, J. Directed movements and selected adhesion of embryonic amphibian cells. *J. exp. Zool.* **128**, 53–120 (1955).

Turner, R. S. and Burger, M. M. Involvement of a carbohydrate group in the active site for surface guided reassociation. *Nature* (*Lond.*) **244**, 509–510 (1973).

Weinbaum, G and Burger, M. M. Two component system for surface guided reassociation of animal cells. *Nature* (*Lond.*) **244**, 510–512 (1973).

Weston, J. A. Cell interactions in neural crest development, in L. G. Silvestri (ed.), Cell Interactions. Proceedings of the 3rd Lepetit Colloquim, London, pp. 286–292, North-Holland Publishing Co., Amsterdam (1972).

Wilson, H. V. On some phenomena of coalescence and regeneration in sponges. *J. exp. Zool.* **5**, 245–258 (1907).

Wilson, H. V. On the behavior of the dissociated cells in hydroids, alcyonania and asterials. *J. exp. Zool.* **11**, 281–338 (1911).

Wolpert, L. Positional information and the spatial patterns of cellular differentiation. *J. theor. Biol.* **25**, 1–47 (1969).

Yen, P. H. and Ballou, C. E. Composition of a specific intercellular agglutination factor. *J. biol. Chem.* **248**, 8316–8318 (1973).

Periodic Cyclic-AMP Signals and Membrane Differentiation in *Dictyostelium*

G. Gerisch, D. Malchow, W. Roos, U. Wick and B. Wurster

Biozentrum der Universität Basel, 4056 Basel, Switzerland

INTRODUCTION

During the development of *Dictyostelium discoideum*, cyclic-AMP acts both by mediating communication between aggregating cells and by stimulating cell differentiation from the growth phase stage to aggregation competence. In the aggregation process cyclic-AMP fulfils the two-fold function of a chemotactic agent (Konijn *et al.*, 1967) and a transmitter in signal propagation from cell to cell (Robertson *et al.*, 1972; Shaffer, 1975; Roos *et al.*, 1975) Pulses of cyclic-AMP are propagated over an aggregation territory in the form of either concentric waves or spirals of chemotactic activity. For signal propagation it is important that the cells are able to regulate the release of cyclic-AMP so that periodic pulses are produced (Gerisch and Wick, 1975). In this chapter we discuss recent results on the periodicity of the signal generating system in *D. discoideum*. One outcome of these studies is the demonstration that short-term activation of adenylate cyclase is a key event in signal generation (Roos and Gerisch, 1976). Another result shows that cyclic-AMP, also in its function as a stimulant of cell differentiation, exerts its maximal effect when applied in repetitive pules (Gerisch *et al.*, 1975a; Darmon *et al.*, 1975).

CHEMORESPONSES RECORDED IN CELL SUSPENSIONS

Biochemical studies on the signal system that controls cell aggregation in *D. discoideum* demand a straightforward registration technique for cellular responses. Figure 1 shows that the responses to chemotactic agents are reflected within seconds in the light-scattering of cell suspensions (Gerisch and Hess, 1974). These optical changes are probably due to changes in cell shape. The light-scattering changes are very sensitive and discriminatory probes for the species specificity of chemotactic agents, and also for differences in the response patterns between developmental stages, and between wild type and mutants (Fig. 1). In the following account, light-scattering changes are used not only to record the cellular responses to extraneous stimuli, but also to record the autonomous generation of chemical signals.

Light scattering responses in cell suspensions

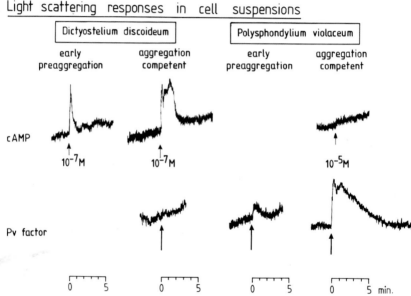

Fig. 1. Responses in cell suspensions to chemotactic agents. *Left*: Cells of *Dictyostelium discoideum* are able to respond to cyclic-AMP by chemotaxis. In cell suspensions, the same agent causes transient decreases of light scattering (*top*). For aggregation-competent cells duplicate response peaks are characteristic. Early during the interphase between growth and aggregation competence, only a single peak is found. Certain non-aggregating mutants do not proceed from this early pattern to the double peak response (not shown). No response of *D. discoideum* to the chemotactic factor of *Polysphondylium violaceum* is observed (*bottom*). *Right:* The responses of *P. violaceum* to their own chemotactic factor show a similar dependence on the developmental stage as the responses of *D. discoideum* to cyclic-AMP. The factor is most probably a peptide (Wurster *et al.*, 1976). No response to cyclic-AMP is observed in this species.

PERIODICITY OF CYCLIC-AMP SYNTHESIS

In shaken suspensions, cells of an appropriate developmental stage syn-chronize their periodic activities. Cyclic-AMP concentrations can oscil-late in these suspensions in a strongly non-sinusoidal mode. At 23°C, cyclic-AMP pulses are formed every 6 to 9 min. The pulses are not accompanied by significant changes of ATP, the substrate of adenylate cyclase (Fig. 2.).

By activating cell surface receptors, a pulse of extracellular cyclic-AMP can induce a pulse of endogenous cyclic-AMP. This is shown: (1) by adding cyclic-AMP to oscillating cells 2 to 3 min in advance of a spontaneous pulse. In this case a precocious pulse is induced, followed by a permanent phase shift of the oscillating system. (2) In cells which

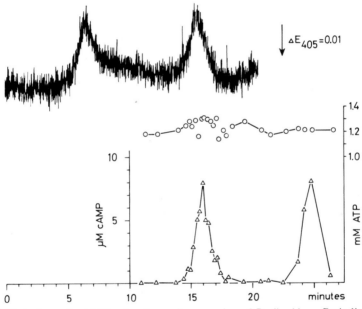

Fig. 2. Periodic cyclic-AMP pulses in a cell suspension of *D. discoideum*. Periodic light-scattering changes (*top*) were measured in phase with spontaneous cyclic-AMP pulses (△), but no concomitant ATP changes were detected (○). The peak intracellular cyclic-AMP concentrations were below 1% of the steady ATP levels. Cells: Strain Ax-2 harvested 4.5 h after the end of growth. Cell density 5×10^7/ml. Other conditions as described by Gerisch and Hess (1974). The plotted concentrations of cyclic-AMP and ATP are calculated as mol/l densely packed cells. Although cyclic-AMP is released into the medium, its life-time there is short because of hydrolysis by phosphodiesterase. Therefore, the total cyclic-AMP measured in the cell suspension is mainly intracellular cyclic-AMP. (From Roos *et al.*, 1977.)

have not yet reached the stage of spontaneous pulse formation, a pulse can be elicited by extracellular cyclic-AMP application. It has been demonstrated under these conditions that a pulse is formed via the activation of adenylate cyclase (Fig. 3). A considerable portion of the synthesized cyclic-AMP is transported into the extracellular space (Gerisch and Wick, 1975, and Fig. 5) where it is able to stimulate other cells.

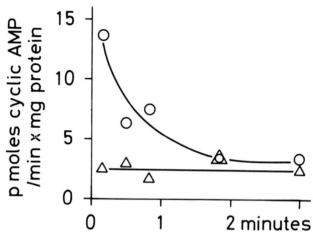

Fig. 3. Receptor-mediated activation of adenylate cyclase. Cells were stimulated by a pulse of 2×10^{-7}M cyclic-AMP. 30 s later the cells were sonicated and, with a delay of not longer than 15 s, transferred to an incubation mixture for the assay of adenylate cyclase. Cyclic-AMP synthesis was assayed according to Salomon *et al.* (1974). After stimulation of the cells and subsequent homogenization, the adenylate cyclase was obtained in a short-lived activated state ($\bigcirc$). Within 1 min the cyclase returned in the incubation mixture to the ground state. After homogenization of non-stimulated cells, a constant basal activity was ovserved ($\triangle$). Cells: Strain Ax-2 harvested 4 h after the end of growth. Cell density 2×10^8/ml. Data are replotted from Roos and Gerisch (1976).

SIGNAL PROCESSING FROM CELL SURFACE RECEPTORS TO INTRACELLULAR TARGETS

Cell surface receptors for cyclic-AMP are believed to mediate the following responses: chemotaxis, activation of adenylate cyclase, phase adjustment of the oscillating system which leads to synchronization of suspended cells, and cell differentiation to the aggregation-competent stage. Various possibilities of receptor coupling to the target systems are shown in Fig. 4. Models (1) and (2) imply that all types of responses elicited by extracellular cyclic-AMP pulses are mediated by one type

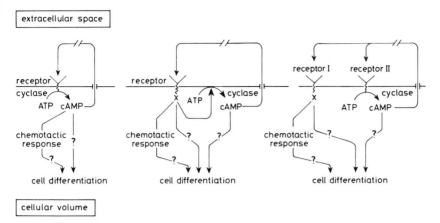

extracellular space

cellular volume

Fig. 4. Hypothetical coupling of cell surface receptors to intracellular targets. Within seconds after cyclic-AMP binding to the receptors, two responses are observed: chemotaxis and adenylate cyclase activation. Cell differentiation is a long-term response which requires repetitive stimulation of the receptors over hours. The simplest assumption is one type of receptor mediating one type of primary response, the activation of adenylate cyclase (*left*). In a more complicated scheme the assumption of an intermediate X is made, which is responsible for both chemotaxis and cyclase activation (*centre*). Complete separation of the signal channels, leading to either chemotaxis or cyclase activation, would imply the existence of two different receptor types (*right*), as suggested by Green and Newell (1975). No specific assumption is made about cell differentiation. Darmon *et al.* (1975) have suggested that the chemotactic response as such is a prerequisite for cell differentiation.

of receptor. Model (3) says that chemotaxis and adenylate cyclase activation are mediated by different receptor types (Green and Newell, 1975).

If the signal processing pathways branch behind the cell surface receptors, as in models (1) and (2), the question arises up to which step these pathways are connected. Already 5 s after cyclic-AMP administration the extension of pseudopods into the direction of the cyclic-AMP gradient is detectable (Gerisch *et al.*, 1975c). Therefore, only reactions that occur during that short interval can be considered to couple the chemotactic response to receptor activation. The interval between stimulation and the peak of intracellular cyclic-AMP is usually longer; it is in the order of 1 to 2 min. However, the increase of intracellular cyclic-AMP begins earlier; and in some cases a fast, short peak of intracellular cyclic-AMP has been found in advance of the main peak (Fig. 5). It is therefore possible that a fast intracellular increase of cyclic-AMP links the activation of cell surface receptors to a local response of the contractile system, which then results in a chemotactic

Fig. 5. Intra- and extracellular cyclic-AMP changes in response to an extraneous cyclic-AMP pulse. At the time indicated by the arrow, cells were stimulated by 5×10^{-8} M cyclic-AMP. A fast and a slower increase of intracellular cyclic-AMP (●) could be distinguished in this experiment. The extracellular cyclic-AMP concentration (○) increased after a delay of half a minute. It should be mentioned that the temporal response patterns obtained in different experiments varied considerably, so that the example shown should not be generalized. Cells: Strain Ax-2 harvested 6 h after the end of growth. Cell density 2×10^{8}/ml. Methods as described by Gerisch and Wick (1975).

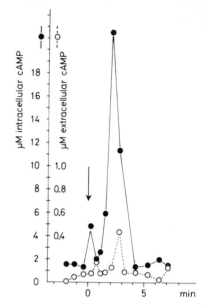

reaction. Another obvious candidate for a common initial step of branched signal processing pathways is Ca ̈ release from intracellular vesicles into the cytoplasmic space. Cell differentiation to aggregation competence, a cyclic-AMP stimulated process, has been found to be accelerated by the Ca ̈ ionophore A 23187 (Brachet, 1976).

BIOCHEMICAL REACTIONS THAT ACCOMPANY CYCLIC-AMP OSCILLATIONS

As outlined above, the periodic generation of cyclic-AMP signals is based on the oscillatory regulation of adenylate cyclase activity in the absence of significant changes of the ATP concentration. Although the mechanism of adenylate cyclase regulation remains unknown, it may be worthwhile to discuss the co-oscillating biochemical reactions which are known. The first observation was an absorption change at 430 nm during a pulse, which was thought to indicate a shift of cytochrome b towards its reduced state (Gerisch and Hess, 1974). An estimate of the total amount of cyclic-AMP synthesized during a pulse indicates that less than 20% of the steady state level of ATP is consumed for cyclic-AMP synthesis (Gerisch and Malchow, 1976). If no other energy-requiring processes are activated during pulse formation, the increase of respiration necessary to compensate the excess ATP consumption should be small. The CO_2 output from oscillatory cells

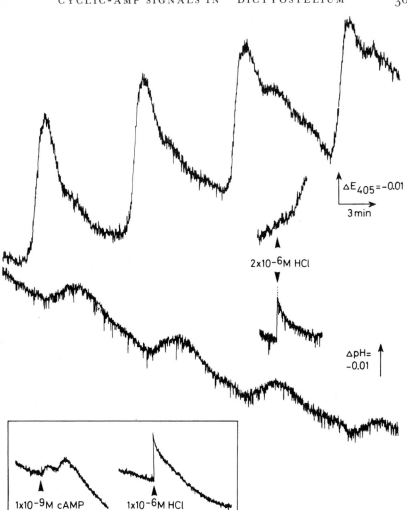

Fig. 6. Extracellular pH changes in suspensions of signalling cells. Spontaneous periodicity was monitored by light scattering (*top*). The extracellular pH oscillated with the same frequency (*below*). For calibration of the pH amplitudes HCl was added (*right*) which had no influence on light scattering. *Inset:* Under similar conditions an extraneous cyclic-AMP pulse induced two separable proton peaks. For comparison a 1000-fold higher HCl pulse was given. At the molarity of the cyclic-AMP pulse, HCl caused no detectable shift. Conditions: 2×10^{-7} Ax-2 cells per ml. The medium was Bonner's unbuffered salt solution containing NaCl, KCl and $CaCl_2$. The extracellular pH of the cell suspension was about 7.4.

suggests an increase of the rate of respiration during a pulse by about 3%. It is conceivable that the observed cytochrome b changes are associated with low-amplitude oscillations of the electron transport rate.

The extracellular pH is another oscillating variable (Malchow and Nanjundiah, 1976). During a pulse, the proton concentration shows a transient increase (Fig. 6). The same is observed after experimental administration of cyclic-AMP. In this case, the pH response is often resolved into a fast and a slow peak (Fig. 6, inset). Possible reasons for the pH shifts include: (1) cyclic-AMP hydrolysis to 5'-AMP, (2) transport of cyclic-AMP through the plasma membrane in the non-ionized state, (3) proton transport, (4) increased CO_2 production, (5) synthesis of another weak acid which penetrates the cell membrane as an undissociated molecule.

Both the fast and the slow pH changes are higher than expected from the hydrolysis of the cyclic-AMP added as a stimulant. However, the hydrolysis of the cyclic-AMP which is released from the cells during a pulse could account for the pH shifts. To test this, phosphodiesterase was inhibited by dithiothreitol. The slow response was still obtained under these conditions, indicating that it was caused by another reaction. It is possible, although not established, that the fast pH peak measures the hydrolysis of the cyclic-AMP which is released from the cells during a pulse.

It is too early to attribute any specific function in the signal system of *D. discoideum* to either these pH changes or the reactions that give rise to them. In principle, the fast response could be part of the chain of reactions from receptor activation to either pulse formation or chemotaxis, if it would not only indicate hydrolysis of the released cyclic-AMP. The slow component of the pH response may coincide in time with the refractory phase, during which the ability of the cells to form a pulse in response to cyclic-AMP is reduced.

EFFECTS OF FOLIC ACID PULSES ON THE OSCILLATING SYSTEM

Folic acid is another chemotactic agent for *D. discoideum* cells, which, in contrast to cyclic-AMP, has a pronounced effect on growth phase cells (Pan *et al.*, 1972). If cells are repeatedly stimulated by folic acid pulses, beginning at 2 h after the end of growth, they start oscillations about 2 h earlier than controls (Würster, 1976). In certain respects folic acid simulates the action of cyclic-AMP, but in others it shows interesting differences. Both cyclic-AMP and folic acid are maximally active when applied in pulses rather than in a continuous flow (Fig. 7). Both

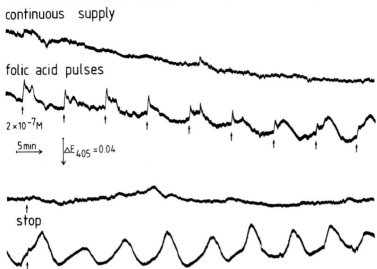

continuous supply

folic acid pulses

2×10^{-7}M

5min

$\Delta E_{405} = 0.04$

stop

Fig. 7. Folic acid stimulated oscillations. A suspension of 5×10^7 Ax-2 cells per ml was divided at 1.5 h after the end of growth into two cuvettes (sample and reference), transferred into an Aminco DW-2 spectrophotometer and kept under identical conditions. Starting at 2 h after the end of growth, folic acid pulses of 2×10^{-7} M amplitude were added in intervals of 8 min to the sample cuvette. To the reference cuvette the same quantity of folic acid per time was supplied in a continous way. The stop of folic acid supply is indicated by the arrow marked "stop".

chemicals accelerate the onset of autonomous oscillations and other developmental processes (Gerisch *et al.*, 1975a; Darmon *et al.*, 1975; Wurster, 1976). The action of cyclic-AMP is distinguished, however, from that of folic acid by the increase of sensitivity of the cells during development from the end of growth to the aggregation competent stage. During oscillations small extracellular cyclic-AMP pulses cause phase shifts, and continuous flow suppresses autonomous oscillations (Gerisch and Hess, 1974). In contrast, the cells are highly sensitive to folic acid in the pre-oscillation state. This sensitivity decreases as soon as autonomous oscillations occur (Würster, 1976). The mechanism of the oscillation-linked decrease is unknown. Its clarification will hopefully provide insights into both the control of receptor function and the biochemical implement of the oscillating system.

   Because they do not interact with established oscillations, small pulses of folic acid can be better used than cyclic-AMP pulses to study the interrelationship between the experimental pulse frequency and the frequency of the induced autonomous oscillations. The autonomous period proved to be about 8 min, independent of the period of folic acid application which was varied from 7 to 11 min (Würster, 1976). The same experiments provide a preliminary answer to a related question: is the frequency of folic acid pulses critical for the induction of precocious oscillations? Periods of 7 and 11 min are both efficient. These results abandon the hypothesis that each folic acid pulse induces a cycle of the length of an autonomous period, and that each consecutive pulse has to fall in a favourable phase of the cycle in order to stimulate. The pulsatile character of the stimuli rather than their frequency seems to be important. A supposed explanation for the pulse effect focusses on the signal processing characteristics of the receptor systems (Gerisch *et al.*, 1975a). Various pieces of evidence indicate that the chemoreceptor systems of *D. discoideum* are constructed similar to those of chemotactic bacteria (Berg and Brown, 1972; Spudich and Koshland, 1975), such that they process changes of effector concentration in time, rather than absolute measures of steady concentrations (Gerisch *et al.*, 1975b).

CELL DIFFERENTIATION AS COUPLED TO THE OSCILLATING SYSTEM

Various cell surface markers are available for the quantitation of molecular changes which are associated with cell differentiation from the growth phase to the aggregation-competent stage. Both the number of cyclic-AMP receptor sites and the activity of cell surface phosphodiesterase increase during that period. Cyclic-AMP pulses enhance this increase (Gerisch *et al.*, 1975c). The control system, therefore, represents a network: cyclic-AMP receptors and phosphodiesterase are involved in the control of cyclic-AMP as well as being controlled by cyclic-AMP.

Other cell surface sites controlled by cyclic-AMP pulses are believed to function in cell-to-cell adhesion of aggregating cells. These "contact sites A" can be assayed by an immunogical technique as well as by their activity: to enable the formation of EDTA-stable cell contacts (Beug *et al.*, 1973). Again, like cyclic-AMP, folic acid has a stimulating effect on contact site A formation, and both agents exert their maximal effect when applied in pulses rather than continuously. It is conceivable that the folic acid acts indirectly by stimulating oscillations, and that autonomous cyclic-AMP pulses are the more direct stimuli of contact site A formation. However, no evidence is yet available that the periodic rise of the intracellular cyclic-AMP concentration is by itself the stimulating factor. Any co-oscillating metabolite would be a likewise plausible candidate. In any case, it is likely that transcriptional and/or translational controls are involved in the developmental regulation of contact sites A, and it remains an exciting question how the oscillating system exerts its effects on either one of these control levels.

ACKNOWLEDGEMENTS

This work was supported by the Schweizerischer Nationalfonds and the Deutsche Forschungsgemeinschaft.

REFERENCES

Berg, H. C. and Brown, D. A. Chemotaxis in *Escherichia coli* analyzed by three-dimensional tracking. *Nature (Lond.)* **239**, 500–504 (1972).

Beug, H., Katz, F. E. and Gerisch, G. Dynamics of antigenic membrane sites relating to cell aggregation in *Dictyostelium discoideum. J. Cell Biol.* **56**, 647–658 (1973).

Brachet, Ph. Differenciation cellulaire. Effet d'un ionophore sur l'agregation de *Dictyostelium discoideum. C.R. Acad. Sci. (Paris)*, Sér. D **282**, 377–379 (1976).

Darmon, M., Brachet, Ph. and Pereira da Silva, L. H. Chemotactic signals induce cell differentiation in *Dictyostelium discoideum. Proc. nat. Acad. Sci. (Wash.)* **72**, 3163–3166 (1975).

Gerisch, G. and Hess, B. Cyclic-AMP-controlled oscillations in suspended *Dictyostelium* cells: their relation to morphogenetic cell interactions. *Proc. nat. Acad. Sci. (Wash.)* **71**, 2118–2122 (1974).

Gerisch, G. and Wick, U. Intracellular oscillations and release of cyclic AMP from *Dictyostelium* cells. *Biochem. biophys. Res. Commun.* **65**, 364–370 (1975).

Gerisch, G., Fromm, H., Huesgen, A. and Wick, U. Control of cell-contact sites by cyclic AMP pulses in differentiating *Dictyostelium* cells. *Nature (Lond.)* **255**, 547–549 (1975a).

Gerisch, G. Hülser, D., Malchow, D. and Wick, U. Cell communication by periodic cyclic-AMP pulses. *Phil. Trans. B* **272**, 181–192 (1975b).

Gerisch, G., Malchow, D., Huesgen, A., Nanjundiah, V., Roos, W., Wick, U. and Hülser, D. Cyclic-AMP reception and cell recognition in *Dictyostelium discoideum.* in D. McMahon and C. F. Fox (ed.), Developmental Biology, ICN-UCLA Symposia on Molecular and Cellular Biology, Vol. 2, pp. 76–88. W. A. Benjamin Inc., California (1975c).

Gerisch, G. and Malchow, D. Cyclic AMP receptors and the control of cell aggregation in *Dictyostelium. Adv. Cyclic Nucleotide Res.* **7**, 49–68 (1976).

Green, A. and Newell, P. C. Evidence for the existence of two types of cAMP-binding sites in aggregating cells of *Dictyostelium discoideum. Cell* **6**, 129–136 (1975).

Konijn, T. M., van de Meene, J. G. C., Bonner, J. T. and Barkley, D. S. The acrasin activity of adenosine-3',5'-cyclic phosphate. *Proc. nat. Acad. Sci. (Wash.)* **58**, 1152–1154 (1967).

Nanjundiah, V. and Malchow, D. pH-Oscillations and cAMP induced pH changes in aggregating slime mould cells. *Hoppe Seylers Z. physiol. Chem.* **357**, 273 (1976).

Pan, P., Hall, E. M. and Bonner, J. T. Folic acid as second chemotactic substance in the cellular slime moulds. *Nature New Biol. (Lond.)* **237**, 181–182 (1972).

Robertson, A., Drage, D. J. and Cohen, M. H. Control of aggregation in *Dictyostelium discoideum* by an external periodic pulse of cyclic adenosine monophosphate. *Science* **175**, 333–335 (1972).

Roos, W. and Gerisch, G. Receptor mediated adenylate cyclase activation in *Dictyostelium discoideum. FEBS Letters* **68**, 170–172 (1976).

Roos, W., Nanjundiah, V., Malchow, D. and Gerisch, G. Amplification of cyclic-AMP signals in aggregating cells of *Dictyostelium discoideum. FEBS Letters* **53**, 139–142 (1975).

Roos, W., Scheidecker, C. and Gerisch, G. Adenylate cyclase activity oscillations as signals for cell aggregation in *Dictyostelium discoideum. Nature (Lond.)* **266**, 259–261 (1977).

Salomon, Y., Londos, C. and Rodbell, M. A highly sensitive adenylate cyclase assay. *Analyt. Biochem.* **58**, 541–548 (1974).

Shaffer, B. M. Secretion of cyclic AMP induced by cyclic AMP in the cellular slime mould *Dictyostelium discoideum. Nature (Lond.)* **255**, 549–552 (1975).

Spudich, J. L. and Koshland, D. E., Jr. Quantitation of the sensory response in bacterial chemotaxis. *Proc. nat. Acad. Sci. (Wash.)* **72**, 710–713 (1975).

Würster, B. Stimulation of cell development in *Dictyostelium discoideum* by folic acid pulses, in Meeting on Dictyostelium, pp. 52. Cold Spr. Harb. Lab. (1976).

Würster, B., Pan, P., Tyan, G. G. and Bonner, J. T. Preliminary characterization of the acrasin of the cellular slime mold *Polysphondylium violaceum. Proc. nat. Acad. Sci. (Wash.)* **73**, 181–182 (1976).

Developmental Interactions Studied with Experimental Teratomas derived from Mutants at the *T/t* Locus in the Mouse

Dorothea Bennett, Karen Artzt, Terry Magnuson and
Martha Spiegelman*

*Department of Anatomy, Cornell University Medical School,
New York, New York 10021, USA*

INTRODUCTION

The set of mutations at the *T/t* locus in the mouse that affects the development of early embryos has provided important analytical tools for analysing developmental mechanisms in mammals. At the present count various mutations in this region have been found when homozygous to produce at least 10 distinctly different lethal syndromes (Bennett, 1975a, for review). Although each of the 10 classes of embryonic impairment can be readily distinguished from any other, the defects produced by *T/t*-locus mutations share some significant features: they all arise during the important first half of development during which blastocyst formation, gastrulation and initial axial organization take place; they all appear to arise from specific blocks in the differentiation of particular components of the ectoderm; and they all seem to have their primary roots in some breakdown in normal mechanisms of cell-to-cell interaction (Bennett, 1975b). It is striking to note that although each *T/t* mutation initially appears to impair the development of only a very specific cell type, and often one that represents only a quantitatively minor portion of the embryo, they usually cause embryonic death within not more than 2 or 3 days, often less, after the appearance of the first visible defect. This suggests that, in mammals at least, the

* *Authors' present address:* Sloan Kettering Institute for Cancer Research, New York, New York 10021, USA.

maintenance of any form of continuing development in early stages is dependent on normal interrelationships among components of the young embryo that are at that time simple and few in number.

The defects are restricted to limited cell types, yet the embryo as a whole dies shortly after each defect occurs. In this sense, the mutations at the T/t locus can be construed as "organizational lethals" which cause lethality by interfering with some pattern of organization that, still unknown to us, is necessary if intra-uterine development is to continue. If this organization were not necessary, we would expect that the embryos would be able to continue developing in a way that was monstrous, perhaps even malignant because of failure of differentiational control mechanisms, but still capable of intra-uterine survival.

This situation contrasts strongly with that seen in the experimental teratomas and teratocarcinomas that occur when embryos are transplanted to extra-uterine sites such as the testis or kidney. Under these conditions the embryos are clearly subjected to traumatic mechanical disorganization, and the growths that they form are grossly distorted yet their cells are capable of surviving and differentiating into a wide variety of adult-type tissues and organs (Stevens, 1970; Damjanov *et al.*, 1971).

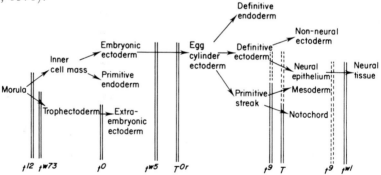

Fig. 1. A diagrammatic representation of developmental transitions blocked in homozygous T/t-locus mutants. ‖, known effects; ⁞⁞, probable effects.

Studies of experimental teratomas and teratocarcinomas also provide evidence that there is a requirement for normal initial organization of the intact embryo if subsequent differentiation is to proceed properly. This fact is best revealed in the sharply differing responses to transplantation of pre-primitive streak and post-primitive streak embryos. Embryos that have securely embarked on primitive streak formation ($7\frac{1}{2}$ days and older in the mouse) produce benign teratomas which differentiate fully, cease to grow, and cannot be serially transplanted. In striking contrast, pre-primitive streak stages will produce teratocarcinomas that contain an uncommitted pool of embryonal carcinoma cells as well as differentiated derivatives of all three germ layers. The embryonal carcinoma component of these tumours is potentially transplantable, and therefore by definition malignant (Damjanov and Solter, 1974, for review). These results can be interpreted as meaning that until primitive streak formation is completed the embryo contains a portion of undifferentiated, uncommitted cells which have the potentiality for uncontrolled growth, and that this potentiality is lost during the definitive organization of the primitive axis of the embryo.

In addition to defining this very sharp difference between pre-and post-primitive streak embryos, the detailed analysis of experimental teratomas has been used to explore the developmental potentialities of the germ layers. The elegant experiments of Skreb *et al.* (1976) have shown, for example, that in the rat at primitive streak stages the ectoderm alone is capable of giving rise to tissue derivatives of all three germ layers, whereas at later stages it has lost this potentiality. In these experiments mesoderm transplanted alone, however, gave rise only to brown fat; this gives a surprising hint that the differentiation of mesoderm is severely restricted in the absence of other germ layers. A host of observations on the types of differentiated somatic derivatives found in teratomas and teratocarcinomas (reviewed in Damjanov and Solter, 1974) can also be used as a gauge of the stringency of the requirements, presumably based on cell–cell interactions, that are necessary to produce any given somatic tissue or organ. For example, elements that occur frequently are cartilage, bone, muscle, neural tissue, and skin; it can be assumed that these elements require relatively simple or widely available interactions. Likewise, simple cysts with a considerable variety of epithelia occur in high frequency, while true organoids such as gut are usually less common. Organs whose requirements of interaction are known on other criteria to be highly specific (Grobstein, 1967), like lung and liver, are very rarely seen.

Because the use of embryos to produce experimental teratomas and teratocarcinomas seemed to be on the one hand a way of releasing embryos from dependence on the constraints of normal organization, and on the other hand a way of assaying for developmental potentiality, we have attempted to use this approach to study a series of several T/t-locus mutant homozygotes.

CHARACTERISTICS OF THE MUTANTS

The mutations we chose for this purpose were:

t^0: homozygotes die shortly after implantation, having failed to achieve organization into embryonic and extra-embryonic ecto-dermal components (Gluecksohn-Schoenheimer, 1940). It is likely that in these embryos the trophectoderm is defective and the ectoplacental cone fails to generate normal extra-embryonic ectoderm (L. J. Smith, personal communication).

t^{w5}: the egg-cylinder formed by homozygotes initially differentiates normally, but subsequently the embryonic ectoderm cells become pyc-notic and die. The extra-embryonic portion of these embryos seems normal at the time when the embryonic ectoderm degenerates; never-theless, the extra-embryonic portions also degenerate and die within days after the disappearance of the embryonic ectoderm (Bennett and Dunn, 1958).

t^9: homozygotes form an abnormal primitive streak, from which only sparse numbers of morphologically abnormal mesoderm cells emerge. Possibly because of membrane defects, these mesoderm cells fail both to migrate extensively and to proliferate normally (Spiegelman and Bennett, 1974); the embryos die at 9–10 days of gestation probably because of inadequacies in the circulatory system.

T: homozygotes produce an abnormally massive primitive streak which, however, fails to undergo normal regression. Furthermore, all axial structures in these embryos, namely neural tube, notochord, somites and gut, differentiate abnormally and form inappropriate associations with one another (Spiegelman, 1976). These embryos die at 10.75 days of gestation because the disorganization of their posterior region precludes the formation of normal placental connections.

EXPERIMENTAL TERATOMAS: THEIR PRODUCTION AND INTERPRETATION

Our intention was to analyse the growths produced by the various homozygotes with special attention to the following questions: (1) whether the embryonic cells were able to form viable growths at all—if they did not, it could be taken as evidence that a particular mutation acted as a cell lethal; (2) whether the histiotypic pattern of growth might reflect particular defects in certain germ layers or cell types; and (3) whether apparently malignant cells of a specific cell type might pre-dominate in these grafts, as was already known to be the case with re-spect to t^9/t^9 teratoid tumours (Artzt and Bennett, 1972).

Matings between appropriate heterozygotes were made, and the ges-tational age of embryos was timed by the vaginal plug method. In the case of t^0 and t^{w5}, embryos were dissected from their deciduae at either 6 or 7 days gestational age; they were freed from yolk sac but were not classed as to developmental stage or genotype before transplantation. In the case of t^9 and T, embryos were dissected at 7, 8 and 9 days of gestation, and classed for developmental stage and characterized as normal or abnormal before transplantation.

Transplants were made into the testis of adult mice as previously described (Artzt and Bennett, 1972). Transplants of t^0 and t^{w5} were made into genetically non-identical, non-H-2 compatible hosts of the BTBRTF/Nev strain. Transplants of t^9 and T had essentially syngeneic H-2 compatible hosts. All embryos were transplanted whole with the exception of those in the T series—where only the posterior third of the embryo was grafted. Controls for the t^0 and t^{w5} transplants consisted of normal litters from the BTBRTF/Nev strain, while controls in the t^9 and T sets were provided by normal littermates. Hosts were killed at approximately 2 or 4 weeks of growth, and their testes removed and prepared for routine histological examination.

Table I presents a summary of the data obtained from grafts of 6- and 7-day embryos in litters segregating for t^0 and t^{w5}. Since these grafts were made from embryos whose genotype was not known, the analysis is necessarily a retrospective one. In this context it should be remembered that lethal t-mutations distort their own transmission through males (Bennett and Dunn, 1971), and that therefore roughly 40–50% of the embryos in these experiments can be expected to be homozygous for the t-lethal in question. The data are presented in terms of embryonic germ layer derivatives rather than histological tissue elements simply as a means of revealing the absence of specific components. No examples of missing germ layer components were found, and all the growths represented were typical experimental teratomas.

Several interesting points emerge from this table. First of all, with respect to t^0, it seems very likely that t^0 homozygotes are capable of undergoing normal (for experimental teratomas!) growth and differentiation when they are released from dependence on their defective extra-embryonic supporting system. Only 3 of 17 grafted 6-day embryos failed to grow, which is quite compatible with the failure rate in controls; part of the growth failures are in any case undoubtedly due to technical difficulties. With 7 day t^0/t^0 embryos, furthermore, more than half failed to grow, confirming that the homozygous population is dead or too moribund at that time to be rescued by transplantation.

With respect to t^{w5} however, the situation is quite different. The overall survival rate is less than 50% for both 6- and 7-day embryo transplants, suggesting that t^{w5} homozygotes may be unable to generate a viable population of embryonic cells capable of forming a teratocarcinoma. Presumably, the embryonic ectoderm dies off, just as it does

TABLE I

Characteristics of growths obtained from embryos from litters segregating t^0 and t^{w5} homozygotes transplanted at 6 and 7 days

Genotype	Duration of graft	Total grafts	(1) None or trophoblast and haemorrhage only	(2) Rejected; can't analyse	(3) Endoderm	(4) Endoderm and mesoderm	(5) Embryonic cells and/or primitive neural cells only	(6) Ectoderm and mesoderm	(7) Ectoderm mesoderm and endoderm	Proportion of (3)–(7) with primitive neural cells and/or embryonic cells (= teratocarcinoma)
						6-day embryos				
t^0	28 days	17	3	—	1	1	—	2	10	9/14
t^{w5}	28 days	8	5	—	—	—	—	1	2	2/3
+/+ Controls	28 days	15	2	—	—	—	—	1	12	9/13
						7-day embryos				
t^0	14 days	25	15	—	1	—	3	5	1	9/10
t^0	28 days	10	6	—	—	—	—	1	3	4/4
t^{w5}	14 days	19	12	—	—	—	4	1	2	7/7
t^{w5}	28 days	15	10	2	—	—	—	2	1	2/3

in the intact embryo. The fact that no growths appear which contain only extra-embryonic structures implies that these elements cannot persist; this suggests in turn that such structures depend in some way on the embryo for their continued maintenance and proliferation. Gardner (1975) has already provided evidence for one instance of this kind, namely that trophectoderm cells lose the capacity to divide unless they remain in intimate contact with the inner cell mass.

One further general point should be noted, namely that virtually all grafts of these 6- and 7-day embryos contain, at both 14 or 28 days post-transplantation, proliferative primitive neural epithelium and embryonic cells, which often represent a large part of the tumour. Since the "fertilization age" of these grafts makes them at least the equivalent of a new-born mouse, these cells are clearly inappropriate to their developmental stage, and probably represent a self-perpetuating stem cell pool.

Table II presents a summary of the data obtained from grafts of early 8-day embryos in litters segregating for t^9 and grafts of late 8-day or early 9-day embryos in litters segregating for T.

The tumours obtained from T homozygotes versus their normal littermates were entirely unremarkable; the two sets were in every respect completely comparable. The fact that T/T embryos were, upon transplantation, able to form normal tissues derived from all three germ layers was already known (Ephrussi, 1935). However, we had suspected that the lack of normal organization of axial structures in the region of the posterior primitive streak might indicate developmental immaturity, and that therefore T homozygotes might be still capable of producing teratocarcinomas at later stages than normal embryos. No embryonal carcinoma or primitive neural cells were found though, thus refuting that hypothesis.

One point of some developmental significance emerges from these results. Although all grafts were made from the posterior third of the embryo, several contained ciliated pseudostratified epithelium, and two actually had histologically normal tracheobronchial structures. Thus at 8 and 9 day stages, well after lung bud formation has begun in the anterior of the embryo, the posterior endoderm still retains the potentiality for giving rise to respiratory elements.

With respect to the t^9 series there is a clear-cut difference between the developmental potential of homozygous t^9 embryos and that of their normal littermates. Whereas control teratomas ($+/t^9$ and $+/+$) all contain derivatives of the three germ layers, t^9/t^9 growths have virtually no mesodermal derivatives. In addition, organoids such as skin or gut which require mesodermal interaction are entirely lacking. In fact,

TABLE II

The distribution of selected tissues and organs in experimental teratomas and teratocarcinomas derived from T and t^9 homozygotes and their littermates

Genotype of embryo	t^9/t^9 8 day	+/+ or +/t^9 8 day	T/T 8–9 day	+/+ or +/T 8–9 day
Number of growths studied	20	22	6	10
Ectoderm				
Primitive neural tissue	17	7	0	0
Lens	13	1	0	0
Retinal pigment	20	8	0	0
Mature nervous tissue	2	21	4	8
Stratified squamous epithelium	19	22	6	7
Mesoderm				
Bone	1	21	4	5
Muscle	0	22	5	4
Fat	2	19	5	9
Endoderm				
Mucous secretory epithelium	5	14	5	8
Ciliated pseudostratified epithelium	10	11	2	2
Cuboidal epithelium	not graded	not graded	4	4
Transitional epithelium	not graded	not graded	2	2
Organoids				
Gut	0	9	3	8
Skin	0	11	5	9
Trachea-bronchus	not graded	not graded	1	1
Acinar glands	not graded	not graded	3	6
Mammary glands	not graded	not graded	1	2

most of the t^9/t^9 tumours are histologically malignant, and consist mainly of a proliferating pool of neuroectoderm which has a proclivity for differentiating into a limited number of ectodermal derivatives such as retinal pigment, lens, and stratified squamous epithelial pearls. While the last element is common in both t^9/t^9 and control growths, primitive neuroepithelium is present in almost all t^9/t^9 growths (17/20), but it is uncommon in control growths (7/22). Likewise, retinal pigment is ubiquitous in t^9/t^9 (20/20) and rarely seen in controls (1/22). Most striking is the virtual absence of mature nervous tissue in t^9/t^9 (2/20) in contrast to its presence in almost all controls (21/22).

These results of t^9/t^9 embryo transplants correlate nicely with the embryological effects of the gene, i.e. its failure to send cells successfully through the primitive streak and generate normal mesoderm. When rescued in an extra-uterine environment, the t^9/t^9 embryo appears to be developmentally "mesodermless" as if the gene had dissected the embryo for the experimenter. Those inductive interactions that depend on mesoderm are absent, and no organs or organoids such as skin, gut, glands, etc., are ever found. Another element that is found only very rarely, and then in small quantity, is mature neural tissue. It is not clear whether the absence of this element reflects a necessity for mesodermal interaction, or whether the same genetic defect that precludes the formation of normal mesoderm also impairs the ability of primitive neural tissue to differentiate further. It is interesting to note in any case that neural epithelium that appears normal in the embryo, but with many of its normal developmental interactions cut off, develops into a malignant embryonal tissue.

CONCLUDING REMARKS

Experimental teratomas and teratocarcinomas produced from genetically defective embryos can thus be a useful tool in analysing development, since they can free the investigation from the constraints imposed by the necessity of the intact embryo to develop as an integrated whole. Furthermore, they may provide, as in the case of t^9/t^9 transplants, a means of obtaining tumours of very homogenous cellular composition that may be useful for studies of biochemical or antigenic parameters.

REFERENCES

Artzt, K. and Bennett, D. A genetically caused embryonal ectodermal tumor in the mouse. *J. nat. Cancer Inst.* **48**, 141–158 (1972).
Bennett, D. and Dunn, L. C. Effects on embryonic development of a group of

genetically similar lethal alleles derived from different populations of wild house mice. *J. Morph.* **103**, 135–158 (1958).

Bennett, D. and Dunn, L. C. Transmission ratio distorting genes on chromosome IX and their interactions, in A. Lengerova and M. Vojtiskova (ed.), Proc. Symp. Immunogenetics of the H-2 System, pp. 90–103, Karger, Basel (1971).

Bennett, D. The *T*-locus of the mouse: A review. *Cell* **6**, 441–454 (1975a).

Bennett, D. *T*-locus mutants: Suggestions for the control of early embryonic organization through cell surface components, in M. Balls and A. Wild (ed.), Brit. Soc. devel. Biol. Symp. Mammalian Early Development, pp. 201–218, Cambridge University Press, London (1975b).

Damjanov, I. and Solter, D. Experimental teratoma. *Curr. Top. Path.* **59**, 69–130 (1974).

Damjanov, I., Solter, D. and Skreb, N. Teratocarcinogenesis as related to age of embryos grafted under the kidney capsule. *Wilhelm Roux' Arch. Entwickl.-Mech. Org.* **167**, 288–290 (1971).

Ephrussi, B. The behaviour *in vitro* of tissues from lethal embryos. *J. exp. Zool.* **70**, 197–204 (1935).

Gardner, R. L. Origin and properties of trophoblast, in R. G. Edwards, C. W. S. Howe and M. H. Johnson (ed.), The Immunobiology of Trophoblast, pp. 43–65, Cambridge Univ. Press, London (1975).

Gluecksohn-Schoenheimer, S. The effect of an early lethal (t^0) in the house mouse. *Genetics* **25**, 391–400 (1940).

Grobstein, C. Mechanisms of organogenetic tissue interactions, in Cell, Tissue and Organ Culture, US Nat. Cancer Inst. Mongr. 26, pp. 279–299 (1967).

Skreb, N., Svager, A. and Levak-Svager, B. Developmental potentialities of the germ layers in mammals, in K. Elliott and M. O'Connor (ed.), Embryogenesis in Mammals, Ciba Foundation Symposium 40, pp. 27–38, Elsevier, Amsterdam (1976).

Spiegelman, M. Electron microscopy of cell associations in *T*-locus mutants, in K. Elliott and M. O'Connor (ed.), Embryogenesis in Mammals, Ciba Foundation Symposium 40, pp. 199–220, Elsevier, Amsterdam (1976).

Spiegelman, M. and Bennett, D. Fine structured study of cell migration in the early mesoderm of normal and mutant mouse embryos (*T*-locus: t^9/t^9). *Embryol. exp. Morph.* **32**, 723–738 (1974).

Stevens, L. C. The development of transplantable teratocarcinomas from intratesticular grafts of pre- and post-implantation mouse embryos. *Develop. Biol.* **21**, 364–382 (1970).

Self-recognition Phenomenon in Acquired Immunity

O. Mäkelä and Saija Koskimies

*Department of Bacteriology and Immunology, Helsinki University,
SF-00290 Helsinki 29, Finland*

INTRODUCTION

Selective contacts between like cells are probably an important element of differentiation. The likeness can be of three main types: *organ-specific* as has been discussed by Moscona (1956) and others, or it can be *genotype-specific* (Oka, 1970) or *species-specific*. Another example of the *organ-specific* contacts is the retino-tectal connectivity (Gottlieb *et al.*, 1976) and they have mainly been observed between cells of an evolutionally advanced individual—most studies have been conducted with vertebrates. An example of species-specific contacts is the fruiting-body formation in slime moulds (Gerisch *et al.*, 1975), and an example of genotype-specific contact formation is colony formation in tunicata (Oka, 1970). We shall suggest in this chapter that at least the genotype-specific recognition can take place in mammalian immune responses.

Selective contacts between cells are probably brought about by complementary molecules. There are at least three possible mechanisms illustrated in Fig. 1. Mechanisms A and B would probably require that the species maintains two gene loci, one for the "female" and the other for the "male" partner of the interaction (receptor and antireceptor of Rosemann, 1970). The genes must be closely linked since they must segregate together to maintain the function. A one-locus explanation can be constructed by postulating that the female structure in mechanisms A and B is an enzyme, e.g. sugar transferase and the male structure is its final product.

Such a recognition may explain species-specific contacts in cellular slime moulds (Chang *et al.*, 1975). Carbohydrate binding is much less

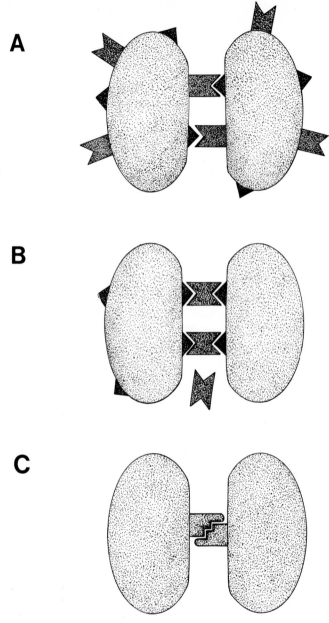

Fig. 1. Possible mechanisms of self-recognition by cells.

likely to explain cell interactions for vertebrate differentiation where there is a demand for a strict regulation of cellular social behaviour (Edelman, 1976). This leaves protein–protein interaction which requires the two gene loci.

Mechanism C of Fig. 1 requires only one self-complementary protein molecule and thus only one gene locus. The best examples of self-complementary proteins are polymers of identical subunits such as concanavalin A and glutamate dehydrogenase.

EFFECT OF SELF-RECOGNITION IN THE TRIGGERING OF B LYMPHOCYTES

The central cell in this discussion is a lymphocyte of the B class. It is a small non-dividing, slowly metabolizing cell, but contact with the correct antigen triggers it to become a metabolically active large and dividing cell (Fig. 1). Some members of the developing clone differentiate further into secretory cells—plasma cells—that produce antibody against the triggering antigen.

Besides the antigen usually another cell is required for this differentiation step of the B lymphocyte. The cell is another type of lymphocyte called the T lymphocyte. The exact mechanism in this cell interaction is unknown, but also the T lymphocyte must recognize the antigen molecule. For the recognition both lymphocytes use surface-associated antibody molecules which are called receptor antibodies.

There are antigens, however, than can induce B lymphocytes into the plasma cell differentiation without the T cell help. They are called T-independent antigens. Syngeneic erythrocytes coupled with a hapten are an example of T-independent antigens, they induce B lymphocytes that are specific for such a hapten without T cell help (Naor et al., 1975) Our recent evidence suggests that a genotype-specific recognition enhances this induction (Koskimies and Mäkelä, 1976).

Our antigen was a synthetic compound NNP (Brownstone et al., 1966) that was covalently coupled to erythrocytes of various types. The coupled erythrocytes were injected into mice, and the induction of NNP-specific lymphocytes was monitored by counting anti-NNP-producing plasma cells in their spleens 4 days later. Since we wanted to study the T cell independent induction of B lymphocytes we used T cell deficient mice. We either injected the mice with anti-thymocyte serum or used mutant mice of the nude type.

By using several combinations of an erythrocyte donor strain and recipient strain we made the general finding that conjugates of syngeneic cells induced a stronger anti-NNP response than conjugates of

TABLE I

The anti-NPP-PFC response (PFC per 10^6 spleen cells) to syngenic, allogenic or xenogenic NNP-erythrocyte conjugates in T cell deficient mice. Data from Koskimies and Mäkelä (1976)

Recipient	Fully syngenic donor	Allogenic donor	H-2 compatible but allogenic donor	H-2 incompatible but otherwise very similar donor	Xenogeneic donor (SRBC)
	I	II	III	IV	V
ATS-CBA	478	248	435		
ATS-C57BL/6	427	137	303	137	137
C57BL/6 nu/nu	448	174			
ATS-(CBA×C57BL/6)F$_1$	389	146			

Mice were immunized i.p. with 0.2 ml of 25% NNP-MRBC. Non-nude mice received a simultaneous injection of ATS (0.3 ml). Nude mice had mainly C57BL/6 background. The donor strains in column IV were B10Br and B10D2. The donor strains for column III were B10Br (for CBA) and LP/J (for C57B1/6). Direct anti-NNP-PFC on day 4 are given (geometric means). One to 5 experiments were pooled for each number. Each recipient strain produced significantly more anti-NNP-PFC in response to syngenic than allogenic, xenogenic or "only H-2 incompatible" conjugates ($p < 0.005$). Each recipient strain produced significantly more anti-NNP-PFC in response to allogenic but H-2 compatible than against allogenic, xenogenic or "only H-2 incompatible" conjugates ($p < 0.01$). The number of recipients tested was 8–10 for each value except the means CBA I and II (it was 24), C57BL/6 I (it was 15), C57BL/6 nu/nu II (it was 4) and C57BL/6 V (it was 5). PFC/10^6 spleen cells against sheep red blood cells were 7.6, $\times \div$ 1.17. ATS = ATS treated.

allogenic erythrocytes (Table I). Syngenic erythrocytes came from another mouse of the same inbred strain and allogeneic erythrocytes from another strain of mice.

There was one exception to the above rule, however; if an allogenic donor strain had the same H-2 allele as the recipient strain the anti-NNP response was of the stronger syngeneic type. This suggested that the H-2 chromosomal region, the major histocompatibility region, is critical for the described phenomenon. A study of congenic strains confirmed this suggestion (Table I). A pair of H-2 congenic strains is genotypically identical except for the H-2 genes and closely linked genes, yet they behaved in our immunizations as allogenic combinations, the response was low. A xenogenic combination also induced a low response (Table I).

In the experiments that we have discussed so far homozygous erythrocytes and homozygous recipients were used. We later immunized F_1 hybrid mice with conjugates of parental erythrocytes or vice versa. The results indicated that parental conjugates injected into F_1 had an intermediate strength as NNP antigens. They were significantly stronger than allogenic conjugates but significantly weaker than syngenic conjugates (Table II). Conjugates of F_1 hybrid erythrocytes injected into parental recipients induced a weak response. The rules are thus slightly different from those in colony formation of tunicate (Oka, 1970). A

TABLE II

The anti-NNP-PFC response (PFC per 10^6 spleen cells) in ATS treated parental and $(CBA \times C57BL/6)F_1$ mice immunized either with NNP conjugated parental, allogenic or $(CBA \times C57BL/6)F_1$ erythrocytes. Data from Koskimies and Mäkelä (1976)

Recipient	Donor			
	$(CBA \times C57BL/6)F_1$	CBA	C57BL/6	BALB/c
$(CBA \times C57BL/6)F_1$	389	286	273[1]	146
CBA	152	699	226	
C57BL/6	87	137	475	

Each value is a pool of two experiments and 10 mice if not otherwise indicated. Recipients for syngenic and allogenic MRBC conjugates are included in the results of Table I. For other explanations see Table I. The first value on the first line is significantly higher than the second and the third ($p < 0.025$ or 0.005), which again are significantly higher than the fourth ($p < 0.005$). On the two other lines the syngenic combination (underlined) produced significantly higher values than the other combinations ($p < 0.005$). Other comparisons exhibit no statistically significant differences ($p > 0.05$).

[1] 20 mice, three experiments.

possible explanation for the weak response elicited by F_1 hybrid erythrocytes is that the amount of the self-recognition structures (see below) is low in heterozygous erythrocytes. It may be suboptimal even in homozygous erythrocytes—erythrocytes have low amounts of H-2 structures which may be the self-recognition structures (for references see Koskimies and Mäkelä, 1976)—but at least it is likely to be twice as high as in heterozygous erythrocytes. By this argument we might have expected the response of F_1 mice to parental erythrocytes to be as high as to F_1 erythrocytes. An *ad hoc* explanation of the finding may be that the allelic forms of, for example, H-2K occurred in hybrid molecules, and they were suboptimal for self-recognition.

DISCUSSION

These findings may appear unexpected; we are used to a rule that if a hapten is attached to an autologous protein it induces a weaker response than if it is attached onto a foreign protein. Two points should be considered, however. One is that the role of the foreign carrier protein probably is to provide recognition sites for T helper cells, and these cells were excluded from the reported experiments. We did other experiments with mice that had not been T cell depleted, and then conjugates of allogenic cells induced a stronger anti-NNP response than conjugates of syngenic cells. Furthermore, the response to conjugates of xenogenic cells was strongest of all.

The other point is that we used cells rather than protein molecules as the carriers of the hapten. This may have created an opportunity for cellular interactions which will be discussed below.

The better immunogenicity of syngenic than allogenic conjugates might have been due to longer survival of the former than of the latter in the circulation of the recipients. Alloantibodies perhaps cleared allogenic conjugates rapidly. This was *a priori* an unlikely possibility since both syngenic and allogenic conjugates were cleared rapidly, within few hours, but a formal proof against this possibility is that conjugates of parental cells in F_1 hybrid recipients were less immunogenic than conjugates of F_1 hybrid cells. The recipients in the former case could not have had antibodies against the parental erythrocytes.

Another trivial explanation could have been that the weaker anti-NNP response to allogenic than against syngenic conjugates was due to antigenic competition. Since the immune system of the recipient was busily making alloantibodies it had less opportunity for making anti-NNP. Against this argue the experiments with H-2 compatible strain combinations (Koskimies and Mäkelä, 1976). Strains B10Br and

CBA differ in many non-H-2 antigens, yet CBA mice responded strongly to B10Br conjugates. Stronger evidence against antigenic competition comes from our unpublished experiments where a limited number of A/J mice were immunized with NNP conjugates of A/J, CBA (only the K-end of H-2 is shared with A/J), BALB/c (only D-end of H-2 is shared with A/J), or C57BL/6 erythrocytes. In these experiments identity in either end of the H-2 complex was enough to make conjugates behave like syngenic conjugates. This was also true of A/J erythrocyte conjugates injected to either BALB/c or CBA mice.

There are at least two ways of explaining why conjugates of syngenic erythrocytes caused a stronger anti-NNP response than conjugates of allogenic or xenogenic erythrocytes (Fig. 2). One explanation is that the receptor antibodies of at least some responding B lymphocytes are complementary to a structure that combines the hapten and a part of the H-2 protein. The other possibility is a dual recognition. The hapten is recognized by a receptor antibody and the self-recognition takes place independently by one of the three mechanisms of Fig. 1.

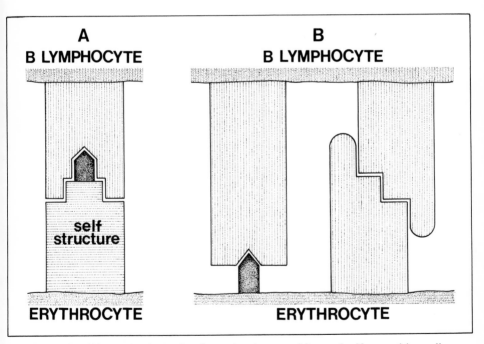

Fig. 2. Possible mechanisms whereby antigenic recognition and self-recognition collaborate in the triggering of lymphocytes. In Alternative A one combining site of the lymphocyte recognizes both the antigen (black) and a self-structure. In Alternative B there is a separate molecule on the lymphocyte for the two recognitions.

We believe that mechanism B of Fig. 2 was operative in our experiments and this is based on two types of evidence. In one approach we tried to exclude close associations of the hapten and H-2 structures by using a protein spacer between the hapten and the erythrocytes. This should have disturbed mechanism A if it was the cause of the syngenic preference. The finding was that syngenic preference still existed, and this suggests that mechanism B was the cause. Another type of test was based on the fact that the humoral antibodies produced by a clone of plasma cells has exactly the same specificity as the receptor antibodies of the maternal B lymphocyte, the initiator of the clone (Klinman and Press, 1975). Our argument was that if the syngeneic preference was based on mechanism A the resulting anti-NNP antibodies would exhibit a preference of conjugates of syngenic erythrocytes over allogenic and certainly over xenogenic erythrocytes. No such preference was observed (unpublished experiments of Koskimies). Anti-NNP antibodies induced by NNP-MRBC lysed NNP-MRBC and NNP-SRBC equally as did anti-NNP antibodies induced by NNP-OA.

There are several lines of evidence that mechanism A often is operative when T lymphocytes recognize the antigen. For example T lymphocytes can act as killer cells which eliminate virus-infected cells. This is a two-step phenomenon: in the first step a virus-infected cell induces a resting T lymphocyte to activate and divide, and in the second step the activated T cells kill the virus-infected cells. Many studies have demonstrated that killing only takes place if the second-step target cells have the same H-2 structures as the stimulator cells (Zinkernagel, 1976; von Boehmer and Haas, 1976).

Another line of evidence comes from studies of delayed type hypersensitivity (DHS), another type of T cell immunity. DHS is also a two-step phenomenon, and macrophages seem important for the presentation of the antigen to the T cells both in the induction phase and in the second step which is called the elicitation step. Again it seems necessary that the macrophages of the first and the second step share H-2 structures (Miller and Vadas, 1976). This suggests that the T cell receptor antibodies recognize a structure which combines elements of the antigen and the H-2.

A further case where lymphocytes seem to recognize H-2 structures is collaboration of T and B cells in the induction of the latter. The mechanism whereby this happens is unknown and even the data do not agree in all points (Kindred and Schreffler, 1972; Katz et al., 1973; Waldman et al., 1975; von Boehmer et al., 1975; Heber-Katz and Wilson, 1975; Katz et al., 1975; Bechtol et al., 1976). We do not know whether mechanism A or B is operative.

CONCLUDING REMARKS

Self-recognition appears to be a factor in mammalian acquired immunity in two ways. One way was demonstrated in anti-hapten antibody formation when T cell deficient mice were immunized by syngenic, allogenic or xenogenic erythrocytes coupled with the hapten. Evidence was presented suggesting that the self-recognition in this case was independent of antigenic recognition, and the H-2 locus was responsible for the molecules participating in the former (mechanism B in Fig. 2).

We propose that the self-recognition alone is not sufficient to induce the lymphocyte, and the antigenic recognition only causes a weak induction. Together the two induce a strong induction.

This self-recognition may be a successor in the evolution of the evolutionally old allotype-specific self-recognition that has been described in the colony formation of compound ascidians (Oka, 1970). The biological role of this recognition in immunity is unknown.

The other type of "self-recognition" has been demonstrated in reactions of T lymphocytes, most clearly in delayed hypersensitivity reactions and so-called cellular cytotoxicity. Mouse T lymphocytes seem to have an unexpectedly large proportion of cells that recognize modified H-2 structures. The modification can be caused at least by a virus infection or chemical haptenation. The hapten and the H-2 structure together seem to form the antigenic determinant. The background of this phenomenon may be the high reactivity of T lymphocytes with all H-2 structures of the species; H-2 structures are privileged antigens. The T cells that are specific for autologous H-2 structures are normally eliminated by self-tolerance but chemically modified autologous H-2 structures again belong to the privileged antigens.

A hypothesis has been put forward that tries to explain why H-2 structures are privileged (Mäkelä et al., 1976).

REFERENCES

Bechtol, K. B. and McDevitt, H. O. Antibody response of CH3↔ (CKB×CWB) F$_1$ tetraparental mice to poly-L (Tyr, Glu) -poly-D, L-Ala-L-Lys immunization. J. exp. Med. **144**, 123–145 (1976).

von Boehmer, H. and Haas, W. Cytotoxic T lymphocytes recognise allogenic tolerated TNP-conjugated cells. Nature (Lond.) **261**, 141–142 (1976).

von Boehmer, H., Hudson, L. and Sprent, J. Collaboration of histoincompatible T and B lymphocytes using cells from tetraparental bone marrow chimeras. J. exp. Med. **142**, 989–997 (1975).

Brownstone, A., Mitchison, N. A. and Pitt-Rivers, R. Chemical and serological studies with an iodine-containing synthetic immunological determinant 4-hydroxy-3-iodo-5-nitrophenylacetic acid (NIP) and related compounds. *Immunology* **10**, 465–479 (1966).

Chang, C.-M., Reitherman, R. W., Rosen, S. D. and Barondes, S. H. Cell surface location of ascoidin, a developmentally regulated carbohydrate binding protein from *Dictyostelium discoidem. Exp. Cell Res.* **95**, 136–142 (1975).

Edelman, G. M. Surface modulation in cell recognition and cell growth. *Science* **192**, 218–226 (1976).

Gerisch, G., Fromm, H., Huesgen, A. and Wick, U. Control of cell-contact sites by cyclic AMP pulses in differentiating *Dictyostelium* cells. *Nature (Lond.)* **255**, 547–549 (1975).

Gottlieb, D. I., Rock, K. and Glaser, L. A gradient of adhesive specificity in developing avian retina. *Proc. nat. Acad. Sci. (Wash.)* **73**, 410–414 (1976).

Herber-Katz, E. and Wilson, D. B. Collaboration of allogeneic T and B lymphocytes in the primary antibody response to sheep erythrocytes *in vitro. J. exp. Med.* **142**, 928–935 (1975).

Katz, D. H., Graves, M., Dorf, M. E., Dimuzio, H. and Benacerraf, B. Cell interactions between histoincompatible T and B lymphocytes are controlled by genes in the I region of the H-2 complex. *J. exp. Med.* **141**, 263–268 (1975).

Kindred, B. and Shreffler, D. C. H-2-dependence of cooperation between T and B cells *in vivo. J. Immunol.* **109**, 940–943 (1972).

Klinman, N. R. and Press, J. L. The B cell specificity repertoire: Its relationship to definable subpopulations. *Transplant. Rev.* **24**, 41–83 (1975).

Mäkelä, O., Koskimies, S. I. and Karjalainen, K. Possible evolution of acquired immunity from self-recognition structures. *Scand. J. Immunol.* **5**, 305–390 (1976).

Miller, J. F. A. P. and Vadas, M. A. Activation of distinct T cell subsets by different major histocompatibility complex gene products. *Cold Spr. Harb. Symp. quant. Biol.* **41**, in press (1976).

Moscona, A. Development of heterotypic combinations of dissociated embryonic chick cells. *Proc. Soc. exp. Biol. Med* **92**, 410–416 (1956).

Naor, D., Saltoun, R. and Falkenberg, F. Lack of requirement for thymocytes for efficient antibody formation to trinitrophenylated mouse red cells in mice: role for thymocytes in suppression of the immune response. *Europ. J. Immunol.* **5**, 220–223 (1975).

Oka, H. Colony specificity in compound ascidians. *Profiles of Japanese Science and Scientists*, 195–206 (1970).

Roseman, S. The synthesis of complex carbohydrates by multiglycosyltransferase systems and their potential function in intercellular adhesion. *Chem. Phys. Lipids* **5**, 270–297 (1970).

Waldmann, H., Pope, H. and Munro, A. J. Cooperation across the histocompatibility barrier. *Nature (Lond.)* **258**, 728–730 (1975).

Zinkernagel, R. M. Virus-specific T-cell-mediated cytotoxicity across the H-2 barrier to virus-altered alloantigen. *Nature (Lond.)* **261**, 139–141 (1976).

SUBJECT INDEX